THE EQUITABLY
RESILIENT CITY

Urban and Industrial Environments Series

Series editors: Robert Gottlieb, Henry R. Luce Professor of Urban and Environmental Policy, Occidental College

Bhavna Shamasunder, Associate Professor of Urban and Environmental Policy, Occidental College

A full list of titles in this series is included at the end of this book.

THE EQUITABLY RESILIENT CITY

SOLIDARITIES AND STRUGGLES IN THE FACE OF CLIMATE CRISIS

ZACHARY B. LAMB AND LAWRENCE J. VALE

THE MIT PRESS   CAMBRIDGE, MASSACHUSETTS   LONDON, ENGLAND

The MIT Press would like to thank the anonymous peer reviewers who provided comments on drafts of this book. The generous work of academic experts is essential for establishing the authority and quality of our publications. We acknowledge with gratitude the contributions of these otherwise uncredited readers.

This book was set in Stone Sans and Stone Serif by Westchester Publishing Services. Printed and bound in the United States of America.

Library of Congress Cataloging-in-Publication Data

Names: Lamb, Zachary, author. | Vale, Lawrence J., 1959– author.
Title: The equitably resilient city : solidarities and struggles in the
    face of climate crisis / Zachary B. Lamb and Lawrence J. Vale.
Description: Cambridge, Massachusetts : The MIT Press, [2024] |
    Series: Urban and industrial environments series | Includes
    bibliographical references and index.
Identifiers: LCCN 2023048911 (print) | LCCN 2023048912 (ebook) |
    ISBN 9780262549868 (paperback) | ISBN 9780262380942 (epub) |
    ISBN 9780262380959 (pdf)
Subjects: LCSH: Sustainable urban development. | City planning—
    Environmental aspects. | Urban ecology (Sociology)
Classification: LCC HT241 .L358 2024 (print) | LCC HT241 (ebook) |
    DDC 307.1/416—dc23/eng/20240205
LC record available at https://lccn.loc.gov/2023048911
LC ebook record available at https://lccn.loc.gov/2023048912

10 9 8 7 6 5 4 3 2 1

CONTENTS

PREFACE AND ACKNOWLEDGMENTS

Finding equitable paths forward in the face of escalating climate crises, urbanization, and systemic inequality is both daunting and essential. The research underlying this book spanned five continents, entailed interviews in nine languages, and took place during a global pandemic. By design and by necessity, the project required working as a team. Specifically, as two white cisgender male faculty from elite US universities, we have sought to maintain a posture of humility as we engage diverse communities beset by long-standing structural inequities. These efforts have profited enormously from contributions by many collaborators with diverse linguistic and cultural competencies. We have endeavored to craft a book that confronts systemic injustice without reducing the communities we write about to an accumulation of damage, victimization, and loss. Instead, we emphasize areas of possibility and local agency without minimizing the burdens of structural injustice. In documenting an array of partial successes, we have remained wary of both self-serving boosterism and self-defeating despair. Finding this balance involves building trust with divergently situated informants who have helped us understand the complexities of their worlds.

FRAMEWORKS, CASES, AND METHODS

Our book features a four-part framework for conceptualizing "the equitably resilient city," linked to a dozen in-depth illustrative core cases. Over the course of a decade of work with MIT's Resilient Cities Housing Initiative, we generated principles for equitable resilience, drawing on prior research on urban disasters, climate change, housing and inequality. We then tested and refined those principles by analyzing interventions around the world, identifying four key dimensions: environment, security, livelihoods, and governance. For each principle, we developed a set of central questions to assess progress at multiple scales. Rather than resort to numerical scoring that can too often be incomplete, unproductively complex, and time bound, we have sought to operationalize the notion of equitable resilience as a multidimensional set of processes and outcomes.

Our twelve case studies share a few central characteristics, which enabled us to test and refine our framework iteratively in actual places. First, each core case includes recently realized built environment interventions. Efforts that are purely process or advocacy oriented and those that are solely focused on changing governance or finance regimes can be transformative, but they are not our focus here. Because we concentrate on resilience in the face of climate change, a crisis that is reshaping the terrain of design and planning in real time, we focus on recent and ongoing intervention rather than projects from the distant past or those that are still "on the drawing board." Assessing recent cases provides powerful insights into contemporary practices but is also limiting. We can only see the progress of a project to date—with the ever-present possibility that promising practices could be stymied by new barriers or be transformed by changing contexts.

We only feature cases that include built environment interventions, but our core cases each treat built environments as integrally related to social structures and institutions. In some instances, this means that a project's innovations may be largely invisible, including changes to governance, ownership, and decision-making processes. We spotlight cases that treat built environments as integrally tied to institutional processes because intervening in built environments without understanding social and organizational repercussions has so often undermined major urban projects in earlier generations, from Haussmannization in Paris to urban renewal in

American inner cities. By centering interventions that take social and institutional contexts seriously, we draw upon insights from decades of urban scholarship, demonstrating how social processes shape the production and functioning of urban space.

Finally, every core case must seek to improve conditions for disadvantaged populations. By contrast, as we discuss throughout the book, many other high-profile projects undertaken in the name of "urban resilience" are either blind to equity concerns or actively harm disadvantaged residents—or both. While equity-blind interventions can incidentally benefit disadvantaged people, the inequity-exacerbating track record of climate adaptation and other major urban interventions demands sustained attention to proactively reducing inequality through transformative climate action.

We do not claim that our case selection is representative in any way. Rather, we chose our sites and stories to achieve maximum range across several dimensions, including geography, hazard types, and scale and type of intervention. Before arriving at our twelve central cases, we gathered a long list of nearly two hundred possibilities from diverse sources, including peer-reviewed literature, popular press accounts in multiple languages, and consultation with regional experts. We conducted exploratory research into potential cases to see if they met our criteria for inclusion and to gauge whether they advanced equity in one or more of our four dimensions of equitable resilience.

To document our cases, we have relied on literature reviews, interviews, and observational site visits. To bring greater depth and nuance, we engaged native speakers to assist in identifying and translating scholarly literature, gray literature and reports, and popular media in English, French, Spanish, Portuguese, Mandarin, Marathi, Hindi, Kiswahili, and Thai.

For each case, we conducted interviews with key actors to assess the projects according to the four central dimensions of environment, security, livelihoods, and governance. We considered impacts across three scales: individual/household, project/community, and city/region. For each case, the team interviewed a range of respondents, including project founders and leaders, operational employees, members of impacted communities, planners and designers, government agencies that interface with the projects, and potential critics. By triangulating across these multiple perspectives and inquiring into the unfolding of initiatives over time, we attempted to reveal successful tactics and strategies, as well as shortcomings and barriers. Our team conducted interviews in each interviewee's preferred language and, whenever authorized and possible, recorded these conversations for coding and analysis.

Despite some COVID-19 restrictions on travel and in-person meetings, the authors or other members of the research team completed observational site visits and in-person interviews in key project locations for all twelve of the core case studies. Supplemented by maps and aerial imagery, we viewed interventions in situ, investigated projects in relation to their contexts, and documented sites in photographs, video, audio, and drawings. This facilitated visceral experiences in places around the globe: huddling over paper maps during downpours in Shenzhen or conducting outdoor interviews in scorching heat on the Pine Ridge Reservation.

Like other researchers who live and work in wealthy, high-emitting nations, we have faced the ethical challenge of writing about the inequity of the climate crisis while seeking to minimize our own contributions to the problem. Adapting to pandemic-related travel restrictions had an unanticipated benefit: it forced us to work with interviewers based near project sites, reducing travel-related greenhouse gas emissions and enforcing a measure of carbon discipline that will inform our work going forward.

This approach also helped us to build an international team—mostly from among our past, present, and future UC Berkeley and MIT students—that could work in nine languages. We coordinated via multicontinental weekly Zoom meetings, with no time zone left behind.

OUR TEAM

This book has benefited enormously from several research assistants. Of these, we offer special thanks to Mora Orensanz for spending four years with us, producing or overseeing all of the book's maps and drawings. Mora also conducted our interviews in Bolivia. Our team in São Paulo greatly enriched our capacity to understand Paraisópolis, notably the incisive work of Ava Hoffman and Lilian Teves's interviews with fellow favela residents, assisted by Ana McIntosh and Prof. Patrícia Cezario. Our discussion of projects in Kibera gained considerably from Kiswahili translations by Elly Wanyoni, case analysis by Patricia Cafferky, and site visits by Ronald Kmau, Haleemah Qureshi, and Jonathan Tarleton. Smriti Bhaya skillfully conducted multilingual interviews in Pune, building on contributions from Ipshita Karmakar and earlier thesis work of Smita Rawoot. Eakapob Huangthanapan ably led our interviews in Bangkok and contributed drone images and drawings. Alessandra Fabbri adeptly directed our interviews in Paris, as did Jesús Contreras for the ROC USA case near Houston, building on work by Stephanie Silva and Sarah Atkinson. Supplementing our own visits to Shenzhen, we are grateful for the interviews conducted by Prof. Gan Xinyue and her students, Lu Yuqi and Peng Ziyin, as well as additional assistance from Shuyue Chen, Colleen Chiu-Shee, Fu Na, and Hongru Cai. Amanda Barnett contributed to our cases in New Orleans and Pine Ridge and conducted site visits in Portland. Finally, we are indebted to the work of Gabriela Zayas del Rio in San Juan, and consultation with Deepak Lamba-Nieves.

In addition, we appreciate the formative contributions of the early leaders of MIT's Resilient Cities Housing Initiative, especially Shomon Shamsuddin and Kian Goh, who helped devise the organizing framework for the book a decade ago—defining "equitable resilience" as a composite of livelihoods, environment, governance, and security ("LEGS"). Further valued support for the early stages of our book came from Carmelo Ignaccolo and Ben Walker, as well as Miriam Wahid, Emily Levenson, and Nadine Eichenlaub. More broadly, we gained significantly from the expertise and support of Berkeley colleagues, including Teresa Caldeira, Daniel Aldana Cohen, Dan Chatman, Renee Chow, Stephen Collier, Zoé Hamstead, Kristina Hill, Justin Hosbey, Ben Metcalf, Carolina Reid, and Danielle Rivera. Similarly, at MIT, we thank Rosabelli Coelho-Keyssar (and the MIT-Brazil Program), Reinhard Goethert, Ken Goldsmith, Janelle Knox-Hayes (and others from the Equitable Resilience Framework team), Janine Marchese, Miho Mazereeuw, Larisa Ovalles, Hashim Sarkis, Jim Wescoat, Sarah Williams, Lizzie Yarina, and Chris Zegras. We also truly appreciate the commitment to supporting open access publication demonstrated by the libraries at Berkeley and MIT through their generous joint underwriting of this book's dissemination.

Beyond our home institutions, we gained greatly from Isabelle Anguelovski, Kazi Khaleed Ashraf, Neil Brenner, Antarin Chakrabarty, the late

Saleemul Huq, Luna Khirfan, Liz Koslov, Carlos Martín, Nate Millington, Rob Olshansky, Kasia Paprocki, Linda Shi, Jason Spicer, and many project-specific key informants, including Oji Alexander, Paul Bradley, Aron Chang, Celina D'Cruz, Edson Elito, Bomee Jung, Hayden Shelby, Paula Santoro, and Maria Sitzoglou.

The team from the MIT Press and its partners has been enormously helpful in shepherding this complex project through each stage. We are particularly grateful to Beth Clevenger and Anthony Zannino for their able management and to Emma Martin, Luane Hutchinson, Emily Simon, and Rashmi Malhotra for copy editing and production support. Erika Millen developed a thorough and insightful index. The book benefited tremendously from the close reading and attentive comments from three anonymous reviewers.

Finally, we are ever grateful to multiple generations of each of our families for their patience and support during years of fieldwork travels, predawn team meetings, and late-night writing sessions.

INTRODUCTION: URBANIZATION, INEQUALITY, AND CLIMATE CRISIS

Adapting to climate change is necessary, urgent, and—all too often—profoundly inequitable.

In Dhaka, the fast-growing capital of Bangladesh, thousands of residents of centrally located but low-lying informal settlements saw their homes bulldozed to make way for new stormwater retention ponds and green spaces. Meanwhile, the government allowed a powerful industry group to remain nearby in their illegally built high-rise headquarters for more than a decade. Across the planet in New Orleans, following the collapse of the city's levees during Hurricane Katrina, planners and politicians called for the conversion of several majority-Black[1] neighborhoods into stormwater-absorbing green infrastructure, while similarly flood-prone majority-white neighborhoods were to be spared and later received more generous rebuilding support.

In both examples, project proponents argued that interventions would reduce their city's flood vulnerability. In each instance, planning and design proposals called for erasing communities inhabited by disempowered residents. Leaders presented the projects as logical and desirable steps to improve urban resilience for cities facing mounting threats from climate change.

Yet, pursuing climate resilience need not come at the expense of disadvantaged people. Residents of canal-side informal settlements in Bangkok gained collective ownership of land and worked with community-oriented architects to redesign their neighborhoods, reducing flood vulnerability and upgrading housing and infrastructure. Facing deadly heat waves, officials in Paris have remade walled-off asphalt schoolyards across the city, enabling new forms of landscape-based experiential learning for students and creating cooling oases for neighborhood residents.

Like the examples in Dhaka and New Orleans, the initiatives in Bangkok and Paris aimed to adapt settlements to climate change. Unlike the first pair of examples, these latter cases actively sought to improve conditions for otherwise disadvantaged people. The difference between these two pairs of scenarios is the difference between resilience and equitable resilience, between pursuing the resilient city and more ambitiously aiming for the equitably resilient city.

Since the beginning of the twenty-first century, resilience has become a prominent organizing concept for governance, planning, and design in cities around the world. Urban resilience initiatives have been driven by the confluence of two major processes: accelerating urbanization and disruptive climate change. Settlements have intensified and expanded into geophysically unstable territories, from flood-prone river deltas to drought- and fire-exposed wildland–urban interface zones. At the same time, climate change has upset long-established climatic patterns, bringing more violent extreme events and disrupting the lives, livelihoods, and built environments of people around the world.

While the convergence of urbanization and climate change has highlighted the need for people, property, and infrastructure to adapt, there is a third global process that is often overlooked in discussions of urban resilience: mounting socioeconomic inequality. The same processes of industrialization and global capitalist expansion that have brought on both urbanization and climate change have also created increasing inequalities across and within societies, generating unfathomable wealth for a few and leaving many urban residents unable to meet their basic needs. With growing inequality, climate change vulnerabilities are also increasingly unevenly distributed. People facing the largest structural disadvantages frequently bear the heaviest burdens from climate change. As the Intergovernmental Panel on Climate Change's (IPCC) Sixth Assessment Report notes, "Cities and settlements are crucial for delivering urgent climate action," in part because of their "concentrated inequalities in risk."[2]

Confronted with this confluence of urbanization, climate change, and inequality, philanthropies, governments, private companies, and advocacy organizations have adopted the goal of improving urban resilience, even as the meaning of "resilience" remains disputed. Those who embrace resilience as a framing concept and a normative aim have used the term to promote an array of physical, institutional, and financial interventions impacting urban settlements. Other more critical observers often regard resilience with great skepticism. The widespread deployment of the term, including by institutions from the World Bank to the US Army and President Donald Trump's Department of Energy, has led critics to reject resilience as a concept. They see it as too easily co-opted by neoliberal schemes to privatize risk, undermine collective action, and further disinvest from public institutions.

Discussions of urban resilience that are dominated by these two polar positions of credulous boosterism and knee-jerk critique have significant blind spots. This book is an attempt to bridge the chasm. In what follows, we develop and illustrate a new framework for what we call "the equitably resilient city."

The book is motivated by three central propositions. First, any attempt to advance urban resilience must link geophysical and social dimensions of vulnerability. Reducing vulnerability to hazards such as floods, fire, and drought is essential but not sufficient. Our understanding of resilience and vulnerability follows from decades of social science research on hazards and socio-ecological relations. Since vulnerability to hazards is fundamentally shaped by inequality, efforts at urban resilience must improve conditions for those who have been rendered vulnerable, or vulnerabilized, across multiple dimensions that shape their lives.

Second, resilience is rooted in the relationships between institutions and the spatial and material configurations of settlements. Too often, resilience research and action focus on one or the other of these domains without recognizing their intrinsic links. Interventions that do not directly reshape the built environment, including insurance, finance, and broad community planning strategies, can shape climate vulnerability and enhance resilience, but they are not our focus.[3] Ultimately, adapting urban settlements to climate change requires changes in the spatial and material configurations of built and natural environments. Even so, we do not center purely physical hazard-mitigation fixes. Rather, our definition of urban resilience links changes in built environments to innovations in social practice, from participatory design and community governance to alternative land and housing ownership structures and livelihood strategies.

Finally, we define resilience not as a stable end state but rather as a complex pursuit that can only be understood in relation to the particular challenges of specific places and times. Accordingly, we ground our discussion of equitable resilience in case studies across a range of scales, geographies, approaches, and hazard types. Too often, both proponents and critics of urban resilience dwell in abstractions—reframing, redefining, and parsing again and again the purported virtues and dangers of resilience. Given the complexity and dynamism of urban processes, all equitable resilience successes are partial, and each incremental improvement is susceptible to loss or retrenchment in the face of changing conditions.

Building resilience is typically regarded as a matter of reducing vulnerability to environmental hazards, but this focus is insufficient. Reducing environmental vulnerability does little for disadvantaged communities if not accompanied by enhanced security, enabling people to stay in place. In turn, the right to remain safely is inconsequential if people cannot satisfy their basic livelihood needs. Finally, equitable resilience requires commitment to the principle of enhanced self-governance. Without the ability to shape decisions affecting one's community, improvements to environment, security, and livelihoods remain precarious. As we use terminology about abstract concepts like environment, security, livelihoods, and governance, we ground these concepts in visceral experiences and basic human questions: Is my family protected from climate-induced disaster? Can I afford to live here? Will I be forced from my home by eviction? Do I have a say in how my community is developed and managed?

Our view of the equitably resilient city is equity centered, socio-spatially focused, and grounded in real-world cases—in all of their messiness, incompleteness, and contingency. These, in short, are stories of solidarities and struggles to pursue equitable resilience in the face of climate crisis.

CITIES, CLIMATE CHANGE, AND RESILIENCE

Settlements are vulnerable to the impacts of climate change because the landscape and climate conditions that have shaped urban space and life over generations are changing with unprecedented speed.[4] Sea level rise and storms are flooding areas previously considered safe.[5] Hotter and drier conditions, combined with ill-conceived land management practices, are producing ferocious wildfires that threaten fast-growing communities at the wildland-urban interface and spread dangerous air pollution across broad territories.[6] Cities in historically temperate areas face prolonged periods of intense heat, devastating communities where behavior, building, and urban space are not well adapted to such conditions.[7]

The costs of climate change for cities and residents are enormous and projected to grow substantially. Between 1998 and 2017, an estimated 91 percent of all disasters globally were attributed to extreme weather events such as those linked to climate change. In that period, floods and storms killed an estimated 375,000 people and caused nearly US$2 trillion in recorded economic losses.[8] In the US alone, the National Climate Assessment estimates that coastal real estate worth approximately US$1 trillion is threatened by sea level rise and storm surge.[9] Extreme heat killed an estimated 166,000 people globally between 1998 and 2017. Recent studies estimate that 37 percent of global heat-related human health impacts between 1991 and 2018 are attributable to climate change.[10] Climate change is also increasing the incidences of some diseases, spreading maladies such as malaria into new territories.[11]

Such statistical catalogues of human misery caused by climate change have grown familiar in recent years. The absolute magnitude of these impacts is alarming. However, the impacts of climate change are especially troubling because this misery is distributed so unevenly. Across the globe and within individual cities, climate change brings massive disruption to some people while barely impacting, or even benefiting, others. A recent study from the National Bureau of Economic Research found enormously uneven impacts from increasing heat, with "today's cold locations benefiting and damages being especially large in today's poor and/or hot locations."[12] Poorer nations are more likely to be in hot regions, which are experiencing some of the direst consequences of climate change. Residents of cities in poorer regions are also less likely to be able to undertake costly adaptations, collectively and individually.[13]

International inequalities in climate impacts and adaptation are especially problematic when viewed in relation to historic greenhouse gas emissions that drive climate change. Many of the poor countries suffering the greatest harms from climate change have contributed very little to historical emissions.[14] According to data from the Global Carbon Project and the Carbon Dioxide Analysis Center, between 1751 and 2017, the US contributed approximately 25 percent of global greenhouse gas emissions, while India, with a vastly larger population, contributed approximately 3 percent. In 2018, the World Resources Institute estimated that the average American produces more than seven times more greenhouse gas emissions than the average Indian.[15] Even as past and present residents of Mumbai or Kolkata are responsible for far fewer emissions than their counterparts in Denver or Dallas, urban dwellers in India are more likely to face flooding, heat, drought, and other climate impacts. They are also less likely to have the resources to move away from trouble, invest in air conditioning or other household adaptations, or benefit from collective adaptation such as upgraded drainage infrastructure.

While these international climate injustices are important, increasingly dramatic wealth gaps within countries, regions, and cities may play an even more powerful role in determining who suffers most from climate change. Hazard vulnerability research has long shown that poor and marginalized people are especially at risk when extreme events strike.[16] In the terminology of hazard and climate change studies, disadvantaged people are especially vulnerable to hazards from climate change because of their high "exposure" and "sensitivity" and their low "adaptive capacity."[17] The urban poor may be more exposed to hazards because they are forced by economic and social circumstances to live in floodplains, steep slide-prone hillsides, or other hazardous places. They may be more sensitive because of less durable housing or poorer infrastructure. Finally, the urban poor disproportionately suffer because of their depressed adaptive capacity, meaning that they have fewer economic, social, and political resources to prepare for and recover from disruptions caused by extreme events.

The social dimensions of vulnerability to climate change and other hazards are often discussed in terms of "social vulnerability", a composite phenomenon that can include an array of characteristics, including poverty, race, gender, immigration and language status, age, and disability.[18] Whether in New Orleans or New Delhi, the same structural drivers of inequality that make life difficult for some urban residents on a day-to-day basis also tend to make those people vulnerable to extreme events such as heat waves and floods. The stark racial and class inequalities in the suffering and recovery from the flooding that followed Hurricane Katrina in New Orleans led some researchers to proclaim that "there is no such thing as a natural disaster."[19] Viewed this way, hazard vulnerability results not simply from the direct causes of an impact but from structural processes that instill and uphold vulnerabilization.

Just as the impacts of climate change are distributed unevenly, so too are the benefits and costs of adapting to climate threats. Research has shown that urban climate change adaptation can harm disadvantaged populations through both "acts of omission" (e.g., insufficient investment in adaptation benefiting less privileged groups) and "acts of commission" (e.g., displacement of disadvantaged groups by infrastructure upgrades).[20] One study found that adaptation tends to exacerbate inequality through four mechanisms: enclosure of public and collective assets, exclusion of marginalized groups from accessing resources, encroachment upon previously protected areas, and entrenchment of preexisting inequalities.[21] Some observers have warned that resilience interventions motivated by climate change might generate new forms of "eco-apartheid,"[22] wherein privileged groups enjoy "premium ecological enclaves" created through "secure urbanism and resilient infrastructure."[23]

Various forms of climate change adaptation can deepen or reduce preexisting urban inequalities. Infrastructure interventions can improve conditions for the urban poor, leave them behind, or drive "climate gentrification."[24] Planned relocation or "managed retreat" can enable disadvantaged communities to control their own resettlement or it can contribute to "adaptation privilege" and "adaptation oppression" by benefiting already well-off people and further marginalizing disempowered groups.[25] Similarly, post-disaster recovery and reconstruction can expand preexisting disparities by enabling "disaster capitalism" and gutting social safety net programs or reduce disparities by providing targeted support to disadvantaged residents.[26]

Climate change is widening already yawning gaps between empowered and disadvantaged groups within cities in multiple ways: through the distribution of climate change vulnerability, the uneven costs and benefits of adaptation, and the disparate burdens and lasting policy changes undertaken in the wake of major disruptions.

THE EQUITABLY RESILIENT CITY

We titled this book *The Equitably Resilient City* with full acknowledgement of the complexity and theoretical debates surrounding each of the four words of the title (including even the "the"). The title does not signal a blind rejection of those contestations. Rather, we welcome those debates and hope that our use of these terms will both link this project to previous work on the resilient city and enrich our collective understanding of the real and urgent issues at hand. Before proceeding to describe the details of our framework and the structure of the book, we consider the four terms that are at the core of this project: "city," "resilient," "equitable," and "the."

ON THE CITY AND HUMAN SETTLEMENTS

The term "city" may, on first examination, appear straightforward. That said, the word can bring to mind an array of images, from the gleaming skylines of Dubai or Shanghai to informal settlements in Rio or Mumbai to the grand boulevards of Paris or Barcelona. Outside the bustle of center cities, defining what is and is not a city or "the city" becomes more complicated. Are ever-expanding suburbs connected by ribbons of highway part of the city? What about the forests, quarries, and farms from which city dwellers acquire their food and building materials? Can our definition of the city encompass mines, oilfields, solar farms, war zones, and "sacrifice zones" associated with the energy production that keeps city lights burning and cars running? Or, going still further afield, what about the orbiting satellites and trans-ocean cables that carry the data that is so crucial to the "command and control" function of global cities?[27] While urban sociologists have long defined "urbanism" with respect to characteristics of size, density, and social heterogeneity,[28] spatial and scalar definitions of the urban or the city are increasingly areas of debate. Critical urban scholars have argued not only that binary distinctions between city and country or between urban and rural are fraught with methodological and conceptual fuzziness, but also that these distinctions obscure the human and ecological costs of urbanization.[29] In reaction to both celebratory and anxious pronouncements about the dawning of the "urban age" in which a majority of the world's population resides in cities,[30] critical scholars have challenged simplistic and inconsistent definitions of "the urban" and presented "planetary urbanization" as an alternative spatial imaginary that includes the much broader socio-ecological processes described above.[31]

In discussing the resilient city, we fully embrace this more complex understanding of urbanization as a dynamic, variegated, and multiscalar process that challenges static definitions. We do not draw hard binary distinctions between city and non-city. The projects that inform and illustrate our framework are drawn from a range of settlement types, from booming East Asian megacities to small Native American settlements in the hills of South Dakota. By using the term "resilient city," we do not mean to suggest that it is possible to improve or even to understand fully the conditions of an entire urban settlement. The planning and design interventions that we discuss here operate across scales, from individual parcels to entire national territories. They are the product of shifting and complex coalitions of actors, including not just government actors but also civil society groups, private-sector actors, and grassroots groups of residents. While we embrace more inclusive terms such as "human settlements" throughout the book,[32] we have chosen to use "city" in the title of the book to link our work to previous research and practice.

THE RESILIENCE OF "RESILIENCE"

"Resilience" is perhaps the most complex and contested of the terms in our title. A recent review of the literature on urban resilience arrives at the following definition: "The ability of an urban system—and all its constituent socio-ecological and socio-technical networks across temporal and spatial scales—to maintain or rapidly return to desired functions in the face of a

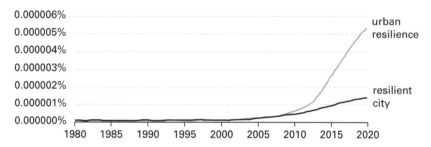

0.1 Dramatic increase in the frequency of the use of the terms "urban resilience" and "resilient city" in the period from 1980 to 2019. *Source:* Mora Orensanz after Google Ngram.

disturbance, to adapt to change, and to quickly transform systems that limit current or future adaptive capacity."[33] The complexity of this definition speaks to the comprehensive ambitions of urban resilience as a normative aim of planning and design. Applying the term "resilience" to cities or to urban systems, people, or places is a relatively recent phenomenon. Since the beginning of the twenty-first century, resilience has proliferated across many domains of urban planning, design, and policy. Graphs of the use of terms such as "urban resilience" and "resilient city" in books show a remarkable growth since 2000 (figure 0.1).

Before 2000, "resilience" appeared frequently in contexts outside of urbanism and planning. In engineering and materials science, "resilience" describes the ability of a material or structure to return to its original form after some perturbation.[34] The term has been used in psychology to describe the capacity of individuals and groups to withstand and recover from stresses.[35] Beginning in the 1970s, ecologists began using "resilience" to describe the tendency of ecological communities to react to and recover from disturbances, whether from natural events such as a fire or human impacts such as logging.[36] Over the following decades, researchers applied the term not only to ecological systems but also to social-ecological systems, considering the relationship of nonhuman species and ecological dynamics to governance, institutions, and other dimensions of human social organization.[37] This social-ecological resilience framing has been applied to a range of activities, including environmental regulation, agriculture, and forestry.

Scholars have fruitfully examined these multiple origins and trajectories for resilience elsewhere.[38] The multiple conceptual origins of resilience have contributed both to the term's utility and to the challenges of applying it in urban planning and design.

As we will discuss in greater depth in the next chapter on the environmental dimensions of equitable resilience, research in human geography, planning, and design has examined how the organization of human settlements shape vulnerability to hazards long before resilience came to prominence in these domains. Instead of resilience, planning researchers discussed "disaster risk reduction" and "hazard mitigation," building upon frameworks from the "natural hazards" school of human geography[39] to consider how strategies such as land-use planning and building codes could reduce vulnerability.[40] In pursuing similar aims, urban designers and landscape architects argued for new modes of planning and design based on analysis of landscape and climatic processes, including hydrology, wind patterns, and soil dynamics.[41]

While this work in planning and design was broadly resonant with the social-ecological framing of resilience, resilience did not make its way into common usage in these domains until after two massive urban shocks: the 9/11 attacks of 2001 and Hurricane Katrina in 2005 (coupled with the devastation from the Indian Ocean tsunami the previous December). These events brought a new sense of urgency to questions of how communities, governments, and built environments cope with and recover from major disruptions. Following the 9/11 attacks and the 7/7 bombings in London four years later, invocations of resilience by planning and design researchers and practitioners focused on securitization and the vulnerabilities of built environments to terrorism and other deliberate human destruction.[42] *The Resilient City: How Modern Cities Recover from Disaster*, an edited volume released in 2005, featured global case studies of cities disrupted by war, attack, and earthquakes.[43] The massive destruction wrought by Hurricane Katrina and the subsequent levee and pump failures that swamped New Orleans brought renewed attention to the vulnerabilities of urban settlements to extreme weather. After these cataclysms, resilience increasingly became the dominant framing for discussing how planners, designers, and decision makers might reform built environments, infrastructures, and institutions to reduce vulnerabilities. A 2006 theme issue of the journal *Built Environment* called for "Learning from Urban Disasters: Planning for Resilient Cities," and the titles of more than a dozen books and countless articles adopted versions of the same "urban resilience" phraseology.

Concerns about urban climate change vulnerability became even more prominent a few years after Katrina when Hurricanes Irene (2010) and Sandy (2012) brought massive flooding to the highly populated northeastern US. Since that region was not widely considered to be vulnerable to such storms and because the region looms large in setting agendas in policy, media, academia, and philanthropy, these hurricanes altered urban climate change discourse in the US and beyond. Such climate-linked disasters made urban resilience an increasingly central concern across sectors. Universities started graduate programs in "risk and resilience" (e.g., Harvard's Graduate School of Design). Philanthropic organizations launched high-profile urban resilience initiatives, including the Rockefeller Foundation's 100 Resilient Cities (100RC) program. Cities hired "chief resilience officers" and generated resilience plans.[44] A community of practice emerged among consultants from engineering, planning, design, and other fields, sometimes critically labeled as a "global urban resilience complex."[45] Several groups created practice-oriented frameworks and indicators to assess resilience, including Arup's "City Resilience Wheel" created for 100RC, the C40 and World Resources Institute's *Inclusive Planning Playbook*,[46] the Assessment Framework on Public Spaces and Urban Resilience,[47] Bruneau's framework for assessing community resilience,[48] and the Urban Land Institute's 10 Principles for Building Resilience.[49] The onset of the global COVID-19 pandemic set off another spasm of writing, research, and reconceptualization of urban resilience, this time adapting the concept to the threat from viral contagion.[50]

While resilience has spread quickly to become a central concept in many domains, it has also proven to be a polarizing construct, embraced by many and critiqued and caricatured by others. As described above, many philanthropic institutions, government agencies, consultants, and planning and design practitioners enthusiastically took up the banner of resilience. Progressive advocates and grassroots organizations have also framed their work as building community resilience.[51] Some have argued that resilience is a useful "boundary object" around which to organize multidisciplinary work because it has been adopted by a diverse array of disciplines, institutions, and actors.[52]

Some observers see the wide applications of resilience as a virtue, but others regard this same characteristic as a troubling vagueness that, at best, renders resilience a "conceptual dead end" and, at worst, makes the concept open to co-optation by regressive interests.[53] Many have criticized the use of resilience as suggesting that it is desirable to "bounce back" to a pre-disturbance equilibrium without questioning whether that previous state was just and without addressing the underlying processes that generate uneven vulnerability in the first place.[54]

A related critique argues that resilience fetishizes "self-regulating systems" in ways resonant with conservative economic thinkers such as Friedrich Hayek. In this framing, invocations of resilience are part of a broader movement to abandon planning for socio-ecological systems in favor of allowing purportedly self-regulating mechanisms to work.[55] Critiques of resilience that frame the concept as promoting neoliberal governance link resilience to mid-twentieth-century complex systems theory and cybernetics.[56] As such, some critics regard resilience as a "powerful technology of contemporary governance and neoliberal rule,"[57] arguing that the concept naturalizes uneven vulnerability, obscures structural causes of inequality, blames victims for their vulnerability, privatizes risk, and justifies the retrenchment of state institutions.[58] These critiques resonate with the words of Tracie Washington, a New Orleans attorney and activist who responded to calls for Louisiana residents to be resilient following a massive oil spill in the Gulf of Mexico, saying: "Stop calling me resilient. Because every time you say, 'Oh, they're resilient,' that means you can do something else to me. I am not resilient."[59] Linking resilience with conservative or neoliberal ideology has led some to ask if resilience is "a right winger's ploy?"[60] In fact, resilience has been espoused by many institutions and actors to justify regressive neoliberal policies and practices. Industry groups and military actors have adopted the language of resilience in promoting supply-chain continuity and preparedness—for instance, in the US Army's "ready and resilient" initiatives.[61] The Federal Energy Regulatory Commission under President Trump promoted policies to subsidize polluting coal-fired power plants to improve "grid resilience."[62] While the mobilization of resilience to advance conservative and neoliberal policy priorities is troubling to many progressive researchers and advocates, recent accounts from the US political press observe that resilience has become "a politically safe word for 'climate change,'" potentially creating space for policy action in the face of growing climate threats.[63]

EQUITABLE RESILIENCE AND TRANSFORMATION

While some critical scholars regard resilience as inextricably bound to neoliberal governance, others have proposed modifying and differentiating forms and practices of resilience to address specific concerns. Recognizing what they regard as the problematic conservative valence of "bounce-back" framings of resilience, many advocates and researchers have postulated alternative "bounce-forward" notions, labeling them alternately "progressive resilience"[64] or "evolutionary resilience."[65] Others have called for modifying resilience with "just" or "equitable."[66]

By modifying the word "resilient" with the adverb "equitably" in our title, we acknowledge that the term has been used to justify a range of damaging practices but contend that such outcomes are not inherent in the pursuit of resilience. In the cases that follow, we construct an alternate trajectory through which the pursuit of climate change resilience can advance rather than damage social equity. In this project, we embrace the idea that equity describes the fair but different treatment of different groups on the basis of their differing need.[67]

Research on hazard and climate vulnerability has long demonstrated that uneven vulnerability is inseparable from broader patterns of marginalization and exploitation within and between social groups.[68] Drawing on the environmental justice movement, which highlighted the uneven burdens of pollution suffered by communities of color and other marginalized groups around the globe,[69] the climate justice movement has reframed climate change as a matter of justice and equity. Pursuing climate justice means redressing the disproportionate harms that climate change is visiting upon already disadvantaged groups, including along fault lines of class, race and ethnicity, immigration status, and gender.[70] Planning and design for equitable resilience therefore must acknowledge and counteract preexisting structures of exploitation and marginalization—the vulnerabilization that has created disparate climate change vulnerabilities.

Some researchers have embraced a three-part framework for social equity in urban resilience planning: (1) distributional equity (who benefits and who is harmed by climate change and adaptation efforts); (2) procedural equity (who makes decisions about priorities and action); and (3) recognitional equity (how differences, including differences in historical marginalization and contemporary need, are acknowledged).[71]

In responding to structural inequalities and seeking transformative potentials, we draw on scholarship articulating progressive, or even emancipatory, potentials in the concept of resilience for planning and design.[72] Kevin Grove argues that resilience might provide a "foothold for subversive action" for reshaping socio-ecological systems through the use of design as a form of understanding.[73] Studies analyzing post–Hurricane Sandy planning in New York City also made the case that resilience is tied to design as a means of analyzing and acting within complex socio-ecological systems.[74] Like Grove, Stephanie Wakefield positions resilience as a possibility-laden alternative to infrastructure planning rooted in stark city/nature separation.[75]

These more nuanced treatments of resilience take seriously concerns that the concept can promote regressive agendas while still holding that resilience can also enable radical action. This approach aligns with mounting calls for "transformational resilience,"[76] "transformative climate adaptation,"[77] and transformative "climate urbanism."[78] In solidarity with such approaches, geographer Mark Pelling poses a central question: "Can adapting to climate change . . . be a mechanism for progressive and transformational change that shifts the balance of political or cultural power in society?"[79] We, too, share in that quest.

Understood in these terms, transformative resilience in the face of climate change requires more than attention to immediate distributional and procedural equity. It also demands that we face deeper questions about how historical and ongoing patterns of marginalization and exploitation have shaped "who has power to act" in the face of climate change.[80] This focus on the distribution of agency within structures of historical and ongoing inequality resonates with Amartya Sen's framing of human well-being as rooted in a context-dependent understanding of human freedoms or "capabilities."[81] If climate vulnerability, resilience, and transformation are understood as fundamentally tied to who has the capability or the power to act, pursuing equitable resilience can never be a wholly technical process of vulnerability reduction. Rather, working toward equitable resilience is inevitably a value-laden political act. Processes, projects, and policies undertaken in the name of resilience can deepen preexisting structural inequalities, but they can also advance transformation and emancipation.

We use "resilience," modified by "equitably," in our title both to put this work in conversation with previous research on urban vulnerability and resilience and to shape ongoing debates about what constitutes planning and design for resilient settlements. In the context of outrageous socioeconomic inequality, any attempt to improve the resilience of a human settlement to climate change must be substantially focused on improving conditions for disadvantaged people. It must do so, moreover, in ways that respect the ways transformation is often led *by* those who have been marginalized rather than enacted on their behalf.

THE "THE"

Finally, it seems worth noting that the first word of our book's title—a seemingly ignorable three-letter definite article—is also a conscious choice. In some ways, it would be simpler, if less ambitious, to call this book "Equitably Resilient Cities." Doing so would reflect the fact that we are presenting a limited number of examples—a subset of sites of equitable resilience initiatives, chosen from many possible cases. To add the "the," however, entails making a bolder and more conceptual claim. Although we do not view equitable resilience as a universal and achievable end state, we deploy the "the" to highlight the commonality of the aspiration. The "the" underscores that this volume seeks to be more than a compilation of chronicles secerned by significant differences. We want to uncover more about what people share when, prodded by an increasingly harsh and unpredictable climate, they must cope with existential challenges to their homeplaces.

In the face of much debate and controversy over its constituent terms, we embrace the phrase "the equitably resilient city" because the label expresses four core underlying assumptions of this project: first, that the climate crisis is bringing enormous suffering around the world; second, that this suffering is distributed unevenly, with some people adapting with relative ease or even profiting from climate chaos and others bearing enormous costs to their health, economic, and psycho-social well-being; third, that the spatial and material configurations of human settlements play a crucial role in mediating these uneven burdens of climate change; and, finally, that these burdens and efforts to overcome them take on particular discernable forms that reveal commonalities across cases.

We seek to demonstrate that the design and planning disciplines can make significant (although not unlimited) contributions in addressing these uneven burdens. Pairing an equitable form of resilience with the idea of city is a reminder that spatial location and the manipulations of space and material matter and that those with skills in these domains—including professional planners and designers—have important roles to play.

THE FOUR LEGS OF EQUITABLE RESILIENCE

Many scholars and policymakers have attempted to define and measure resilience, often yielding frameworks that are either narrowly focused on a single dimension such as hazard mitigation[82] or so expansive and complex as to seemingly encompass all elements of human life. In seeking a tangible and actionable middle ground, we have developed a four-part framework that is both holistic and bounded, linking environmental safety and vitality, security from displacement, enhanced livelihoods, and self-governance. The framework is holistic in addressing a range of human needs whose satisfaction requires not just spatial and material interventions to reduce vulnerability but also the more complex construction of human agency, institutions, and solidarities that make improved environments worth inhabiting. At the same time, our framework is bounded, with each of its components filtered through an equity lens.

In our view, building equitable resilience for human settlements includes four central principles: livelihoods, environment, governance, and security. For largely mnemonic purposes, we refer to these principles collectively as the four LEGS of equitable resilience. The book is structured in four main parts, one for each of the four LEGS. As the case studies demonstrate, there is no universal logic to the priority or sequencing of the principles in practice. Groups pursuing equitable resilience combine and layer activities across these dimensions as circumstances demand. The cases demonstrate that there can be both productive synergies and challenging tensions between the different LEGS components of equitable resilience.

Each of the book's four parts includes three main components. First, a framing chapter discusses the particular LEGS concept, tracing relevant lineages in planning and design and exploring links to other fields. We illustrate problematic patterns as well as promising dimensions, using examples from around the world. Each framing chapter is followed by three case studies analyzing interventions that have achieved partial successes in the domain of interest and in other LEGS dimensions. Each part includes a brief conclusion.

We selected the twelve core cases featured in the book to represent a range of pathways for pursuing equitable resilience. We include cases from the Global North and South from different settlement types, from dense urban core areas to smaller and more isolated settings (figure 0.2). We include cases in places facing many different threats, from drought and water stress in high mountain deserts to flooding in lowland deltas. The production of inequality in each case varies. Some communities featured are disadvantaged because of immigration or migration status. Others have faced intergenerational racism and Indigenous land theft. In some cases, gendered power imbalances have created elevated vulnerability to climate change and other threats. The projects featured in the cases also vary in the composition of their central actors. Some efforts started with government initiatives. Others emerged from grassroots movements. Most are reliant on coalitions joining public and private actors across scales and sectors.

Given our central concern with climate adaptation, we devote the book's first main part to discussing environmental dimensions, with subsequent parts focusing on each of the other three principles.

OVERALL BOOK STRUCTURE

PART I: ENVIRONMENT

Most discussions of climate and hazard resilience foreground reducing environmental vulnerability, often to the exclusion of other concerns. Our framing of the equitably resilient city begins with, but is not limited to, the need to reduce the vulnerability of residents, especially socioeconomically disadvantaged residents, to environmental risks and stresses.

Reducing environmental vulnerability is necessary, but not sufficient, to enable equitable resilience. In the environment realm, advancing equitable resilience also means supporting the ecological vitality on which life depends and enhancing the capacity of residents, especially the most vulnerable, to understand the landscape and ecological processes that contribute to hazard vulnerability.

We situate our discussion of the environmental dimension of equitable resilience with respect to other traditions, including the human geography framing of natural hazards and urban political ecology. The framing chapter for the environment part outlines shifting discourses and practices in planning and design related to how settlements respond to climatic and landscape processes and delineates types of climate adaptation in housing, infrastructure, and settlement patterns.

The core case studies in this part include the Gentilly Resilience District (GRD; case 1), an attempt to reform how one New Orleans neighborhood relates to stormwater; an effort to rehouse residents from flood-prone areas of the Paraisópolis favela in São Paulo (case 2); and the Paris OASIS project (case 3), which is transforming schoolyards to enable experiential education and create cooling refuges for nearby residents. In these case studies, we highlight both successes and ways in which the projects have fallen short or struggled to address critical issues.

Following the core cases, a brief conclusion to part I recounts key components of the environment dimension and discusses areas for potential transformative action.

PART II: SECURITY

Reducing environmental vulnerability delivers little benefit to disadvantaged people if they are threatened with displacement. Accordingly, part II focuses on security. Equitable resilience requires that interventions enhance the security of residents in the face of threats of displacement.

Climate change destabilizes and disrupts geophysical conditions in cities: melting permafrost undermines buildings and infrastructure; sea levels rise and claim coastal lands; rivers swell, eroding their banks and displacing communities. The violence and displacement of these geophysical processes is often accompanied by violence and displacement in human relations.

Communities impacted by hazards frequently find themselves insecure with respect to their tenure rights as public and private entities use the chaos of disasters to take lands without adequate processes or compensation.

Interventions to enhance urban resilience, like other major projects, frequently displace the poor, both directly and indirectly. City officials clear hillside favelas in the name of slope stabilization. Planners push for new green infrastructure investments in parks and open space to improve conditions in low-income neighborhoods. However, these interventions may also push property values higher, driving displacement of residents with fewer resources.

The framing chapter for part II begins by considering how hazard exposure and adaptation can undermine the security of communities and individuals by weakening their rights to stay in place. We highlight several recent projects in which security is central to design and planning for resilience across scales, including projects that nurture a sense of belonging and connection at the individual and household level, strengthen informal cohesion, and seek external recognition from state and non-state authorities to resist displacement and other forms of violence.

The core cases in this part include Pasadena Trails, a cooperative manufactured home park (MHP) in Texas that mobilized to reduce flood vulnerability (case 4); Comunidad María Auxiliadora (CMA), a women-led community in Cochabamba, Bolivia, where collective land tenure enabled improved water infrastructure and housing security (case 5); and Baan Mankong, a program in Thailand in which community-engaged design and collective land tenure enables low-income residents to control infrastructure upgrades and rehousing along flood-prone waterways (case 6). Part II closes with a discussion of cross-cutting themes and possibilities for transformation in the area of security.

PART III: LIVELIHOODS

Even if a resilience-focused intervention reduces the hazard vulnerability of disadvantaged residents and promotes security against displacement, it can still harm those residents by undermining their ability to live affordably and decently. In framing part III, centered on livelihoods, we discuss how interventions undertaken in the name of resilience can threaten the ability of disadvantaged people to meet their basic needs. Conversely, promoting equitable resilience means supporting stable and dignified livelihoods for residents.

The history of planning, design, and development is littered with examples of infrastructure installations, housing projects, and other interventions that are nominally intended to improve conditions for poor communities but instead disrupt or destroy the ability of those very people to earn livelihoods through paid labor, informal work, or subsistence activities. New seawalls cut off fishing communities from the shore. Resettlement housing shifts low-income residents to outlying areas far from jobs and affordable transportation. Leaders use post-disaster urgency as a pretext for suspending fair labor standards.

The core cases featured in the livelihoods part include Living Cully, an initiative in a low-income neighborhood of Portland, Oregon, that trains local residents to advance green infrastructure and energy justice to improve neighborhood conditions and fight displacement (case 7); Yerwada, an in situ redevelopment of informal settlements in Pune, India, that preserved residents' access to center city livelihoods (case 8); and Dafen, a village engulfed by the urbanization of Shenzhen, China, reinvented as a thriving community of art producers (case 9). Part III closes with consideration of cross-cutting themes and a discussion of transformative potentials in resilient livelihoods.

PART IV: GOVERNANCE

Finally, in part IV, we argue that the quest for equitable resilience involves more than environmental quality, heightened security, and economic stability; it is also rooted in the personal efficacy and community empowerment to shape how places are designed, developed, and governed. Residents need to be able to play meaningful roles in making decisions that structure their lives. The imposition of arbitrary decisions by outside powers strips residents of the agency necessary to live according to their own priorities and values. As such, equitable resilience should empower residents through self-governance.

The framing chapter for part IV recounts how climate change challenges existing institutions and norms governing urban settlements.

While ambitious urban resilience projects are necessary, leaders in many cities have used the imperative of adaptation as a pretext for weakening democratic decision making. We argue that strengthening self-governance among impacted communities is critical, while acknowledging that devolving control to local communities comes with its own challenges. We examine instances in which democratically legitimated governance has been subverted in the name of resilience and adaptation and explore alternative strategies for advancing self-governance among urban poor communities under environmental threat. Doing so illustrates the multiscalar dimensions of self-governance for equitable resilience, including improving individual self-efficacy; exercising community control over decision-making processes, both during project design and in ongoing management; and establishing effective linkages to external centers of power and resources.

The core cases discussed in part IV include Thunder Valley, a community development initiative led by Lakota people on the Pine Ridge Reservation in South Dakota, which uses job training and house construction as part of a larger process of community regeneration and liberation (case 10); the work of the Kounkuey Design Initiative (KDI) in Kibera, Nairobi—Kenya's largest informal settlement—which uses community-driven design and management to build new public spaces that reduce flooding and accommodate local economic activity (case 11); and a community land trust (CLT) along the Martín Peña Channel (el Caño Martín Peña) in San Juan, Puerto Rico, which is working to ensure that residents benefit from new green space, drainage infrastructure, and public space while avoiding displacement (case 12; figure 0.2 and table 0.1).

Part IV closes with a brief conclusion, tracing points of convergence and divergence between the three cases and pointing toward areas for transformation in the domain of community self-governance for equitable resilience.

SUMMING UP AND LOOKING FORWARD

After the four central substantive parts, the book closes with a concluding chapter that synthesizes the major points from the previous parts, drawing together common lessons in the form of ten axioms of equitable resilience.

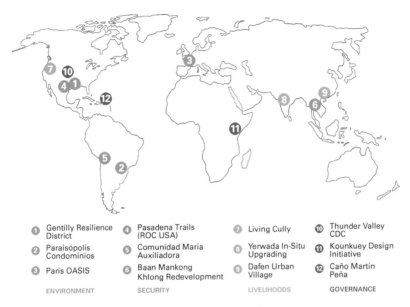

<table>
<tr><td>①</td><td>Gentilly Resilience
District</td><td>④</td><td>Pasadena Trails
(ROC USA)</td><td>⑦</td><td>Living Cully</td><td>⑩</td><td>Thunder Valley
CDC</td></tr>
<tr><td>②</td><td>Paraisópolis
Condomínios</td><td>⑤</td><td>Comunidad María
Auxiliadora</td><td>⑧</td><td>Yerwada In-Situ
Upgrading</td><td>⑪</td><td>Kounkuey Design
Initiative</td></tr>
<tr><td>③</td><td>Paris OASIS</td><td>⑥</td><td>Baan Mankong
Khlong Redevelopment</td><td>⑨</td><td>Dafen Urban
Village</td><td>⑫</td><td>Caño Martín
Peña</td></tr>
<tr><td></td><td>ENVIRONMENT</td><td></td><td>SECURITY</td><td></td><td>LIVELIHOODS</td><td></td><td>GOVERNANCE</td></tr>
</table>

0.2 Location of core case-study sites. *Source*: Mora Orensanz.

LEARNING FROM PARTIAL SUCCESSES: USING THE LEGS FRAMEWORK

The LEGS framework is not a static tally in the mold of the LEED rating system developed by the US Green Building Council to measure the environmental impacts of buildings. There are no universally applicable standards for evaluating equitable resilience. Human settlements, climate change impacts, and adaptation are all marked by place-specific complexity, dynamism, and uncertainty. As such, any universal scorecard for equitable resilience would end up being either excessively specific or vague and overbroad. We do not intend to grade projects, either on binary (success vs. failure) or linear scales of performance with respect to livelihoods, environment, governance, and security. Rather, we intend for the LEGS framework and the questions that it invites to serve as a heuristic for understanding and learning from the inevitably partial successes of interventions. We provide a set of lenses through which to learn from real interventions by bringing together critical inquiry with openness to positive change and transformative potentials, even when successes are inescapably imperfect, incomplete, or fleeting. For every case, we graphically illustrate the directionalities of

Table 0.1

Core case studies

Case Number	Case Name	Principal Focus	Location
1	Gentilly Resilience District	Environment	New Orleans, LA
2	Paraisópolis Condomínios	Environment	São Paulo, Brazil
3	Paris OASIS	Environment	Paris, France
4	Pasadena Trails (ROC USA)	Security	Houston, TX
5	Comunidad María Auxiliadora	Security	Cochabamba, Bolivia
6	Baan Mankong	Security	Bangkok, Thailand
7	Living Cully	Livelihoods	Portland, OR
8	Yerwada In Situ Upgrading	Livelihoods	Pune, India
9	Dafen Urban Village	Livelihoods	Shenzhen, China
10	Thunder Valley Community Development Corporation	Governance	Pine Ridge Reservation, SD
11	Kibera Public Space Initiative	Governance	Kibera, Nairobi, Kenya
12	Caño Martín Peña	Governance	San Juan, Puerto Rico

change, making clear where solidarities have significantly overcome struggles and where they have not.

SCALES OF EQUITABLE RESILIENCE

In considering both impacts and responses to climate change, questions of scale are essential to judging equity. Adaptation actions formulated to protect some areas can, when viewed within a larger scalar framework, be maladaptive.[83] For instance, protecting urbanized areas from flooding can increase danger in adjacent unprotected areas.[84] Scalar interactions can be endlessly complex and difficult to discern. So, we address the question of scale in two distinct ways. First, the cases that we highlight span a range of magnitudes, from small temporary interventions to national programs for infrastructure and housing upgrading. Some cases may include multiple interventions that are part of a larger program. In each instance, we situate the case at the center of our focus within the larger temporal and scalar context, seeking to explain how the core case is emblematic of broader patterns. We indicate where there may be important points of departure, and what these patterns might tell us about processes of scaling up or applying a model across contexts.

The second way that we engage with questions of scale is through a multiscalar analytical framework for examining how each of the LEGS dimensions manifests across three scalar categories: the individual or household scale, the community or project scale, and the broader city or regional context. This three-scale analytic framework is summarized in table 0.2.

INDIVIDUAL AND HOUSEHOLD SCALE

Resilience projects are often critiqued as privatizing risk and placing the burdens of adaptation on individuals and households rather than enabling collective provisioning.[85] While we wholeheartedly agree with this critique, we also recognize that there is critical work to be done by planners, designers, and decision makers in supporting individuals and households to improve their own lives. In the realm of livelihoods, interventions can support residents in acquiring and maintaining the skills, tools, and resources necessary for stable and dignified forms of work. With respect to environmental conditions, projects can help people better understand the hazard risks and ecological dynamics of their landscapes so that they can adapt, respond, and recover more wisely. In the domain of governance, a project might support the capacity of residents to participate in individual and collective problem solving. Finally, in the sphere of security, a project can improve or harm residents' sense of personal safety, stability, and attachment to place.

From our analysis of cases, it is clear that purely individual or household-level interventions are not sufficient. Nor is individual capacity building a suitable substitute for effective collective action through governments or other institutions. Nonetheless, it is disempowering to ignore the role of individual and household level agency in shaping risk. Collective adaptation has implications not only on aggregate social groups. Rather, collective interventions can enable or foreclose upon individual agency. Projects

Table 0.2
The LEGS principles at three scales

	Individual/Household	Project/Community	City/Region
Livelihoods	**Capacity** Does the project support residents in acquiring or maintaining the skills, tools, and capital necessary for stable and dignified livelihoods?	**Accommodation** Does the project create spatial and material conditions that support residents' livelihoods, including home-based and informal work?	**Access** Does the project enable residents to access livelihoods through access to work opportunities and/or affordable transportation?
Environment	**Knowledge** Does the project improve residents' understanding of hazard and ecological dynamics?	**Protection** Does the project reduce the frequency and/or severity of environmental hazard impacts?	**Ecological Vitality** Does the project support the ecological systems on which residents rely?
Governance	**Self-Efficacy** Does the project support residents in developing individual and collective efficacy in addressing problems?	**Control** Does the project allow residents to play a meaningful role in designing interventions and/or ongoing management?	**Linkage** Does the project create pathways for residents to advocate for themselves and others with respect to larger institutions, including governments?
Security	**Stability** Does the project support or maintain residents' sense of personal safety, stability, belonging, and attachment to their place?	**Cohesion** Does the project support or maintain residents' informal security through collective cohesion to resist threats of displacement?	**Recognition** Does the project support residents in gaining formal recognition of their right to protection from displacement?

pursuing equitable resilience cannot require already disadvantaged residents to shoulder a disproportionate burden of adaptation, but nor can they ignore the capacity of all people, perhaps especially those who have been historically marginalized and exploited, to improve conditions for themselves, their families, and their neighbors.

COMMUNITY OR PROJECT SCALE

Many of the projects that we highlight enable individual- and household-level agency, but the projects all have broader aims to support or protect a specific territory or group of people. At the project or community scale, thoughtful design and management of spaces can accommodate and support residents' livelihoods, for example making space for home-based fabrication or storage for work-related materials and equipment. In the environmental dimension, a project can reduce the frequency and severity of disruptions from climate-based hazards—for instance, accommodating stormwater within a landscape or providing relief from urban heat. With respect to governance, community-scale benefits can include enabling residents to shape decisions about both the initial design of a project and its ongoing management. Finally, an intervention can support security at the community scale by building internal solidarity to resist threats of displacement.

Just as there are dangers to focusing exclusively on the individual and household scale in building equitable resilience, there can be serious problems with conceiving of planning and design interventions as discrete "projects." As a means of organizing efforts across institutions and individuals, a project framing can be clarifying, but it can also obscure important processes that happen both within and beyond the bounds of the project. "Projectification" of funding and attention can limit the effectiveness of interventions that require sustained action across broader spatial and temporal scales.[86] For these reasons, project or community-level aspirations need to be paired with attention to both smaller and larger scales of action and analysis.

BEYOND THE PROJECT

Just as a narrow focus on the project scale can obscure smaller-scale impacts on individuals and households, it can also hide important relationships beyond the project that can threaten or strengthen the impacts of an intervention. In analyzing our core cases, we consider how key interventions relate to broader physical, social, and governance processes. In the environmental dimension, an intervention can promote or harm the vitality of ecological systems on which residents and communities rely. The location of projects can enable or restrict access to livelihoods through proximity and access to employment centers and affordable transportation. Interventions can support community self-governance by creating or sustaining links to external sources of power and resources, including government agencies and decision-making processes. Finally, by bridging to scales beyond the discrete project, interventions can support the security of communities by gaining recognition from external authorities to help resist displacement.

ON TEMPORAL DYNAMISM AND CHANGE

Just as pursuing or analyzing equitable resilience requires multiscalar spatial analysis, it also requires a nuanced and complex understanding of temporal dynamism. Equitable resilience is not a stable state that can be definitively and permanently attained. Rather, equitable resilience is linked to a set of normative aims that must adjust with unstable and uncertain socio-ecological conditions. While some interventions may have positive impacts quickly, others can take many years to achieve even a small measure of their goals. Some may produce transformative impacts but prove unstable, collapsing due to pressures from within or outside of a community. Others may gain recognition and external support, lending stability. However, such externally supported staying power may undermine initial radical or transformative practices. Among the case studies that we highlight, there are instances of all of these temporal dynamics—and more. In our narratives and visualizations, we treat core case studies not as abstract or idealized projects but rather as complex, continuously unfolding, variegated, and contingent initiatives that may stabilize, grow, evolve, degrade, or collapse. In some cases, a single project may simultaneously make great strides in improving conditions on one dimension while losing ground on other fronts. Seen collectively, these solidarities and struggles reveal the possibilities for what the equitably resilient city can provide.

I.
ENVIRONMENT

EQUITABLY RESILIENT ENVIRONMENTS: KNOWLEDGE, PROTECTION, AND ECOLOGICAL VITALITY

How can the negative impacts of environmental hazards on disadvantaged urban residents be ameliorated? In this section, we discuss how uneven vulnerabilities are produced—and how planning and design interventions can reduce these vulnerabilities. These chapters also make clear that while reducing environmental harms is necessary, it is not sufficient to advance truly equitable resilience.

As already noted, climate and hazards researchers often view vulnerability in terms of exposure to hazards, sensitivity to disruption, and adaptive capacity to prepare, withstand, and recover.[1] Low-income people face distinct burdens across all three of these dimensions yet also frequently develop creative measures to cope with vulnerability. From hillside favelas in Rio to mobile home parks in river floodplains in Vermont, disadvantaged people are more likely to live in hazard-exposed landscapes as they seek affordable accommodation and are pushed out of safer areas. They are disproportionally sensitive to disruption from hazards due to physical factors such as substandard housing and inadequate infrastructure as well as social challenges, including linguistic, education, and disability barriers.[2] Finally, poverty can reduce adaptive capacity by limiting access to the resources, information, and skills necessary to adapt and recover.[3] Vulnerability to climate change and other threats is shaped not just by poverty but also by structural disadvantages rooted in race, gender, religion, and other factors that can restrict where and how people live.[4]

Researchers and practitioners have long recognized that poverty and other social disadvantages can make people more vulnerable to hazards, but all too often, efforts to promote urban resilience have actually aggravated conditions for disadvantaged groups.[5] Much as the environmental justice movement has exposed the radically uneven impacts of toxic pollution,[6] efforts to adapt settlements to emerging climate threats must reckon with the uneven impacts of both climate hazards themselves and attempts to reduce hazard risks.

PLANNING, DESIGN, AND URBAN ENVIRONMENTAL HAZARDS

Centuries before planners and designers used the label of "resilience," their predecessors considered the relationship between settlements and underlying landscape and climate processes. Ancient Greek philosopher Aristotle[7] and Roman architectural theorist Vitruvius[8] recommended rules for orienting cities and houses to optimize wind and solar exposure for comfort and health. Cities in the Ming dynasty of China and in precolonial Mexico integrated extensive infrastructures of canals, lakes, and earthworks to manage water for flood control, irrigation, and defense against invasion.[9] After Spanish colonial forces colonized much of the Western hemisphere, King Philip II issued the Laws of the Indies in 1573, codifying standards for settler colonial town construction, including recommendations on settlement siting, street widths and orientation, and other factors that shaped the comfort and health of residents.[10]

Meanwhile, much city making in Europe embraced abstract geometries, responding not to specific climatic and landscape processes but rather to purported Enlightenment principles of universal rational spatial order— what Spiro Kostof terms "the grand manner."[11] From the gridded settlements in the New World to designed capitals from New Delhi (India) to Brasília (Brazil) to Astana (Kazakhstan), urbanization patterns frequently ignored ecosystems.[12] In other cities too, advances in engineering and construction enabled urban expansion in response to colonial settlement and industrialization. These urban extensions, enabled by "Promethean projects" of infrastructural modernization, such as dams, aqueducts, levees, and land "reclamation" projects, allowed settlements to spread into areas that had previously been deemed inhospitable to inhabitation.[13] Whether inspired by City Beautiful neo-classicism, Garden City neo-traditionalism, or International Modernism's formal austerity, urban expansions from the late nineteenth century through the mid-twentieth century frequently relied on interventionist engineering to overcome constraints rather than adjusting to dynamic landscape and climate processes.

Though dominant strains of urban planning and design in the nineteenth and twentieth centuries paid little heed to landscape and climatic processes, a parallel lineage of thinkers and designers sought a more accommodating path. Pioneering landscape architect Frederick Law Olmsted argued that urban green spaces could perform infrastructural functions such as channeling stormwater while offering other benefits, including improved health and social uplift across barriers of class and immigration.[14] The Valley Section proposed by Scottish proto-planner Patrick Geddes related landscape formations to the social, political, and spatial order of human settlements.[15] Another Scot, Ian McHarg, transformed Geddes's abstract notions into operational strategies to guide urbanization through "designing with nature."[16] McHarg developed methods for multifaceted "suitability analysis" that layered features including slope, hydrology, soil type, and vegetation to determine where and how settlements should take shape to minimize environmental harm and reduce vulnerability to hazards such as flooding. Subsequent writing reinforced McHarg's idea that landscape and climate should be regarded as dynamic and interlinked processes influencing city development.[17] Kevin Lynch's theory of "good city form" included principles of "vitality," or ecological responsiveness, and "fit," or adaptability to changing conditions.[18] These same principles of ecological responsiveness and adaptability later became central to "landscape urbanism," an influential movement in the early twenty-first century.[19]

The growing centrality of landscape and climate processes to urban design and planning holds both promise and peril for equitable resilience. Increasingly volatile environmental conditions demand that urban settlements be better adapted to landscape and climate conditions to safeguard lives, property, and infrastructure. However, when design strategies are driven by narrow technocratic understandings of environmental conditions, they miss the socio-ecological complexity of hazard risk. Planners may well understand where and how settlements *should* be built to minimize hazard risks, but forging politically viable and equitable strategies for achieving such settlement patterns has proven more difficult.[20]

Addressing the environmental dimension of equitable resilience demands attention to the structural drivers of uneven risk. Everyday inequalities differentiate and deepen social vulnerabilities. Acting equitably entails interrogating the political ecology of hazards—questioning how particular conceptualizations of risk and adaptation harm disadvantaged people.[21]

CLIMATE CRISES AND CITIES

Climate change matters for the design and planning of cities because urban form shapes both greenhouse gas emissions and the distribution of hazard risk.[22] Urban and rural settlements increasingly face extreme heat and cold, drought, fire, and floods. Areas once considered safe now face regular threats. Seasonal patterns shift, rendering monsoon rains unpredictable across South Asia, extending "fire season" in the American West and lengthening "hurricane season" in warmer Atlantic waters.[23] People in every region face threats from climate change, but residents of less wealthy nations are especially vulnerable. The World Meteorological Organization estimates that, between 1970 and 2019, 91 percent of deaths from weather, climate, and water events occurred in countries categorized as "developing economies" by the United Nations.[24]

EXTREME TEMPERATURES

The processes once called "global warming" now get cast as "climate change." Although it is now clear that climate change encompasses many complex threats to settlements beyond "warming," extreme heat can have serious risks, damaging infrastructure, disrupting food supplies, and impacting human health in multiple ways. Heat kills more Americans than all other weather-related causes and increasingly severe heat events have been particularly harmful in South Asia.[25] Social scientists have long recognized the dire and uneven impacts of heat waves on urban poor residents.[26] Yet, heat is often referred to as an "invisible" killer because its impacts are less visible than other climate threats such as coastal storms.[27] Extreme heat may be visually unspectacular, but its long-term negative impact perpetrates what Rob Nixon calls "slow violence" against the poor.[28] The impacts of extreme heat vary enormously both because of physiological differences related to factors such as age, chronic disease, and pregnancy and because people with more resources can more easily take adaptive action, using air conditioning or relocating temporarily.[29] Low-income populations and people of color are also more exposed to the urban heat island (UHI) effect, a phenomenon whereby local temperatures are elevated in areas with extensive pavement, heat-exhausting buildings and vehicles, and less shading and evaporative cooling from trees and open spaces. Studies have found that low-income communities of color face conditions more than 10°F (5.5°C) warmer than surrounding areas.[30] Excessive heat is also associated with housing serving disadvantaged populations. For instance, subsidized housing in California is more likely to be located in areas that experience extreme heat.[31] Residents of manufactured housing or mobile homes are especially vulnerable to extreme heat in many areas of the US.[32]

DROUGHT AND WATER STRESS

Extreme heat is often related to drought. Although drought risk is frequently associated with rural agricultural communities, it can also threaten urban water supplies and imperil urban ecosystems. Despite decades of aggressive water infrastructure development,[33] communities across the American West face water shortages, spurring restrictions on new development, and drastic water-saving practices.[34] In 2018, the city of Cape Town, South Africa, nearly ran out of water after years of prolonged drought.[35] The impacts of water stress are unequal, even within the same urban region. While social movements in the Global North and South have fought for the recognition of a "right to water,"[36] water has become increasingly privatized and commoditized under neoliberal urban governance regimes, leading to rationing and limited access for the poor.[37]

WILDFIRES

Increasing heat and drought contribute to more frequent and severe wildfires in many regions, including the American West, Canada, Australia,

Southern Europe, and Central Africa. Fire is essential to many ecosystems, but extended hot, dry conditions along with decades of fire suppression and increased settlement in the wildland-urban interface have put more people, property, and infrastructure at risk.[38] Recent studies estimate annual costs of wildfire preparedness, suppression, evacuation, and loss in the US at between US$71 and US$341 billion.[39] In the past, fire adaptation and recovery has marginalized low-income communities while facilitating continued settlement and exploitation of resources by powerful groups.[40] Manufactured housing or mobile home communities, an important source of affordable housing in many areas across the American West, are also disproportionately located in fire-vulnerable areas, leading to devastating losses such as those in the community of Paradise, California, during massive wildfires in 2018.[41] In some settings, fires can disproportionately impact the wealthy,[42] but fire also poses serious risks in informal settlements, refugee camps, and homeless encampments, with their extreme density, unregulated construction, limited access, and inadequate infrastructure. The uneven harms of wildfires spread beyond areas threatened with destruction, blanketing whole regions in smoke for extended periods, posing serious health risks, especially for sensitive groups and people without the means to avoid smoke through air filtration and other adaptations.[43]

FLOODING AND STORMS

Flooding and storms are among the most dramatic climate change risks. Like wildfires, increasing flood risk is a function of several factors, including changing climate conditions, increasing settlement in hazard-prone areas, and generations of unwise risk mitigation practices that have reduced the frequency of small floods and amplified the severity of less frequent major ones. Climate change increases flood risks across three categories: pluvial or rain-driven flooding, fluvial or river flooding, and coastal flooding associated with sea level rise and storms.

Extreme rain events are increasingly inundating cities across the globe. In the summer of 2011, nearly six inches of rain fell on temperate Copenhagen in two hours. In the summer of 2021, the eastern Chinese city of Zhengzhou saw nearly eight inches of rain in one hour, trapping drivers and subway riders in flooded tunnels.[44] Climate change is widely blamed for increasing the severity and frequency of such cloudburst events as warmer air transports and deposits enormous quantities of rain in settlements around the world.[45] Heavy rain falling on impervious urban landscapes of pavement and buildings overwhelms storm sewers, canals, and drainage pumps. Such pluvial floods cause disproportionate harm in low-income neighborhoods which are frequently located in low-lying terrain and in areas with inadequate drainage infrastructure. In April 2022, floods from torrential rainstorms around Durban, South Africa, killed at least 448 people, mostly in informal settlements in "locations unsuitable for housing" and yet intensely settled by desperate people in search of opportunity.[46]

In settlements around the world, neighborhoods with topographically descriptive names such as "bottom," "flat," and "hollow"[47] were built by forcing waterways into culverts and sewer pipes to make way for housing for lower-income people.[48] In many cities, aging infrastructure can no longer accommodate intensifying rainfall, leaving communities at enormous risk. The torrential rains associated with Hurricane Ida that flooded New York City neighborhoods in 2021 killed eleven people in basement apartments, demonstrating the confluence of climate change–driven flooding and inadequate housing.[49]

As climate change drives heavier rain and faster snow melt, river flooding is also intensifying. Climate change shifts the timing and volume of stream flows, overwhelming existing levees and other defenses built to keep waterways away from settlements. Interventionist engineering projects undertaken over the course of the nineteenth and twentieth centuries have left urban settlements from Kolkata to New Orleans to Rotterdam at tremendous risk as flood levels rise. To make floodplains and river deltas available for settlement, planners, engineers, and developers diked, filled, pumped, and drained lowlands, confining waters to ever-narrower territories, which can no longer contain floods.[50] Like the lowlands inundated by pluvial flooding, fluvial floods threaten floodplain settlements that are disproportionately home to disadvantaged people. In the US, manufactured home parks are more likely than other settlement types to be located in floodplains.[51] A recent study found that at least 9 percent of public housing units in the US are in designated floodplains.[52] The impacts of river flooding fall unevenly across societies. Research from Bangladesh suggests that low-income women face especially heavy burdens from flooding.[53]

Along with the threat of flooding from heavy rains and rivers, climate change brings additional risks from sea level rise and coastal storms. Low-lying coastal zones are among the most urbanized regions of the planet. These areas are especially vulnerable as sea levels rise. Projections from the US National Climate Assessment indicate that sea level rise could reach as much as 8.2 feet by 2100,[54] while the IPCC estimates that ice-sheet melting could lead to more than sixteen feet of global mean sea level rise.[55] Many low-lying coastal cities already experience "sunny day flooding" even without rain or storms.[56] Sea level rise exacerbates flooding from heavy rains by reducing the drainage capacity of existing infrastructure networks and raising groundwater levels.[57] Sea level rise is projected to displace some thirteen million people within the US alone by the end of the twenty-first century.[58] People often think of urbanized coasts and waterfronts as privileged landscapes for the wealthy. Yet, many areas at extreme risk from coastal flooding are home to low-income people, from waterfront informal settlements in Lagos to Shenzhen's urban villages to public housing in New Orleans, Norfolk, and New York City.

Mounting climate change–driven flooding has brought increased scrutiny to existing risk-mitigation strategies, including flood insurance,[59] and encouraged more comprehensive approaches that integrate infrastructure, land use and building regulation, and other methods to reduce risk.[60]

ENVIRONMENTALLY EQUITABLE RESILIENCE ACROSS SCALES

Planning and design for equitable resilience must confront the environmental dimensions of climate change and other threats across scales. In our assessment, an intervention advances the environmental dimension of equitable resilience if it improves conditions for disadvantaged people at three linked scales: (1) community or project scale—reducing the frequency or severity of disruptions from hazards; (2) city or regional scale—improving the vitality of ecological systems on which settlements rely; and/or (3) individual or household scale—increasing residents' actionable knowledge with respect to the landscape and climatic processes that underlie hazard risk.

This scalar analysis is not meant to provide a checklist or sequenced hierarchy. An intervention can make significant contributions to equitable resilience without being attentive to all three scales.

PROTECTION: ENVIRONMENTAL RESILIENCE AT THE
SCALE OF "THE PROJECT"

Hazard mitigation in housing, infrastructure, and settlement form has long been the core focus of urban resilience and disaster planning. Unfortunately, if equity is not a central focus, such efforts can deepen existing inequalities. Therefore, interventions aimed at improving equitable resilience should reduce the frequency or severity of hazard disruptions for disadvantaged people.

Protective Housing   New and retrofitted housing can reduce hazard sensitivity by making buildings more resistant to damage from hazards. From the Great Chicago Fire of 1871 to Hurricane Andrew's devastation of South Florida in 1992 to the Champlain Towers condominium collapse in Surfside, Florida, in 2021, urban disasters often spur new building regulations meant to reduce the susceptibility of buildings to shocks.[61] Hazard mitigation–focused building codes can prohibit flammable construction in fire-prone settings, require elevated floors and equipment in flood zones, or mandate reinforced structures in high wind zones. While mitigating hazards for housing can reduce harms to people and property, these efforts can also increase building costs, contributing to gentrification and displacement.[62]

Protective Infrastructure   Many efforts to protect against climate hazards include upgrading infrastructure. Infrastructural protections can include "gray" infrastructures such as levees to hold back floodwaters and reservoirs to cope with drought. In response to problems with conventional gray infrastructure, many climate adaptation initiatives call for alternative and hybrid approaches that go by many names, including "green infrastructure," "nature-based solutions," and "ecosystem-based adaptation."[63] These approaches use landscapes and ecosystems to reduce hazard risk while delivering other benefits. Popular strategies include stormwater-retaining bioswales, coastal wetland restoration to buffer against storm surge, and street trees to reduce UHI effects.

Cities around the world are experimenting with new hazard-mitigation infrastructures to cope with climate change threats. Flood mitigation has been a particular area of focus. The East Side Coastal Resiliency Project in Manhattan combines gray and green infrastructure to protect adjacent public housing and neighborhoods from flooding.[64] The Benthemplein water square in Rotterdam and Taasinge Square "cloud burst park" in Copenhagen each retrofitted existing open spaces to manage stormwater while creating recreational and ecological amenities.[65]

Infrastructural protections are often essential to urbanization, but both gray and green infrastructure can also create uneven impacts. For example, the devastation of New Orleans following the failure of the city's flood defenses during Hurricane Katrina dramatically illustrates the "levee effect,"[66] whereby flood protections create an unwarranted sense of security, encouraging risk-blind settlement that places more people and property at risk when infrastructures are overwhelmed. Structural flood defenses such as levees do not eliminate flood damage; they simply shift flooding from protected to unprotected areas.[67] New protective infrastructure can have other inequitable impacts, including direct displacement for construction or "green gentrification," when investments in green infrastructure lead to rising property values.[68]

From the Eko Atlantic infill off Lagos, Nigeria,[69] to the "Great Garuda" plan for Jakarta, Indonesia,[70] some cities facing grave climate vulnerability have proposed to fund adaptation infrastructure by building new urban districts on "reclaimed" land at the edges of existing urban areas. Less grandiose versions of such development-driven resilience proliferate as leaders in climate-vulnerable places look to new luxury real estate development to increase tax revenues and, in some cases, provide physical barriers to protect existing settlements. A developer of new high-end waterfront housing in Boston alluded to this dynamic, saying, "large scale waterfront developments are becoming the first lines of defense for protecting low-lying and vulnerable communities."[71] Using new urban development to fund protective infrastructure for existing settlements is not inherently inequitable, but it does raise concerns that these new projects could lead to inequitable conditions in which a privileged few enjoy relative environmental security, while less fortunate groups suffer under worsening climate chaos.

Protective Settlement Patterns   If conventional protective infrastructure such as floodwalls are meant to keep hazards away from people and property, regulating settlement patterns and land use through mechanisms such as zoning and floodplain development restrictions reduces risk by keeping people away from hazards.[72] These tools can help guide development away from climate hazards, but they do little to address growing climate vulnerabilities in existing settlements. Worse, realigning settlement patterns in response to climate change can deepen existing inequalities. As residents with resources move out of harm's way, they may lay claim to less hazard-exposed lower-income neighborhoods, leading to climate gentrification.[73]

The inequity and inefficiency of unplanned relocation has encouraged interest in "managed retreat."[74] Here, too, early evidence suggests that climate change–driven settlement realignment can replicate the inequities of earlier mass relocations, from "slum clearance" under urban renewal to community relocations for dams, highways, and other infrastructures.[75] Studies of relocation in vulnerable neighborhoods, from post-Katrina New Orleans[76] to post-Sandy New York[77] to coastal Alaska,[78] suggest that people with more resources, power, and privilege are better positioned to take advantage of government programs than poorer residents. Moreover, even when disadvantaged communities do voluntarily use buyout programs to leave vulnerable sites, it can be hard for them to find safe, affordable, and welcoming "receiving communities."[79]

Leaders often justify displacement in the name of building green infrastructure using the rationale that lowland territories are not suitable for habitation. The infamous "green dot" proposal for post-Katrina New Orleans favored new water-retention parkland over older Black communities[80]; São Paulo's Tietê River Valley Park displaced forty thousand poor people[81]; and the construction of Dhaka's Hatirjheel stormwater retention lakes displaced entire informal settlements.[82] When disadvantaged people are removed to make way for green infrastructure, claims of resilience fall short of *equitable* resilience.

ECOLOGICAL VITALITY: ENVIRONMENTAL RESILIENCE AT LARGER SCALES

Beyond the scale of individual projects, planning and design interventions can promote environmental resilience by restoring or retaining the function of the ecological systems on which urban settlements rely. In opposition to various place-based and Indigenous forms of environmental knowledge,

dominant strains of planning and design have long treated urban settlements as separate from "natural" landscape and climate processes.[83] Now,
however, dramatic climate disturbances demonstrate the extent to which
human settlements rely on the health of wider environmental systems.
Deforestation of hillslopes can cause flash floods and mudslides. Destruction of wetlands exposes coastal settlements to storm surge. Climate
change places additional stresses on the environmental systems necessary
for human settlement. Urban hazard-mitigation interventions can have
negative or positive impacts on ecosystem function, restoring hydrological
flows or impeding them, enriching biodiversity or further impoverishing
ecosystems.

Examples of hazard-mitigation efforts that harm ecosystems abound.
Levee construction, stream channelization, and culverting can reduce
nuisance flooding, but these strategies treat waterways as purely technical systems and destroy ecosystem functions both within streams and in
adjoining landscapes. As they contemplate adaption to rising sea levels,
leaders in coastal US cities from Boston to San Francisco to New Orleans
have all considered building massive coastal barriers to regulate the flow
of water in and out of nearby harbors, bays, and lakes. The enormous
system of dikes and dams constructed by the Dutch Delta Works program
after deadly flooding in 1953 demonstrates the efficacy of such efforts in
reducing flooding. Unfortunately, it also demonstrates the devastating
ecological costs, destroying coastal estuaries and riverine ecosystems.[84]

Increasingly, designers and planners seek to mitigate climate hazards
through projects that support ecological functioning. Such "nature-based
solutions" include a wide range of infrastructures,[85] including mangrove
restoration, coastal marsh plantation, and living oyster reefs to dampen
the impacts of coastal storms. Proposals in the San Francisco Bay Area call
for "horizontal levees," broad shallow-sloped earthen berms that would
block coastal flooding while allowing ecological communities to adapt
to sea level rise along the slopes gradually.[86] In deltaic settings, including
coastal Bangladesh, the Netherlands, and South Louisiana, pilot projects
have sought to undo the damage done by earlier generations of levees
through "depoldering," piercing, or relocating levees to make "room
for the river" and restore connections to surrounding landscapes.[87] For
decades, designers and planners in cities from Berkeley to Zürich to Seoul

have opportunistically "daylit" urban creeks, removing underground pipes and restoring streambeds.[88] Such efforts can reduce flood risk while improving ecosystem functions, biodiversity, and water quality.[89]

Designers have implemented landscape-based green infrastructures in urban regions around the world, often featuring permeable paving and bio-swales to capture, retain, and infiltrate stormwater into soils rather than directing water as quickly as possible into storm sewers. From Philadelphia's "Green City, Clean Waters" project[90] to China's national "Sponge City" initiative,[91] leaders increasingly embrace alternative approaches to urban stormwater management that deliver other benefits, including aesthetic, ecological, and recreational improvements. While nature-based solutions and green infrastructure projects frequently focus on managing urban stormwater, other efforts—including the Paris OASIS (see case 3) and the Barcelona Superilla (Superblock) initiatives[92]—highlight the potential of landscape-based strategies to mitigate extreme heat while delivering other benefits to humans and other species. Ensuring that the least advantaged people benefit from these efforts is a core challenge of equitable resilience.

LANDSCAPE AND HAZARD LITERACY FOR INDIVIDUALS AND HOUSEHOLDS

Project-scale hazard mitigation and broader ecological vitality are crucial aspects of equitable resilience, but intervention at the scale of individual households or even a single person can also contribute to equitable resilience. At these smaller scales, we focus on how interventions can improve landscape and hazard literacy, increasing residents' understanding of hazard-generating processes and enabling them to make informed decisions about where to live, how to build, and how to mitigate risks. Focusing on individual knowledge and action can be controversial. Much "critical resilience" research has rightfully highlighted the danger of treating hazard resilience as an individual responsibility.[93] However, it can also be danger-ous and disempowering to focus exclusively on structural drivers of vulner-ability, ignoring the agency of individuals and households to shape their own responses to climate risks.[94]

Designers and planners can support "landscape literacy"[95] through "eco-revelatory design,"[96] helping residents make more informed decisions with respect to hazards. For example, Anne Whiston Spirn's long-running West Philadelphia Landscape Project paired local public-school students with landscape architecture graduate students to explore the ongoing legacy of buried floodplains in the low-income neighborhood near their school.[97] Several recent projects use similar strategies to teach children and others about climate and landscape risks, including through school-based programs such as Ripple Effect in New Orleans and the East Harlem Resilience Study in New York City.[98] Aimed at broader publics, the Elevate San Rafael project created for the Resilient by Design competition in the Bay Area of California included a mobile flood education truck.[99]

Other efforts to improve landscape and climate literacy are proliferating in urban regions around the world. The Mahila Housing Trust trains "climate saathis," low-income women from informal settlements in India, to understand climate change vulnerability and intervene through strategies such as cool roofs and rainwater harvesting.[100] Citizen science initiatives aim not only to improve hazard literacy, but also to enroll vulnerable communities in gathering and sharing experiential knowledge of climate impacts. Examples include ISeeChange.org,[101] the West Oakland Environmental Indicators Project,[102] and Peta Jakarta, a crowdsourced flood-reporting platform in Indonesia.[103] These efforts are rooted in long-standing calls from natural hazards researchers to communicate hazard risks more effectively before, during, and after extreme events.[104]

Table I.1 brings together the three scalar dimensions of equitable environmental resilience, pairing key principles with central questions and examples.

PARTIAL SUCCESSES IN EQUITABLE ENVIRONMENTAL RESILIENCE

Improving individual- and household-level landscape and hazard literacy can counteract the increasing alienation of people from landscape processes and improve hazard-mitigation decision making, but such efforts must be paired with collective actions at broader scales. Focusing solely on improving individual knowledge in the hope that people will make rational risk-mitigation decisions does not acknowledge historically embedded

Table I.1

Three scales of equitable environmental resilience

Scale	Equitable Resilience Principle	Question	Examples
Individual/ Household	Literacy	Does the intervention help residents to understand hazard and ecological dynamics?	West Philadelphia Landscape Project; Ripple Effect; Peta Jakarta
Project/ Community	Protection	Does the intervention reduce the frequency or severity of hazard impacts on urban poor residents?	East Side Coastal Resiliency Project; Post–Hurricane Andrew Building Code Revisions; Benthemplein Water Square
City/Region	Ecological Vitality	Does the intervention support the ecological systems on which urban settlements rely?	Urban stream daylighting; "Room for the River"; Green City, Clean Waters

inequalities that restrict the realms of choice. For poor people around the world, the choices of where to live, how to mitigate hazard risks, and when to adapt or move are shaped more by economic, political, and social structures than by individual or household-level knowledge. It is in this sense that they are *vulnerabilized*, not just vulnerable.

The following three case studies illustrate the promise and challenges of intervening to improve the environmental aspects of equitable resilience. The Gentilly Resilience District in New Orleans (case 1) is an ambitious effort to transform public and private space to accommodate stormwater better. While the project has seen some success in small-scale stormwater management, implementation of many of its most ambitious elements has been delayed, hampered by political transitions and questions about their effectiveness at reducing flood risk. Next, we investigate the Paraisópolis Condomínios (case 2), a series of multifamily housing developments serving about a thousand families relocated from the immediately adjacent Paraisópolis favela in São Paulo, Brazil. While partially successful, the Condomínios have also suffered from delays and ongoing struggles to regularize the housing tenure of residents, as well as insufficient scale to meet larger demand. Finally, the Paris OASIS initiative (case 3) is a pilot project transforming paved schoolyards in the French capital for experiential environmental education and community heat resilience. While the initial projects show commendable promise, questions remain regarding their impacts and attempts to scale up.

Case 1

GENTILLY RESILIENCE DISTRICT: AMBITIOUS AIMS AND POLITICAL PERILS IN POST-KATRINA NEW ORLEANS

OVERVIEW

Ten years after the flooding following Hurricane Katrina devastated the New Orleans neighborhood of Gentilly, the city won a federal grant for a suite of projects demonstrating a new approach to "living with water," including extensive investments in green infrastructure, public arts, and job training. The Gentilly Resilience District (GRD) proposal aimed to advance equitable resilience by reducing flood vulnerability in a middle- and working-class Black-majority neighborhood and creating a new green infrastructure workforce to ensure that residents benefit from the investments. Eight years after the initial award, few GRD projects had started construction, delayed by shifting leadership and COVID-19. Drawing on site visits as well as interviews with participants and critics, we assess the GRD's ambitious agenda and implementation struggles, revealing lessons about the possibilities and limitations of competition-driven, top-down infrastructure projects.

INTRODUCTION: KATRINA IN GENTILLY

Gentilly is a large neighborhood, spread across nearly nine square miles in north-central New Orleans (figure 1.2). The area that is now Gentilly

1.1 Rendering of green-blue street infrastructure to be completed as part of the Gentilly Resilience District. *Source*: Stantec.

1.2 Gentilly Resilience District location within New Orleans. *Source*: Mora Orensanz. 1. University of New Orleans; 2. Dillard University; 3. City Park; 4. London Avenue Canal; 5. Audubon Park; 6. French Quarter; 7. Central Business District; 8. Industrial Canal; 9. Louis Armstrong International Airport.

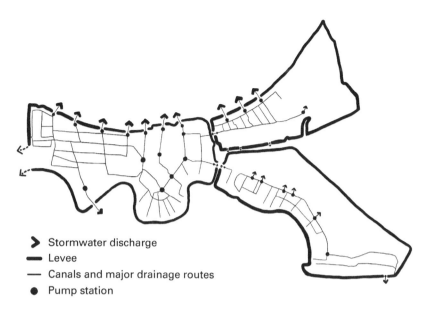

> Stormwater discharge
━ Levee
— Canals and major drainage routes
● Pump station

1.3 New Orleans's urbanization is dependent on a network of levees, floodwalls, canals, and pumps. *Source*: Mora Orensanz.

was once swampland dominated by cypress and tupelo trees. Gentilly's lands were drained and subdivided, its streets were laid out, and most of its buildings were built in the middle decades of the twentieth century after the wetlands were "reclaimed" by the Orleans Levee Board and the New Orleans Sewerage and Water Board with their levees, pumps, canals, and storm drains.[1] These infrastructures converted the flooded forest into relatively dry land for urban expansion, accommodating New Orleans's growing population and its appetite for suburban-style residential neighborhoods (figure 1.3). While the popular image of New Orleans features shotgun houses, wrought-iron balconies, and antebellum mansions, Gentilly is characterized by modest craftsman bungalows and brick ranch houses. Gentilly is largely a middle- and working-class neighborhood. While it is a majority-Black neighborhood, Gentilly's sub-neighborhoods are racially and economically diverse, ranging from Pontchartrain Park in the east, which is more than 95 percent Black, with nearly 25 percent living in poverty, to Lake Terrace and Lake Oaks in the north, which are 56 percent white, with only 10 percent living in poverty.[2]

When New Orleans's levees and floodwalls collapsed during Hurricane Katrina in August 2005, relatively small topographic differences largely

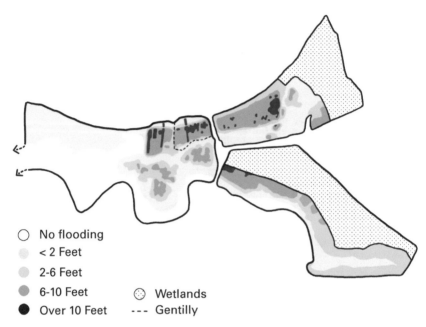

○ No flooding
  < 2 Feet
  2-6 Feet
● 6-10 Feet   ⊙ Wetlands
● Over 10 Feet   --- Gentilly

1.4 Flooding following Hurricane Katrina was most extreme in low-lying neighborhoods such as Gentilly. *Source*: Mora Orensanz.

determined flood depths. The iconic French Quarter and other old neighborhoods on the relative high ground along the Mississippi River were largely spared, but Gentilly and other newer low-lying neighborhoods sat under as much as twelve feet of flood water for more than two weeks (figure 1.4).[3] Shaped by inequalities of wealth, privilege, and political power, recovery from the floods was profoundly uneven. Neighborhoods that were submerged beneath the same deep murky waters faced divergent reconstruction processes.[4] When the Bring New Orleans Back Commission released the first official planning proposal after Katrina in January 2006, now-infamous maps showed green dots over several heavily flooded neighborhoods, indicating places where residential areas would be replaced with stormwater-absorbing green space. While Gentilly and several other majority-Black neighborhoods were green dotted, on the other side of City Park, Lakeview, a wealthier and whiter neighborhood that was similarly devastated by the floods, was to be spared. The so-called green dot map was quickly scrapped after overwhelming resident opposition.[5]

Although they escaped the displacement envisioned in the green dot map, residents in Gentilly and other Black-majority neighborhoods continued to suffer from structural disadvantages as they rebuilt their homes and lives. The Louisiana Road Home program and other state and federal reconstruction programs stacked the deck against neighborhoods such as Gentilly, providing more support for homeowners than renters and greater compensation in areas with high property values. While post-disaster construction costs made rebuilding everywhere more expensive, residents in Gentilly received less reconstruction aid than those in Lakeview because the "pre-storm value" of their homes was lower. Many residents pulled together private funds, insurance settlements, and public and philanthropic support to rebuild. Nonprofit housing developers built hundreds of houses for low- and middle-income people in the neighborhood. And yet, many of those new and rehabbed homes sat next to vacant lots covered in head-high weeds. Gentilly's population in 2019 was 13 percent lower than in 2000 before Hurricane Katrina.

In 2015, the City of New Orleans submitted an ambitious proposal to the National Disaster Resilience Competition (NDRC), sponsored by the Department of Housing and Urban Development and the Rockefeller Foundation. The city received US$141 million to implement what they called the Gentilly Resilience District. The GRD aimed to demonstrate a different approach to managing stormwater and flood risk. Where New Orleans had, since the early eighteenth century, adopted an adversarial approach to its watery landscape, keeping water away from settlements with levees, floodwalls, drainage pumps, pipes, and canals,[6] the GRD promised an approach described as "living with water."[7]

The GRD emerged from sustained efforts over the decade following Hurricane Katrina. With the threat of flooding in New Orleans growing due to intensifying storms, sea level rise, and coastal wetland loss, leaders from local government, advocacy organizations, and the design and planning community conducted a series of visioning processes that aimed to reorient the city's relationship to water to avoid future flooding catastrophes. Designers and planners from local firms such as Waggonner and Ball and Dana Brown and Associates championed collaborations with Dutch flooding experts and designers who were engaged in parallel discussions about reimagining levee-protected settlements in the Netherlands. First, in a series of "Dutch Dialogues" starting in 2008,[8] and later, in the multivolume Greater New Orleans Urban Water Plan,[9] these groups laid out a vision for accommodating water in the city. Rather than pumping rainwater out of the city's bowl-shaped levee-protected areas as quickly as possible, the plans called for slowing stormwater to reduce pressure on New Orleans's aging drainage infrastructure. This approach is also intended to slow subsidence, which occurs when moisture-laden soils, dried through pumping, become compacted, causing the ground to sink, compounding flooding threats.

Chastened by the popular rejection of the green dot map, local officials and designers changed tack in these later green infrastructure proposals.[10] Rather than calling to replace neighborhoods with green infrastructure, a prospect that critics had labeled "ethnic cleansing,"[11] the GRD called for accommodating stormwater in public parks, street rights-of-way, and vacant land.

After a decade of post-Katrina planning, the GRD is the first large-scale attempt to realize this "living with water" vision in New Orleans. While many of its most ambitious aims have yet to be implemented, the unfolding case of the GRD offers insights for planners and designers pursuing equitable resilience.

GENTILLY RESILIENCE DISTRICT AND EQUITABLE RESILIENCE

The GRD proposal includes an ambitious suite of projects and programs, most of which are focused on accommodating stormwater within the neighborhood's built environment (figure 1.5). The GRD proposal aims to ensure that project benefits are broadly shared and benefit New Orleans's low-income residents. As an effort to pursue equitable resilience, the GRD primarily focuses on improving environmental conditions with less robust attention to other dimensions.

ENVIRONMENT: REINTEGRATING WATER INTO A DELTA CITY NEIGHBORHOOD

In the words of one city official, the GRD is focused on "transforming water from a threat into an asset in the public realm."[12] Most of the projects funded by the NDRC grant focus on retrofitting Gentilly landscapes to

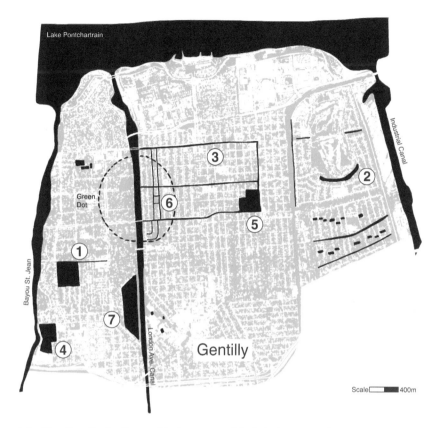

1.5 The Gentilly Resilience District proposal included green infrastructure projects throughout the neighborhood. *Source*: Mora Orensanz. 1. Mirabeau Water Garden; 2. Pontilly Neighborhood Stormwater Network; 3. Blue and Green Corridors; 4. St. Bernard Neighborhood Campus; 5. Milne Campus; 6. St. Anthony Green Streets; 7. Dillard Wetlands.

retain stormwater to reduce pressure on the city's drainage pumps, reduce flooding, and slow subsidence. The GRD includes projects to advance flood resilience across our three scales of analysis.

MITIGATING FLOOD RISK AT THE PROJECT SCALE

Nine of the twelve core components of the GRD focus on stormwater management through landscape interventions. Projects vary in scale and approach but are unified in slowing the path of rainwater to the city's pumps and delivering a range of recreational, mobility, and environmental co-benefits.

The GRD includes proposals for new large open spaces designed to retain water, in some cases creating flooded forests that echo the area's precolonial landscapes. Among these new parks are the Mirabeau Water Garden, a twenty-five-acre parcel formerly owned by a Catholic religious order, and the Dillard Wetlands, a twenty-seven-acre forest tucked between the London Avenue Canal and Dillard University, a historically Black university. Other projects will integrate stormwater retention into more intensively used facilities, including the St. Bernard Neighborhood Campus, which will capture and store rainwater under refurbished playing fields. Other GRD projects will create new public spaces with stormwater retention on city-owned canal and street corridors. The Pontilly Neighborhood Stormwater Network will build on previous efforts funded by the Federal Emergency Management Agency to create multifunction public spaces along the Dwyer Canal serving residents of Gentilly Woods and Pontchartrain Park, one of the first suburban neighborhoods in the US built for a growing Black middle class in the mid-twentieth century.[13] The GRD's Blue and Green Streets initiative will transform eight miles of street right-of-way in Gentilly with daylighted drainage canals and other features to retain and convey stormwater visibly, along with traffic calming, pedestrian and bike infrastructure improvements, and new playgrounds and a community center (figure 1.6). In addition to these projects on publicly owned land, the GRD includes US$5 million for the New Orleans Redevelopment Authority's (NORA) Community Adaptation Program (CAP), which provided incentives for low- and middle-income residents to capture stormwater on private lots. The program provided grants up to US$25,000 to more than two hundred households for interventions such as bioswales, rain gardens, and de-paving.

DESIGN FOR WATER AND FLOOD LITERACY AT THE HOUSEHOLD AND INDIVIDUAL SCALE

The GRD calls for reintegrating water into the fabric of Gentilly, both for functional reasons (slowing surface water to relieve pressure on pumps and reduce subsidence) and to shift how residents relate to their watery landscape.

Several GRD projects emphasize public pedagogy, environmental literacy, and community outreach. GRD projects include visible surface water

1.6 The GRD aims to shift New Orleans's stormwater infrastructure from a system that hides stormwater in concrete channels behind floodwalls (above) to reintegrate water into the fabric of the neighborhood, including through Blue and Green Streets (below). *Source*: Mora Orensanz.

and water-themed public art intended to encourage better understanding of landscape and water dynamics. Landscape architect Dana Brown has been central to GRD efforts to build "legible landscapes,"[14] echoing Anne Whiston Spirn, under whom Brown studied, who advocates for improving public "landscape literacy" as a means of reducing hazard vulnerability in disadvantaged communities.[15] This approach is premised on the idea that people can only make wise risk-mitigation choices about where and how they live if they understand their landscapes.

Landscape literacy–focused GRD projects include the CAP initiative, public arts programming, as well as several projects that will increase the visibility of water in Gentilly's landscapes. While much of New Orleans's drainage infrastructure was buried or hidden behind high floodwalls in the twentieth century, the GRD aims to bring water back into view, including through open water features in Blue and Green Streets, water-play features in new playgrounds, and waterside walking and biking paths.[16] Arts New Orleans enlisted local artists to shift public perceptions of water, including through a public arts fellowship program to train artists in water infrastructure and flooding and grants for placemaking along the London Avenue Canal. NORA officials responsible for the homeowner-focused CAP initiative regard public education as a central goal. They reported that participating homeowners became allies in spreading the word about the program and underlying stormwater mitigation goals.[17]

These landscape literacy–focused GRD projects build on other post-Katrina efforts to change how New Orleanians relate to water through education, outreach, and political mobilization, including the school-based experiential water education programs of Ripple Effect; the Water Collaborative's coalition building for equitable flood risk management; and Water-Wise NOLA, an effort by local nonprofits and landscape professionals focused on education and demonstrating green infrastructure strategies. Local designer Aron Chang, who has been central to many of these efforts, says that "water literacy" is essential because "if ninety [out of a hundred] people can't tell you what the relationship between pumping and subsidence is or can't tell you that we spend over US$50 million a year pumping stormwater or can't tell you that the river created this land," it will be impossible to make good policy and planning choices.[18]

SUPPORTING THE HEALTH OF ECOLOGICAL SYSTEMS
BEYOND THE PROJECT SCALE

The GRD aims to demonstrate a distributed model for managing water in New Orleans, collecting geographically distributed projects into a "district." This "living with water" approach has many labels, including "low-impact development" and "sponge city," and can offer benefits beyond flood mitigation. Some of these co-benefits support the health of ecological systems. The ecological benefits of the GRD's stormwater projects include improving water quality in Lake Pontchartrain and other water bodies that receive the city's stormwater, improved habitats for birds and other animals, and reduced urban heat.

While some of these environmental co-benefits may not be daily concerns for residents, others, such as reducing urban heat, are clearly connected to health and well-being. Mary Kincaid of the city's Project Delivery Unit spoke to the potential of the CAP program and other GRD projects to reduce energy burdens for low-income residents, saying that green infrastructure will "help [residents] with household cooling," which is a particular concern because "77% of our days [in New Orleans] are cooling days," and low-income residents "disproportionately spend their income on energy usage."[19] For many households, extreme heat and cooling costs are even more salient problems than flooding.

Some of the GRD's new public spaces, such as the Mirabeau Water Garden and Dillard Wetlands, are large enough to contribute to habitat and biodiversity improvements. However, some project participants question the city's capacity to provide the long-term maintenance and management necessary to realize these benefits. A researcher engaged in the project reports that larger GRD sites could provide adequate space for rich ecosystems that can keep nuisance species such as mosquitos in check. However, they also point out that there is a disconnect between the "static landscape architecture rendering[s]" of the GRD proposal and complex adaptive management necessary to build functional ecosystems, which requires "trial and error and being willing to change things."[20]

The GRD was proposed as a collection of distributed projects that together would demonstrate an alternative approach to managing water, but some project participants worry that the project-by-project approach has led to a lack of coordination that could reduce the impact of the broader initiative. The city government created a group called the Resilience Design Review Committee to coordinate its stormwater-resilience efforts. Nonetheless, one designer working on the GRD reported that the "corralling of everything as a big district idea, actually never happened." Rather, the GRD is "really just a list of projects . . . and so the interrelationships between them were not [realized]. There wasn't a mechanism in the city or . . . in the contracts . . . That didn't exist."[21]

While the GRD projects, however well or poorly coordinated, are most squarely focused on environmental improvements (e.g., water literacy, hazard mitigation, and socio-ecological co-benefits), some elements of the proposal speak to the other dimensions of equitable resilience.

SECURITY: GREEN INFRASTRUCTURE AS A POTENTIAL DRIVER OF DISPLACEMENT

In the realm of security, the GRD has both promising elements and significant challenges ahead. Crucially, the GRD involves no direct displacement, a major departure from the post-Katrina green dot map, which would have dislodged residents of many low-income neighborhoods to make way for green infrastructure.

The GRD's approach to flood hazard mitigation also departs from other functionally focused approaches in making significant investments in public art and placemaking—interventions that can contribute to security by strengthening the bonds between neighbors and between people and their neighborhood. NORA executive director, Brenda Breaux, spoke to the value of the GRD's public art along the London Avenue Canal as an "anchor" to enhance neighborhood social cohesion—a gathering point "to spark conversations and get to know your neighbors."[22] Such placemaking efforts required building project teams that could understand the neighborhoods. As Rami Diaz, a designer on several GRD projects, noted, these projects are "not just solving a stormwater issue." They are trying to "make good design and improve the urban place," which requires building "heavily local teams" that "get in there and design stuff that is very specific and very responsive and not generic."[23]

The GRD aims to support security by avoiding direct displacement and investing in placemaking, but in the years since the initial proposal, changing market dynamics in New Orleans have raised fears that the GRD could contribute to rising housing prices and displace low-income residents. In Gentilly, as in other New Orleans neighborhoods, flood damage and uneven recovery led to significant population decline and vacancy after Hurricane Katrina. One of the four core objectives of the GRD was to "encourage neighborhood revitalization."[24] However, in the years since the NDRC award, Gentilly has seen housing prices rise considerably. Some worried that simply calling the GRD "a resilience district" could be "contributing to gentrification."[25] One project participant reported that while NORA once enthusiastically converted Gentilly vacant lots into "stormwater lots," as housing demand and "the market value of those lots started going way up," authorities could no longer justify such uses of land.[26]

While the GRD represents a shift in water management, accommodating water without displacement and using public art and placemaking to strengthen residents' place-based bonds, there are mounting concerns that the projects and their branding under the banner of "resilience," could contribute to indirect displacement, undermining the security of existing residents.

LIVELIHOODS: RESILIENCE JOBS?

Given New Orleans's long-standing racialized inequality, city leaders have emphasized that efforts to improve flood resilience should also create livelihood opportunities for disadvantaged residents. Jeff Hébert—who variously served as executive director of NORA, Deputy Mayor, Chief Administrative Officer, and Chief Resilience Officer of New Orleans—was a key figure in conceptualizing the GRD. Hébert plainly expressed the anti-poverty aims of the city's climate resilience efforts, saying, "I don't care if you build the biggest [flood] walls in the world; if the people inside those walls are living in poverty, what is the point?"[27]

To direct the benefits from the GRD's resilience investments to disadvantaged residents, leaders earmarked US$3 million of the overall NDRC grant for workforce development. This training was meant to increase the capacity of disadvantaged residents. However, these efforts have been slow to gain traction. The GRD's first green infrastructure workforce training program took place in December 2021, six years after the initial grant award.

Without a full workforce training program, the Community Adaptation Program served as a training ground for a new workforce and an incubator for small businesses hoping to do flood resilience work. Seth Knudsen, who was instrumental in the CAP efforts, observes that NORA's network of smaller projects and contractors was able to "educate and train a large number of workers," preparing them to work on contractor teams for future larger-scale projects.[28]

Marketing materials for CAP include these goals, stating that the program "will help build skilled labor capacity in the 'green' infrastructure industry" and that contractors "must be eager participants to help train and support a new workforce of capable but new recruits."[29] Breaux argues that because Section 3 HUD standards required CAP contractors to hire low-income residents, the program helped "to build an entire industry of green infrastructure individuals who can take this skillset to ultimately develop it, implement, and grow their own businesses."[30]

Although CAP and other GRD programs promised to build green infrastructure capacity and support local livelihoods, some people involved in the projects are concerned that most of the funding is being absorbed by consultants and will not ultimately build capacity or benefit impacted communities. One project leader argued that because "every single project is bid out" to consultants, the government does not actually build internal capacity and "very little of that money" is benefiting vulnerable communities. They continued, "I'm of the belief that we would be far better off investing money in people."[31]

While the GRD's commitment to green infrastructure workforce training represents a potentially transformative new direction for the city, as with other elements of the GRD, progress has been limited, and potential payoffs remain largely unrealized.

GOVERNANCE: CREATIVE ENGAGEMENT WITHIN A LIMITED SCOPE

While the GRD vision includes ambitious efforts to engage and communicate with residents, the GRD also illustrates the challenges of instituting meaningful self-governance practices in the context of infrastructure development, a realm of planning and design that remains overwhelmingly top-down and expert driven.

1.7 Some GRD projects have included creative efforts to engage residents. *Source:* Mora Orensanz.

In part motivated by HUD's requirement to institute what they called "design by engagement" principles, GRD projects have involved a greater degree of community outreach than previous generations of infrastructure development (figure 1.7).[32] Even as the COVID-19 pandemic made in-person community meetings impossible during a critical period of project planning between 2020 and 2022, some GRD project teams instituted creative methods to communicate project visions and receive input from impacted residents. Beyond the scores of online community meetings and neighborhood mailers sent out for various GRD projects, the St. Anthony Green Streets team experimented with several novel ways to engage residents safely in the midst of the pandemic. Between 2020 and 2022, the team held the Gentilly Art Parade and a series of Lantern Walks, during which residents learned about the public art installations and green infrastructure principles underlying the GRD, including through viewing boxes displaying renderings of future interventions.[33]

Despite these efforts at creative outreach, some participants pointed out that the structure of the NDRC competition meant that many key decisions and priorities were already established before any neighborhood consultation. One designer said, "Everything was basically set through the competition process and guided by [The Rockefeller Foundation] . . . I remember the first public meetings. People already knew this. They said, 'Why are these streets chosen? And not those streets? What about my street?' And there were no answers to those questions other than it was already decided . . . when there was a group meeting at NORA headquarters with Rockefeller . . . it was already done."[34]

Given its reliance on international consultants, some critics allege that the GRD simply extends the "global resilience complex," just like previous green infrastructure planning projects in New Orleans.[35] Designer Aron Chang quipped that "you're more likely to have worked on the Urban Water Plan if you lived in Amsterdam than if you lived in Central City" (i.e., in a low-income majority-Black New Orleans neighborhood).[36]

The design competition model used in the GRD can be an effective way to bring together experts from across geographic and disciplinary barriers to forge visions for complex resilience interventions, but the GRD also shows that if projects are not supported by broad-based public legitimacy, shifts in political priorities can threaten the whole model. The mayoral administration of Mitch Landrieu developed the winning NDRC proposal, and Jeff Hébert, an MIT-trained planner and native New Orleanian, established the vision and assembled the team. Hébert marshalled the combined resources of multiple city and state-chartered agencies to build a unified vision for the GRD. But Landrieu was term limited, leading Hébert and virtually every other official responsible for the GRD proposal to join his exit from city government. The next mayor, Latoya Cantrell, and her team embraced the GRD and moved the projects forward, but many people working on GRD projects attribute their slow progress, at least in part, to the reduced capacity and massive turnover in city staff. One GRD designer said that although Mayor Cantrell intended to "implement the Urban Water Plan," the projects have languished because "the entire Office of Resilience and Sustainability basically left, and now has been whittled down to like one or two employees."[37]

The GRD's slow progress illustrates the vulnerability of this kind of top-down, competition-driven model. Because these green infrastructure projects emerged from an expert-driven competition for federal resources rather than from a neighborhood-based or grassroots movement, several GRD project participants expressed concerns that the approach does not have the broad community support necessary to maintain existing projects, much less expand the strategies to other areas. One official said maintenance and future green infrastructure projects can only happen if it is "something that the public wants."[38]

Even though the projects did not emerge from a community movement, some GRD proponents simply presumed that community members would fill the gaps left by insufficient public funds for maintenance. One GRD designer spoke to these unrealistic expectations, saying, "We do engagement and then [there is] this surprise that people don't jump and become stewards . . . Why would somebody start going out in 100°F heat in the summer to pluck weeds from a rain garden that you designed for them? . . . Why would you expect that?"[39] Similarly, another GRD project leader pointed out the need for the city to build capacity and resources within public institutions to ensure the success of green infrastructure, saying, "The only way that it's durable" is to "pay people, lots of people, a good wage and . . . develop that locally rooted technical workforce within public institutions." Nongovernmental organizations (NGOs) may succeed in "pulling in foundation money to implement these green infrastructure projects," they added, "but in the long term, the funders aren't always going to fund that. And they're certainly not going to fund maintenance."[40]

Balancing community input and self-governance with expert knowledge is a challenge in many infrastructure projects, both green and gray. Some GRD projects have demonstrated admirable creativity in engaging impacted communities. Yet, project participants expressed concerns about the governance of the initiative in terms of both its structure—competition-driven, externally funded, project-by-project implementation—and its uneven and slow implementation due to shifting political priorities and personnel. Because green infrastructure requires major shifts in both design and long-term maintenance, these concerns raise doubts about whether the GRD will have the popular legitimacy necessary for long-term success.

CONCLUSION

The projects in the Gentilly Resilience District aim to shift how New Orleans, one of the world's most flood-vulnerable cities, manages water. The GRD has brought together a broad coalition of experts to realize a vision reflecting years of effort following Katrina's devastation. Many

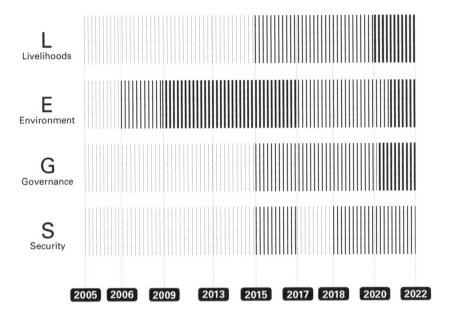

1.8 The Gentilly Resilience District proposal includes activities meant to enhance each of the four dimensions of equitable resilience, with particular emphasis on environment and livelihood aims. Progress on all fronts has been slow. *Source*: Smriti Bhaya. 2005: Hurricane Katrina; 2006: Bring New Orleans Back Commission (green dot map); 2009: Dutch Dialogues; 2013: Greater New Orleans Water Plan; 2015: Resilient New Orleans Plan/ GRD Proposal; 2017: Rain floods overwhelm pumps; 2018: Mayoral transition and staff turnover; 2020: COVID-19 begins; 2022: CAP funds expended, and job training starts.

elements of the GRD are focused on equitable resilience, including its focus on reducing flood risk and subsidence in a neighborhood with a large low- and middle-income population, commitment to implement green infrastructure without displacing residents, workforce training, and limited but creative community engagement in the face of challenging circumstances (figure 1.8).

The city received NDRC funds in 2015, but as of mid-2023, very few projects had started construction much less been completed. Aside from the household incentive-focused CAP initiative, virtually every other component of the GRD was in some stage of planning, design, or bidding, with no significant on-the-ground progress to show after nearly eight years of work.

Even if the GRD projects are completed, some local experts fear they will not adequately address the larger problem: New Orleans is a sinking, flood-vulnerable city in an era of rising seas. One designer argues that much of the change represented by the GRD and aligned projects is discursive and symbolic, asserting that while "every engineering, planning and design firm in town now uses the words of 'green infrastructure' and 'living with water,'" no one is actually "wrestling with the fact that half of the city is still below sea level, that the entire Gentilly Resilience District exists below sea level, and that we're investing $141 million in an area that will flood again."[41]

The GRD has realized some notable successes. Yet, the initiative is just as instructive in illustrating the significant challenges of pursuing equitable resilience through state-initiated, expert-driven, competition-based projects. The GRD proposal won a significant federal grant because a talented team of people within city government, local NGOs, and design and planning firms created a compelling collaborative vision. And yet Gentilly has relatively little to show for the years of effort and millions of dollars spent on consultants. Meanwhile, changing housing and real estate dynamics in New Orleans have shifted Gentilly from being a neighborhood in need of "revitalization" to a site of increasingly intense concern over tenure security and displacement, raising fears that the branding of the "resilience district" could contribute to "green gentrification" or "climate gentrification."[42] As the GRD has struggled to move from design vision into built reality, it has become clear that such projects can be jeopardized by many factors, including shifts in personnel and political priorities, fragmented project-by-project administration and funding structures, inadequate attention to long-term maintenance, and insufficient community support. Finally, even if built and maintained perfectly, the GRD interventions will not protect the neighborhood from another Katrina-like failure of the city's conventional gray infrastructure. While the GRD's US$141 million is one of the largest single investments in green infrastructure in a US city, that investment pales in comparison to the US$14 billion spent to expand conventional flood protections (pumps, levees, floodwalls, and gates) after Katrina.

The GRD shows that even bold efforts to invest in environmental improvements must reckon with the historical legacy of urbanization and infrastructures, requiring robust attention to the other key dimensions of equitable resilience: empowered community self-governance, security of tenure, and enhanced livelihoods.

Case 2

PARAISÓPOLIS: SEEKING ENVIRONMENTALLY SECURE HOUSING IN A SÃO PAULO FAVELA

OVERVIEW

The poorest residents of polarized megacities are often relegated to risky locations subject to threats such as flooding, landslides, fires, and toxins. This case highlights an effort to improve conditions for thousands of residents in the massive Paraisópolis favela in São Paulo, Brazil. By constructing mid-rise apartment complexes on an environmentally safe site at the favela's perimeter, close to flood-prone settlements slated for removal, the project endeavored to promote more equitable forms of resilience. The Paraisópolis Condomínios project, one piece of the city's larger favela urbanization program, engages all four dimensions of equitable resilience: environment, livelihoods, security, and governance. The project was meant to provide environmental protection to favela residents while maintaining affordability and mitigating displacement. It also aims to improve livelihoods, strengthen secure tenure, and enable residents to share in their own governance. Success in all these facets has been far from complete. Interviews with policymakers, Condomínio and favela residents, designers, and researchers, together with site visits and analysis of documents and previous studies, reveal significant achievements and illustrate obstacles that can complicate the pursuit of equitable resilience.

2.1 Paraisópolis Condomínios, with favela and Morumbi towers behind. *Source*: Photograph by Fabio Knoll, archive of Elito Arquitetos Associados.

INTRODUCTION: URBANIZING PARAISÓPOLIS

Favelas in Brazil, including those clinging to the slopes above Rio's elite beachfront redoubts, are iconic, contested, and often misunderstood. Although many favelas formed as peripheral settlements following widespread urbanization, they now offer poor residents the advantages of urban centrality. São Paulo's favelas typically lack the dramatic settings of their counterparts in Rio, but they share similar environmental risks, many of which are exacerbated by climate change. Similarly, favela residents often face immense security challenges from both state violence and criminal gangs, threatening livelihoods and undermining communal self-governance.[1]

Ever since Janice Perlman sought to counter "the myth of marginality" in the 1970s, social scientists have been reevaluating the lifeworlds of *favelados*, appreciating ingenuity and community solidarity while acknowledging the harsh realities of displacement, exploitation, discrimination, and violent repression.[2] Successive Brazilian governments have claimed to improve favela conditions through a variety of interventions, including novel housing policies and intensive policing. Such efforts have improved conditions in some places while also triggering new resentments. Many past favela interventions have forced residents to ever-further urban peripheries, away from the advantages of their previous settlements, including proximity to jobs and social networks. Given this fraught history of vulnerability and vulnerabilization, it is worth seeking out and learning from ambitious efforts such as São Paulo's Paraisópolis Condomínios.

The settlement, sometimes called the Paraisópolis Complex (Complexo Paraisópolis), is composed of three areas: Paraisópolis itself, Jardim Colombo, and Porto Seguro. Paraisópolis is further divided into five zones, delineated by micro watersheds: Antonico, Brejo, Centro, Grotão, and Grotinho. Of these, Antonico, Grotão, and Grotinho are the most precarious areas, with dense settlements facing extreme environmental risk.

As São Paulo's second-largest favela, Paraisópolis houses an estimated hundred thousand residents, although the official population is barely half that. The settlement is familiar to many international audiences, in part because Paraisópolis is ringed by the wealthy subdistrict of Morumbi, whose adjacent luxury towers underscore iconic inequality. Paraisópolis has other prominent neighbors: major TV stations and the Governor's Mansion are within a kilometer, and the settlement is not far from São Paulo's financial center (figures 2.2 and 2.3).

Paraisópolis was first planned in 1921 as a vast housing estate on the former Morumbi farm. With two thousand carefully drawn parcels of five hundred square meters each, it featured a grid of hundred-meter by two-hundred-meter blocks on twenty-meter-wide roads.[3] The promise of "Paradise City" initially attracted higher-income buyers, but most soon abandoned their parcels due to limited access, lack of infrastructure, and steep slopes. The site remained largely vacant and its taxes largely unpaid for decades.[4] As São Paulo's urbanization accelerated in the mid-twentieth century, favelas grew across the city, even as authorities called for their eradication.[5] During the 1950s, many low-income households settled in Paraisópolis. Some new residents acquired land in this still semi-rural area from *grileiros*, sham owners transacting properties under the guise of legality. Meanwhile, the high-income neighborhood of Morumbi grew in the more accessible surrounding area. In Paraisópolis, construction of wooden *barracos* (shacks) proliferated through the 1960s and 1970s, partly driven by demand for construction labor in adjacent neighborhoods.[6] As migrants from Brazil's northeast poured in, the government proposed expropriating parts of the community but did not follow through.

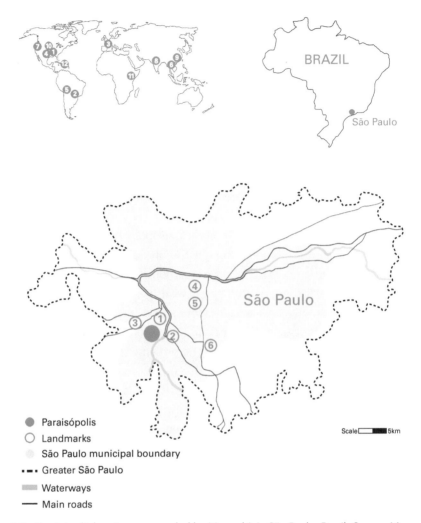

2.2 Paraisópolis location, surrounded by Morumbi, in São Paulo, Brazil. *Source*: Mora Orensanz. 1. Governor's Mansion; 2. TV Globo Headquarters; 3. TV Band; 4. Financial Center; 5. Avenida Paulista; 6. Congonhas Domestic Airport.

In the 1980s, urban policy gradually shifted from favela removal to upgrading.[7] In 1982, wary residents formed the Paraisópolis Union of Residents, which became a potent voice for favela residents, including resistance to growing private-sector development that residents blamed for increasing displacement. The Union initiated dialogues with the city to secure public lighting and running water in some areas of the favela.[8]

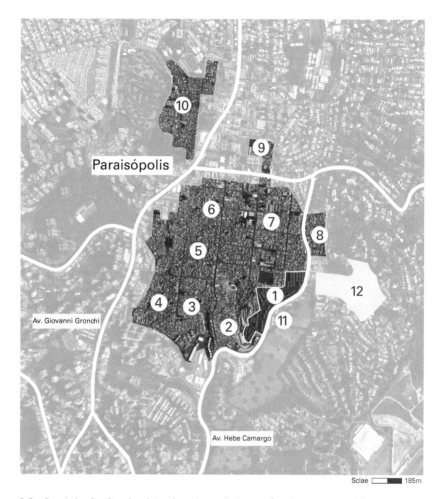

2.3 Paraisópolis Condomínios location relative to favela surrounded by Morumbi. *Source*: Smriti Bhaya. 1. Paraisópolis Condomínios; 2. Grotão; 3. Grotinho; 4. Antonico III; 5. Antonico II; 6. Antonico I; 7 Centro; 8. Brejo; 9. Porto Seguro; 10. Jardim Colombo; 11. Proposed Monorail Station; 12. Parque Paraisópolis.

Even so, as Paraisópolis's population grew to an estimated fifteen thousand in 1986, Mayor Janio Quadros proposed what became popularly known as the "favela removal law" (*lei de desfavelamento*).[9] Election of more pro-poor city and national leaders in the late 1980s and early 1990s reduced displacement threats and initiated infrastructure and service upgrades.[10]

In the mid-1990s, São Paulo leaders proposed the Favela Verticaliza-tion Program (PROVER), widely known as Projeto Cingapura ("Singapore

Project") in reference to Singapore's high-rise public housing. Describing this effort, Maria Teresa Diniz, who coordinated the Paraisópolis Urbanization Program from 2005 to 2012, lamented PROVER's standardized approach: "We have over 22,000 apartments which are exactly the same in the city. You get a piece of land and say, *how many 'Cingapuras' can you stamp on this piece of land?*"[11] Projecto Cingapura, coupled with Programa Guarapiranga, a program of environmental remediation financed by the World Bank and the Inter-American Development Bank, led to large-scale resettlement away from other established favelas, spurring growth in Paraisópolis, especially on the steep slopes of Grotão and Grotinho.[12]

During the 2000s, Paraisópolis began to receive public investments in infrastructure, housing, and environmental improvements,[13] largely driven by Brazil's 2001 City Statute, which required municipal master plans and created the designation of Zones of Special Social Interest (ZEIS) to facilitate land tenure regularization and physical upgrading in favelas across the country.[14] In accordance with the São Paulo Master Plan, city officials designated Paraisópolis as a ZEIS in 2002, requiring "dignified housing" with security of tenure, adequate sanitary facilities, habitable conditions, and "essential public services."[15] Also in 2002, Luiz Inácio Lula da Silva of the Workers' Party was elected president, prompting further investment in housing by a state-owned bank, Caixa Econômica Federal, among other social spending priorities. Paulistanos embarked on housing policies aimed at "transforming favelas into neighborhoods"—starting with Paraisópolis.[16]

São Paulo's Municipal Housing Secretariat (SEHAB) commenced Phase I of the Paraisópolis Urbanization Program in 2006, prioritizing high-hazard risk areas for upgrades, including new public stairs, drainage and road upgrades, creek canalization, new water and sewer lines, and roughly a hundred units of new housing.[17] With funding through the first phase of the federal Growth Acceleration Program (PAC-1), the city started more ambitious interventions, including construction of housing and institutional facilities along the favela's perimeter. Phase II projects, launched in 2008, included the canalization of Antonico Creek, stabilization of steep slopes in Grotão, consolidation of the existing road network, and pavement and drainage improvements.[18] Paraisópolis's dense maze of alleys presented logistical as well as environmental challenges for upgrading. Most residents had no registered addresses and therefore could not receive mail or deliveries. Distant from bus lines, even rideshare drivers refused to enter many parts of the favela.[19]

SEHAB proposed a series of ambitious projects intended to benefit thousands of families across the favela.[20] Unfortunately, as mayoral administrations shifted, projects lagged and sometimes stalled completely often after families had already been displaced. In some cases, people reoccupied dangerous sites that had been cleared but never upgraded, leaving residents exposed to floods, landslides, and fires.[21]

In this context, construction of a thousand new deeply subsidized low-income housing units in the Paraisópolis Condomínios commenced. In tandem with opening an avenue on the favela's southeastern edge, completion of the Condomínios stands as a counterpoint to many broken promises, even though this effort, too, has its detractors. The Condomínios linked improved environmental safety with new housing for favela residents close to their former neighborhoods rather than in distant housing projects (figure 2.4).

THE PARAISÓPOLIS CONDOMÍNIOS AND EQUITABLE RESILIENCE

The seven building complexes, collectively known as the Paraisópolis Condomínios,[22] mark a significant departure in many ways. Given the well-earned distrust of government institutions among favela residents and the scarcity of publicly owned land in Paraisópolis, proponents faced a serious challenge: How could the city relocate residents from environmentally risky areas without displacing them from the community? In response, the city used an atypical, even innovative, approach. Taking full advantage of the political will of Mayor Gilberto Kassab (in office 2006–2012),[23] the government expropriated a privately owned area of "good and expensive" land known as Fazendinha ("little farm") adjacent to the favela.[24] Then, breaking with the formulaic Cingapura style of low-income housing, they linked construction of the Condomínios to other interventions, including drainage and sanitation upgrades, retaining walls and paving, new green spaces, and public services (including a social services center and daycare).[25]

Condomínios A, B, C, E, F, and G were designed by Elito Arquitetos in 2007. The boldly colored concrete-block buildings were built into the

2.4 The Paraisópolis Condomínios offered new and safer housing (bottom) in proximity to residents' former dwellings in precarious areas (top). *Source*: Mora Orensanz.

hillslopes with entrance walkways partway up the buildings, with four floors of apartments above and one to four below, depending on the topography. This mid-level entry arrangement enables residents to access up to nine levels of apartments without exceeding the four flights of stairs allowed without an elevator in the Social Interest Housing regulations. Entrances provide access to stairwells as well as common facilities and accessible apartments for residents with disabilities.[26] Although all apartments feature two bedrooms, nonstructural interior walls enable residents to adapt their homes to suit their needs. Architect Elisabete França, SEHAB's housing superintendent between 2005 and 2012, praises the "windows that open 100 percent outward [to improve] ventilation and lighting, [and] the light-colored roof to avoid problems with [over]heating."[27] As another departure from the previous model, the favela-facing ground floor of the Condomínio D blocks feature commercial spaces. Contractors completed work on the first 783 apartments (Condomínios A, B, C, D, and F) by March 2011, and residents occupied two further phases totaling 284 more units (Condomínios E, G, and H) between 2012 and 2014 (figure 2.5).[28]

In sum, by operating on expropriated vacant land, the project provided more than a thousand units of low-income housing on a highly desirable site, without direct displacement. Accommodating nearly four thousand people,[29] the Paraisópolis Condomínios have been a boon for many residents. For others, however, moving to environmental safety has brought problems, including financial challenges that threaten their livelihoods, lingering insecure tenure, and reduced control over neighborhood governance.

ENVIRONMENT: PROTECTING FAVELA RESIDENTS WHILE MITIGATING DISPLACEMENT

In assessing the Paraisópolis Condomínios as contributors to equitable resilience, we begin with the environmental dimension because reducing hazard vulnerability was a core motivation for the project. As environmental interventions, however, the Condomínios do not stand on their own. This project is linked to other projects in Paraisópolis that displaced residents in the name of hazard mitigation.

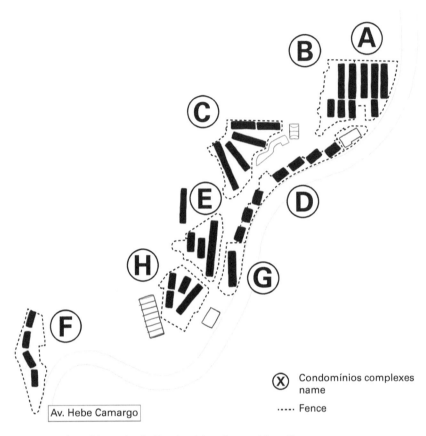

2.5 Site plan of Paraisópolis Condomínios. *Source*: Mora Orensanz.

CONDOMÍNIOS AS PROTECTION FROM
ENVIRONMENTAL HAZARDS

The Condomínios and allied efforts resettled thousands of residents from risky areas, enabling environmental infrastructure improvements, including drainage, water and sanitation, road paving, and landslide mitigation. Project leader Maria Teresa Diniz emphasizes the intervention's geotechnical risk reductions: "You had risk areas for mudslides, for very steep hills— forty meters high [in areas] like Grotão and Grotinho." Houses collapsed into creeks. Mudslides destroyed homes. Most of the favela had inadequate sewage and solid waste management, threatening human health and water quality. Flooding meant more than unwanted water. "It's sewage with solid waste," Diniz observes. "That's why you need to do the works along the rivers and in the lowest points."[30] These infrastructure projects connect directly to housing development: the Condomínios rehoused many residents displaced by environmental upgrades in hazard zones.

Vanessa, a twenty-four-year-old renter in Condomínio A, spent her youth living with grandparents in the favela, conscious of the risks: "In houses where they are like on top of one another, you are always thinking 'Man, what if there's a heavy rain and it collapses?'" Daniel had visceral experience living directly above a flood-prone creek in Sossego Alley, where floodwaters frequently damaged his furniture and possessions. When he eventually moved to Condomínio B in 2010, his family was forced to "get rid of everything because of the mold" (figure 2.6).[31]

The Condomínios offer a stark contrast to more hazard-prone parts of the favela. Joildo Santos, a resident of Condomínio E, gratefully observes that in his new home, these risks simply "don't exist." As a community journalist and former vice president of what is now called the Paraisópolis Union of Residents and Business, he advocates for media coverage of the positive aspects of life in and around the favela. He summarizes several improvements that the Condomínios bring to residents: "individual entrances, waste collection, easy access to streets, and public lighting."[32] For João, a twenty-eight-year-old who moved to the Condomínios from Antonico where he worried about heavy rains coming through his family's tile roof, the new home offered "a very peaceful environment."[33] For Lea, who grew up in *barracos* in the favela and is a new renter in the Condomínios, the benefits are clear: her new abode reduced "the risk of flooding, the risk of [house] collapse."

Persistent environmental vulnerabilities across Paraisópolis underscore the urgency of further intervention, especially given mounting climate threats. In late 2014, following intense flooding in the Antonico sector, where approximately three thousand families live precariously in houses built atop the creek, two thousand people peacefully protested to demand long-awaited canalization and housing upgrades. With little official response and periodic torrential rains, house cave-ins remain common in many parts of the favela.[34] In October 2021, with upgrades still not delivered, flooding caused the collapse of a three-story house. Because this area

2.6 Ongoing struggles in the former Sossego Alley, previous site of Daniel's house. *Source:* Mora Orensanz.

floods regularly, community organizers termed the event an "announced tragedy." The incident resulted in one fatality and at least five serious injuries, and damaged six other houses, displacing nearly a dozen families. In response, residents staged another protest, carrying symbolic coffins through Morumbi, demanding action.[35] Floodwaters and landslides are hardly the only threat. Large fires occur almost every year in Paraisópolis, destroying more than a thousand dwellings in the last decade.[36] While the Condomínio and allied infrastructure projects have reduced hazard vulnerability for many Paraisópolis residents, risks remain.

IMPROVING THE ENVIRONMENTAL PERFORMANCE
OF HOUSING

The Condomínio projects sought environmental improvements through low-cost green building tactics. Edson Elito described sustainability-related features of the project, including planting shade trees, extensive insulation, and reduction of construction waste. To incentivize residents to conserve energy, each apartment has separate monitors for water, gas, and electricity. Units feature water-efficient plumbing fixtures, high-efficiency lighting, and building-wide rainwater capture.[37]

The combined green building and infrastructure upgrades made the Condomínio project the first social housing development in Brazil to earn a gold rating from the Caixa Econômica Federal's Blue Seal (Selo Azul) Awards program (a rating system inspired by the US Green Building Council's LEED assessments).[38] The Paraisópolis Condomínios project met all of the program criteria, including in the "social practices" category, which includes training construction laborers, community participation, employment generation, and "social risk mitigation." In the words of an official from the Caixa Econômica bank, "What set the Paraisópolis project apart were the social practices. Beyond construction workers who benefited from educational and training programs, the project also envisions activities with residents, including their involvement in the project's elaboration."[39]

While the Condomínio and allied projects may support ecosystem health through green building strategies and mitigation of risks from landslides, sewage, and solid waste pollution, the projects' contributions to overall urban ecological health are partial and deeply imperfect. Lingering problems with household waste and sewage dumping in area creeks continue to compromise safety and ecological health. Further, even where upgrades were completed, flood-control efforts frequently confined creeks to hard concrete channels, an approach that can harm water quality and exacerbate flash flooding. Many nearby upgrading projects were also significantly delayed or cancelled only after residents were displaced. Without proper rehousing, many desperate residents reoccupied some of the favela's riskiest areas.

SECURITY: PROMISES AND REALITIES

The quest for security remains a priority for many Paraisópolis residents. Residents suffer from periodic violence at the hands of both drug trafficking gangs and police militias, although those problems are even more acute in other favelas. Many in the community resent the ways that biased media coverage stigmatizes Paraisópolis. Residents have long fought to improve their community, including persistent demands for recognition of their right to remain. Still, while the Paraisópolis Condomínios promised to provide secure housing, these and other recent projects have often fallen short. For those choosing to relocate to one of the Condomínios, the new environment enhanced some aspects of security, but also imposed new costs.

SAFETY WITHOUT SOLIDARITY: CREATING NEW BARRIERS

Given the persistent threats from crime and violence in Paraisópolis, many Condomínio residents supported construction of walls, fences, and security checkpoints to separate themselves from the adjacent favela (figure 2.7). The complexes feature perimeter fences, topped in some places by barbed wire. Edson Elito and the site design team responded to security concerns among prospective residents, noting that they championed physical barriers as affirming their "change of status." The Condomínio buildings' mid-level entries accommodate security stations to control access. In the architect's assessment, this socio-spatial balance between adjacency and separation has worked well: "Today there isn't really great tension between the residents of the Condomínios and the residents of the favela."[40]

Housing researcher Paula Santoro is less certain of this, emphasizing that the Condomínios created a significant barrier, especially from the side facing Hebe Camargo. She observes that the new development created "a new real estate front for the middle class," flanked by a "walled avenue," which separated "the precarious settlement and the [nearby] park." Urbanization planner Maria Teresa Diniz resists that characterization, and sees these spatial divisions as an inevitable outcome of the political processes, given the role of "people from the 'rich side'" in opposing social projects and "push[ing] the avenue closer to the [favela] so that they can keep it further away from them."[41] The resulting streetscape, Santoro reflects, is "a

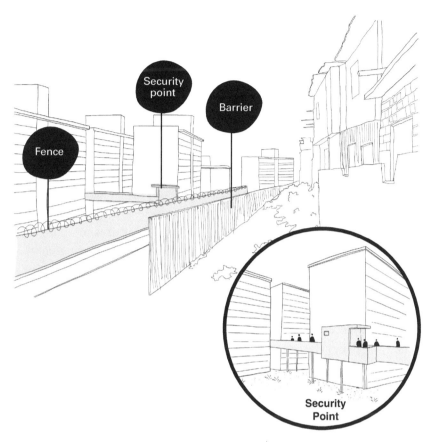

2.7 Paraisópolis Condomínios are adjacent to the favela yet sit behind secure barriers. *Source*: Mora Orensanz, based on photograph by Lawrence Vale.

very barren place, not a very inviting public intervention—especially for women," she comments. "It's not a place where you feel safe."[42] Juliana, a forty-five-year-old seamstress who operates from Condomínio D along Avenida Independência but does not live there, concurs that she feels safe on the Independência side of the complex until evening, but notes that "if I go walking on Hebe Camargo at the wrong time, they say I'll get mugged."[43]

Those who actually live in the Condomínios generally view these as a safe refuge that also retains their sense of connection to the favela community. Daniel observes that residents of the Condomínios still receive the same respect as others from the favela because "they know we are from the community as well." He adds: "We have never seen a case of mugging or any type of violence." Márcia, a thirty-five-year-old woman who now lives in Condomínio C with her mother, reports feeling "a lot safer than where I lived before, which was an alley facing the street." "Here," she continues, "if I open the door, I'm facing the hall, not the street itself." "I love living in Paraisópolis," she comments, "but I feel safer here in these buildings because before you enter the apartment you have to get through the gate."[44]

While many residents report that the formal security measures of the Condomínios increase their sense of personal safety, others prefer the informal security provided by the dense social networks of the favela. For Vanessa, the narrow ways of the favela felt safer because "you know the people better, and here on the main street you don't." It's "all cars [and] strangers." She misses engaging with "the old ladies who take care of everyone else all day long." Márcia describes a similar feeling of loss after moving to the Condomínios. She reported that after her home in Grotinho was judged to be "at risk," "they kicked us out." After four years of rental assistance, a Condomínio unit became available. With one day's notice, the movers came: "We didn't choose; they just gave us the key and said, 'this block, this floor,' and that was it." Although she appreciates the newfound environmental safety of the apartment, she reported missing her former "wonderful" neighbors and the mutual support structure in Grotinho.[45] In short, new forms of security may come with the loss of old forms of solidarity.

The security of the Condomínios themselves is inseparable from the larger insecurities associated with the favela urbanization program. The new settlements produced on the favela periphery also disrupted older homes and communities. This means that those who made it to the Condomínios were, overwhelmingly, families that had previously experienced an extremely complex security environment in the favela itself.[46] "People from the community probably feel safe there," housing scholar Santoro observes, "but it's a sense of security based on regimes of private territorial control using coercive and violent instruments." The fenced territories and camera-surveilled corridors of the Condomínios are efforts to substitute for these other "regimes of control," ranging from "the moral control exercised by the church to the control of drug trafficking gangs" and police militias. Based on her research with the University of São Paulo's LabCidade, Santoro argues that "security" cannot be measured simply by examining "the formal relationship between the sidewalk and the building." Rather, "The question of security is much deeper and has to do with the forms of control that low-income territories have acquired in São Paulo."[47] Closely linked to this quest for security and community, the long history of forced removals underscores the importance of tenure security.

THE INSTABILITY OF SECURE TENURE

Some residents report that moving to the Condomínios has contributed to both physical security and greater security of tenure, but many still struggle. For decades, the number of households displaced from Paraisópolis far exceeded the number of new housing units with secure tenure. In some cases, residents displaced by government action but not rehoused have depended on rental assistance for years on end. Joildo Santos estimates that there are ten thousand such households in Paraisópolis. Researchers found that the increasing number of households on government rental assistance has contributed to an overall increase in rents across Paraisópolis during the period when the new Condomínios were built.[48] Unconvinced that they would ever receive new affordable housing, many households accepted what they termed a *cheque despejo* ("eviction check"), a payment of R$5,000 (currently about US$900). Santos observes that these payments were so inadequate that many residents who took the eviction checks to leave informal self-built homes "ended up building other informal housing in another location—sometimes in even worse conditions than they were in the first place."[49]

For those who declined the *cheque despejo* and attained a place in the Condomínios, the promises of more secure tenure remain only partly fulfilled. City officials told residents that, after twenty years of Condomínio payments, they would become the title holders of new apartments. However, residents still lack full security of tenure. They merely have permission to use these properties but do not yet have formal deeds that could be used to sell their apartments. Some people do sell, Santos explains, "but it's an under-the-table contract." Lacking formal ownership, "those who want to get the apartment off their hands end up creating contracts saying that they sold it to so-and-so, and when it's regularized, it's their property."[50] Even without formal ownership, Marcelo, a forty-two-year-old man who lives with his wife and daughters in an apartment he bought from a former owner, feels confident that he will be able to sell, given that he had "authorization and documentation from the CDHU (State Housing and Urban Development Company), with registration at a notary's office and a lawyer." While informal transactions may suit buyers and sellers, they can present challenges to renters who lack basic protections against displacement. Letícia, who grew up in *barracos* in the favela and is a new renter in the Condomínios, reported feeling insecure because the owner of her unit may return and force her to move at any moment: "It can happen any time because it's theirs."[51]

Even for unit owners who do not intend to sell, staying in their Condomínio unit can be challenging. The apartments are 80 percent subsidized, but residents still face increased living expenses. Social workers assisting with relocation of favela residents into the Condomínios helped residents to calculate their new cost of living. They explained that expenses would include not just the modest monthly acquisition charge of R$100–R$120 (approximately US$20), but also additional payments for water, energy, and Condomínio fees. As Maria Teresa Diniz acknowledges, these were new "costs that they didn't [previously] have in their own houses." Still, she insists, "their overall expenses become much lower than before" if they had paid high rents in the favela.[52] In theory, paying monthly Condomínio installments moved households toward full ownership, but a lack of clarity surrounding the acquisition and titling process made the payments functionally equivalent to monthly rent. Interviews with residents suggest that some Condomínio households, despite the considerable subsidy, fail to pay monthly installments and consequently risk losing their homes. Many found the cost of Condomínio ownership to be unsustainably high, making the question of livelihoods especially important to any assessment of equitable resilience.

LIVELIHOODS: LINKING HOUSING TO LASTING AFFORDABILITY

Moving to greater environmental safety is clearly a substantial gain for many favela households. Yet, interviews with relocated residents reveal significant trade-offs.

MORE THAN HOUSING BUT LESS THAN FULL ACCOMMODATION FOR LIVELIHOODS

Breaking with the previous Cingapura model, the Condomínio projects incorporated nonresidential services and activities, including public schools, a technical school, a kindergarten, physical and mental health services, and a library.[53] The new district includes a unified education center with a theater, cinema, and recreation areas. In 2021, the large Paraisópolis Park opened just opposite Condomínio A.[54] The projects also support livelihoods by accommodating formal and informal work from home. With WiFi access, some work from computers, while others do informal home-based work such as hair cutting, food sales, and artistic endeavors.

Diverse businesses operate from the mixed-use ground floor of Condomínio D—but nowhere else (figure 2.8). Architect Edson Elito explains that including more ground-floor retail "didn't fit well in the Condomínios," since "commercial spaces need a special type of access" and because "the government didn't want to have a lot of commercial spaces."[55] Maria Teresa Diniz explains that including commercial spaces in social interest housing is "a huge taboo in São Paulo because it's not easy to administer afterward." She points to the potential for conflicts caused by "people who open up bars and then people drink all night long and then the residents don't want them." Still, she recognizes that when they relocated households from the favela, they also removed services. In the favela, "a family that had a beauty salon on the ground floor" would live upstairs. "They're going to move into the housing units and how are they going to work? You're also removing their work, so you also need to have resettlement for shops and services." In the case of the Condomínios, only those residents who were already registered as shop owners or service owners received commercial space options. The government left the Condomínio residents to decide what types of businesses would be allowed.[56]

Not surprisingly, some businesses proved incompatible with the standardized spaces provided. Paula Santoro comments that some kinds of businesses simply "didn't fit." She cites "a guy who had a pizzeria with a wood-burning oven" that could not be installed on the ground floor of a Condomínio because it lacked appropriate ventilation. Others objected to placing their business in Condomínio D because they wanted to locate on

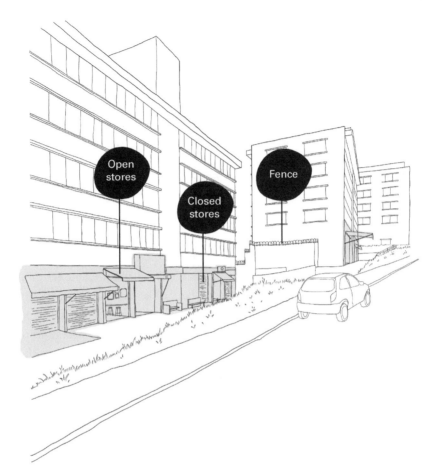

Open
stores

Closed
stores

Fence

2.8 Condomínio D features ground-floor commercial enterprises. Businesses in these spaces have met with mixed success. *Source:* Mora Orensanz.

"the most intense commercial streets" rather than in what they perceived as a peripheral location, distant from most favela customers.[57] Juliana, who migrated to Paraisópolis in the 1980s from the Northeast Region of Brazil, opened a snack bar in Condominío D in 2013. She struggled due to a lack of foot traffic, exacerbated during the COVID-19 pandemic. To make up for lost earnings, she turned to other skills, including tailoring and fortune-telling. Juliana observes, "the restaurants that open last a year, then close," as did a tattoo parlor. Among the businesses that have fared better are multiple hairdressers, an accountant, and a pet shop and dog groomer.[58]

LIVING LIFE ON THE EDGE: ACCESS TO LIVELIHOODS

Interviewees consistently noted that the location of the Condomínios enabled residents to access a variety of livelihoods, both within the favela and beyond. Because the complexes are located on new roadways with regular bus service, residents can readily move across the city for employment and other amenities. A long-planned monorail line to the airport includes a proposed station near the Condomínios, which would further improve accessibility.[59] Conversely, locating the Condomínios on the edge of Paraisópolis means that residents find themselves distant from the favela's existing commercial hubs and from family and friends in their former neighborhoods.[60]

TESTING HOUSEHOLD CAPACITY: THE TRUE COSTS OF SUBSIDIZED CONDOMÍNIOS

While the Condomínio projects support residents' livelihoods through both accommodating income generation within the complex and accessibility to outside employment and services, many have faced challenges to their livelihoods from the higher cost of living that Condomínio residence imposes. Numerous respondents expressed concern about increased utilities costs. Resident leader Joildo Santos noted that improvements to the public areas of Condomínios, such as installation of finished flooring or electronic key access, cause increases in monthly fees. If a basic principle of equitable resilience entails judging a project on its capacity to support stable and dignified livelihoods, then it must be acknowledged that moving to the Paraisópolis Condomínios has actually caused substantial livelihood challenges for some households.

Whether the move to the Condomínio project increased or decreased living expenses varies from one household to another. Márcia recounts that her cost of living "went up significantly." Marcelo's family found Condomínio costs for piped gas and electricity meant higher expenditures. They planned to move closer to relatives who could assist with childcare. Aline, a thirty-eight-year-old receptionist, purchased her apartment from someone who could not keep up with the expenses. She can manage the cost but, like Marcelo, is looking to sell and move closer to her family, who can help care for her daughter. By contrast, Lucas, a nineteen-year-old who has lived with his parents in Condomínio C for more than a decade, says that his family's monthly costs declined in comparison to renting the *barraco* that was razed to make way for the technical school. Looking around his apartment, he observes, "Today we wouldn't have so many things if it weren't for that." Thirty-one-year-old Antônia, who moved from an alley to Condomínio A with her husband and young child, estimated that their reduced rent was balanced by higher utility payments. Daniel, who works as a motorcycle courier, acknowledges higher expenses but praises the practicality of the location and improved "peace of mind" from living in a mold-free "nice and organized place."[61]

In 2016, a local TV news program featured interviews with residents of the Condomínios who could not afford the costs of the new housing units and returned to informal housing in the favela. The reasons varied. A resident named Maria showed her unaffordable gas bill, which reached 1,270.67 reais (nearly US$250) for three months. According to the report, some thirty residents had their utilities cut off when they could not pay. Dona Maria Nilza, with 9,000 reais (more than US$1,700) in unpaid bills, dreaded "being kicked out, living in a place that I don't have the means to pay for." The report estimated that about half of the Condomínio residents were behind on payments and that at least half of original residents had already left (as of 2016).[62] Resident leader Joildo Santos insists that the attrition is smaller. "It's not a lot of people. It happened . . . but it's not significant. Maybe 10% of the families." Raimundo lived in the Condomínios for a decade until his family rented out their unit following the death of his father from COVID-19 in 2020. He observes that, back in 2010, many people sold their apartments quickly, especially if they had a large family that could not squeeze into a two-bedroom forty-eight-square-meter apartment.[63] Most of those choosing to depart more recently appear to be renting out their unit, often for more money than it will cost them to live elsewhere.

GOVERNANCE: REGAINING TRUST

As with livelihoods and security, attempts to advance resident self-governance in the Paraisópolis Condomínio project have also seen marked progress but remain imperfect. Seen most positively, the project fulfilled the longtime demands of favela residents to obtain decent housing close to their community networks. Less auspiciously, many residents viewed participation mechanisms as merely consultative, informing residents of plans without truly empowering them to shape fundamental aspects.

GOVERNING URBANIZATION IN PARAISÓPOLIS: UNEASY LINKAGES

The development of the Condomínios played out amid deeply rooted mistrust of city officials due to a history of repressive policing, neglect, and inconsistent investment in favelas. Often improvements in favela public works initiated under one mayoral administration have been abandoned following political transitions.

While efforts at official participatory governance have been criticized by some, Paula Santoro from LabCidade points out that, in São Paulo, citywide social movements are essential to the success of any individual neighborhood project.[64] Accordingly, Paraisópolis urbanization coordinator Diniz describes a robust participatory process, noting, "Between 2005 and 2012, we had a team of fourteen social workers inside the community every day." She cites the centrality of the favela's thirty-six-member management council, composed of elected representatives—half of whom are from civil society and half public servants—as well as numerous working groups that engage those residents most effected by nearby projects. "In the first four years," Diniz estimates, "we had over a thousand meetings a year with the community. It was insane—weekends, nights. You'd have simultaneous meetings because we had such a large team." "Due to the impossibility of having all entities participating," she adds, their process identified NGOs, landowners, and leaders to represent the community, including the Union of Residents. They also engaged neighbors from the broader community, including a seat for those from the "rich side" of Morumbi.

Given the high stakes for residents, many of these meetings proved contentious. "Every time we would come in," Diniz recalls, "[residents] would lash out on us. In the beginning, we had very difficult fights because they didn't believe us. We said, 'We're going to do the housing project here, we're going to remove you now; we're going to put you into rental assistance.' Residents would say: 'But who guarantees me? This piece of paper is nothing. You're just showing me a nice picture. I only want to move out of my house when I see the building ready.' It was very hard to convince and [achieve] their trust." Diniz learned that being open about shortcomings and delays helped. "This is something very important when you're doing a participatory process that nobody teaches us in architecture school—that they need to trust you and . . . whenever you have bad news to tell, you have to say it . . . If you lie and they find out later . . . you're done—they're not going to believe you anymore."[65]

Despite the many community meetings, Paula Santoro contends that the development of the Paraisópolis Condomínios was "not an example of participatory governance." At base, she observes, the "[ZEIS-required] committee was assembled a posteriori." "The committee approved the plan," she contends, but only after "the project was already being carried out." The plans for the Condomínios "were presented to the community; they were not designed with, or stemming from, the community."[66]

EROSION OF SELF-EFFICACY

In terms of community participation and collective efficacy, resident and community journalist Joildo Santos says, "This process was very tense, but it was gratifying to see some things come to fruition and achieve some victories that had not been previously possible."[67] Interviews suggest that some residents actively participated in meetings while Condomínio construction proceeded and were able to choose where to live, while some others were simply assigned to available apartments. Ultimately, the overall Paraisópolis redevelopment embodies a curious contradiction: democratically enacted displacement. Despite a semblance of participation, one scholar charges, it reproduced "age-old processes of expulsion, removing residents to make way for public works interventions."[68]

THE LIMITS OF RESIDENT CONTROL

Some households left their Condomínio apartments due to concerns about governance. These residents resented Condomínio rules, viewed as imposing problematic "new limits that did not exist in previous places of residence."[69] Some preferred a less encumbered social life. Raimundo grew up in Grotão near many flooded *barracos*. When his neighborhood was demolished and replaced with a technical school, Raimundo observed that moving to a Condomínio caused him to "lose a certain freedom." Previously, he observed, "we could do whatever we wanted," but in the Condomínio, "you don't live in a house anymore, so there are rules." Raimundo found that he lost many friendships from his former neighborhood. "You have to start over with your neighbors," he remarks wistfully. The people in the formal housing were "not the neighbors you can leave your child with, or the neighbor you can ask to borrow a little salt." Unwilling to accept such trade-offs, some returned to environmentally riskier places that gave them more autonomy and access to networks of mutual aid.[70]

Others resisted living in the Condomínios because this implied having "a relationship with the government." Despite the opportunity to shift to "a housing alternative within the realm of legality," many residents "prefer to stay in the 'old' favela."[71] Each Condomínio does things somewhat differently, but all residents had to adjust to a system of governance that operates like a residents' association, delegating considerable authority to managers.[72] Ultimately, life in the Condomínios is shaped primarily by individual households making decisions for themselves and their family. In contrast, favela life is characterized by constant informal social bargaining.

CONCLUSION: THE ONGOING QUEST FOR ENVIRONMENTAL SAFETY AND COMMUNITY CONTROL

With safe and well-located homes for more than a thousand families who previously lived in hazardous conditions, the Paraisópolis Condomínios constitute a hard-fought and substantial achievement. At the same time, many households forced out of homes in the favela have not been rehoused and must make do with meager rent assistance. With few new housing developments, favela residents continue to cope with broken promises and new waves of environmental disaster (figure 2.9).[73]

Persistent tragedies typically occur in the interior portions of the favela, far from public view. This is a reminder that many dramatic architectural statements located on the periphery—whether the Condomínio projects or numerous high-profile design proposals that have generated international publicity only to languish on the drawing board—may mask systemic problems that remain.

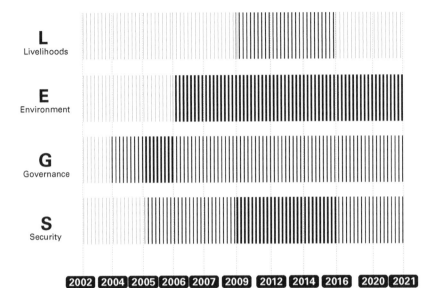

L
Livelihoods

E
Environment

G
Governance

S
Security

2002 2004 2005 2006 2007 2009 2012 2014 2016 2020 2021

2.9 Paraisópolis Condomínios have made significant environmental improvements but slower progress on livelihoods, governance, and security. *Source*: Smriti Bhaya. 2002: Paraisópolis designated as a ZEIS, Lula elected Brazil's president; 2004: José Serra elected São Paulo's mayor; 2005: SEHAB launches Paraisópolis Urbanization Program; 2006: Infrastructure improvements begin in high-risk parts of the favela, Gilberto Kassab elected São Paulo's mayor; 2007: Growth Acceleration Program; 2009–2011: Condomínios A, B, C, D, and F built and occupied; 2012–2013: Condomínios E and G built and occupied; 2014: Vila Andrade G (formerly Condomínio H) opens; 2016–2017: Large fires destroy hundreds of Paraisópolis homes; 2020: COVID-19 pandemic begins; 2021: Rains cause house collapses, Parque Paraisópolis inaugurated.

Ultimately, the creation of well-located nearby apartments can help some Paraisópolis families but falls well short of the community's overall need. Moreover, lack of community input into the design of the Condomínios, the high monthly costs, the loss of valued networks among former favela neighborhoods, and the lingering absence of formal titles to apartments all pose problems for residents. The Paraisópolis Condomínios make a substantial contribution to equitable resilience in environmental terms but achieved less in the realms of support for livelihoods, security of tenure, and empowered self-governance.

Case 3

PARIS OASIS: CO-DESIGN FOR HEAT-ADAPTIVE SCHOOLYARD RENOVATIONS

OVERVIEW

The Paris OASIS pilot initiative included climate-adaptive renovations of ten schoolyards to support experiential education and resilience-enhancing social cohesion (figure 3.2). The schoolyards are also designed to provide publicly accessible cooling refuges for times of extreme heat. Based on our interviews and site visits, the OASIS initiative demonstrates the power of participatory design for adaptive pedagogical landscapes and the potential of school-based strategies to provide distributed scalable climate-resilience intervention. At the same time, the project also carries lessons from its challenges, including limited engagement with lower-income residents.

INTRODUCTION: COPING WITH URBAN HEAT

For several weeks in the summer of 2003, Western Europe suffered unusually hot temperatures. By the time temperatures returned to more temperate historical norms, more than seventy thousand people had died.[1] In France alone, nearly fifteen thousand people perished, more than seven hundred of whom lived in Paris, where temperatures reached 35°C (95°F) for ten straight days.[2] Subsequent European summers have been even hotter.

3.1 OASIS schoolyard at École Élémentaire Maryse Hilsz in Paris's twentieth arrondissement. *Source*: Joséphine Brueder/City of Paris.

Globally, extreme temperatures are less visually dramatic and cause less obvious damage than other climate change threats such as floods and wildfires, but they can be just as deadly.[3] Between 2000 and 2019, nearly 10 percent of global human fatalities resulted from "non-optimal temperatures." Although extreme cold kills more people worldwide than extreme heat, cold events are decreasing in frequency, while deadly heat waves are increasing, especially in regions such as Northern and Eastern Europe, where people are unaccustomed to heat and built environments are not well adapted.[4]

The dangers of heat, like other climate threats, are shaped by both physical factors in the built environment and social factors. The physical materials that make up urban environments absorb and hold more heat than nonurban landscapes, elevating temperatures and reducing nighttime cooling, making it harder for people to recover from hot days.[5] This urban heat island (UHI) effect can cause industrial and urban areas to be more than 10°C–15°C (18°F–27°F) hotter than nearby nonurban landscapes.[6] Paris is especially vulnerable to heat waves because the city has less than 10 percent vegetated green space—markedly less than other large cities

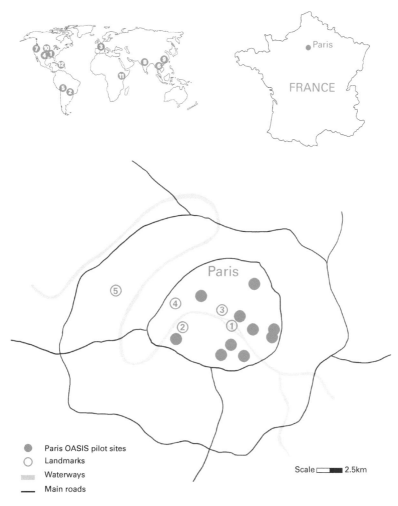

3.2 OASIS pilot project locations within Paris. *Source*: Mora Orensanz. 1. Notre Dame Cathedral; 2. Eiffel Tower; 3. The Louvre; 4. Arc de Triomphe; 5. La Défense.

such as New York City (27 percent) or London (33 percent).[7] Increasing the tree canopy of urban landscapes to 40 percent can reduce summer daytime high temperatures by 5.5°C (approximately 10°F). Reducing impervious surfaces, such as asphalt streets, sidewalks, and parking lots, is especially effective in reducing nighttime temperatures, allowing people and places to cool after hot days.[8] Factors such as age, poverty, race, and social isolation place some urban residents at higher risk during heat waves.[9] Recent studies have shown that urban areas in the US that were disinvested because of racially biased "redlining" lending practices many decades ago still have less vegetated cover and suffer from elevated heat stress.[10]

In 2015, as part of the Rockefeller Foundation's 100 Resilient Cities program, the city of Paris adopted a resilience strategy that included several initiatives to adapt to heat waves such as the one that beset Paris in 2003—and recurred repeatedly in recent years. These adaptation measures included changes to built environments and efforts to promote social cohesion to increase the resilience of vulnerable elderly and isolated populations. One action item in the resilience strategy sought to "transform schoolyards into cooling island 'oases.'"[11] Pursuing this goal, the city government renovated a first round of thirty-one Paris schoolyards, starting in 2017.

PARIS OASIS AND EQUITABLE RESILIENCE

Following these initial schoolyard adaptation projects, a coalition applied for and received a US$5 million grant from the European Regional Development Fund's Urban Innovative Actions (UIA) initiative in the fall of 2018. With UIA support, the schoolyard renovations, which had been a technically focused renovation effort by city engineers, expanded to include government, civil society, and academic actors who, together, developed a pilot program to renovate ten more schoolyards across the city (figure 3.3). The new initiative took the name Paris OASIS, based on an acronym for Openness, Adaptation, Sensitization, Innovation, and Social Ties, reflecting the program's broad ambitions.

The composition of the OASIS team demonstrates an expansive and process-oriented approach to climate change adaptation. The team brought together funders and administrative authorities with technical experts in built environment design, pedagogy, and research on both social and

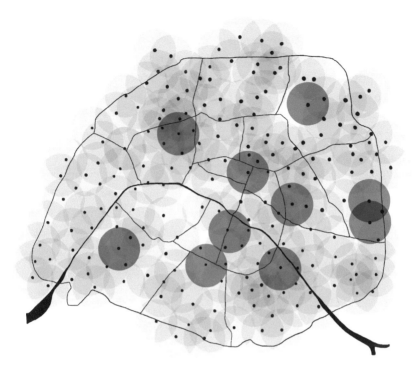

3.3 The OASIS program plans to renovate schoolyards across Paris. Shaded regions indicate 500-meter radii around each school. Darker shaded regions indicate areas around the OASIS pilot sites. *Source*: Mora Orensanz.

environmental outcomes. It included representatives from several departments from the City of Paris, including the Resilience Office. It also included the Paris Councils for Architecture, Urbanism, Environment (CAUE), the Ligue de l'Enseignement (Teachers' League), the Laboratory for Interdisciplinary Evaluation of Public Policies (LIEPP) at Sciences Po, the Interdisciplinary Research Laboratory for Future Energies (ESIEE), and the French meteorological service, Météo France.

The OASIS interventions in the ten schoolyards included intensive co-design processes and substantial adaptive renovations spread across ten different arrondissements (districts) over three years. These pilot projects were to test strategies for converting all of the city's 761 schoolyards into neighborhood-cooling oases by 2050, which, if completed, would increase green space in the city by some 20 percent.[12] Because most Paris residents

live within two hundred meters of a schoolyard, these oases would provide a network of naturally cooled social spaces throughout the city. The
OASIS schoolyard renovations were intended to improve resilience, not
just by physically cooling the environment and increasing neighborhood
social cohesion but also by providing stimulating environments for school
children to learn about natural landscape and climate processes. By the end
of 2022, a total of one hundred schoolyards had been renovated using a
similar set of design processes and strategies.[13]

While the OASIS projects successfully combine built environment adaptation with participatory design and experiential education, there are also
lessons to be learned from its struggles, including challenges in creating
inclusive participation. As with many other projects launched immediately
before the COVID-19 pandemic, some aspects of the OASIS initiative suffered from restricted activities and delays.

CLIMATE REFUGES AND LEARNING OASES: ENVIRONMENTAL
DIMENSIONS OF RESILIENCE

The OASIS initiative is intended to deliver multiple environmental benefits across scales. Proponents initially conceived it as a means of adapting
the built environment to mitigate heat risk for schools and their neighborhoods. The project embraced other environmental goals, including
supporting experiential environmental education and improving ecosystem health beyond the schoolyards. In pursuit of these aims, the program
adopted three central principles: diversification of spaces and materials,
abundant natural materials and vegetation, and sobriety of design, including using low-tech, locally manufactured, and reused materials. All of the
OASIS pilot projects share these three principles and link heat-adaptive
open-space renovations with experiential education. Within this framework, site-specific co-design processes determined the particular features
and form of each intervention (figure 3.4).

COOLING OASES: MITIGATING CLIMATE RISK AT
THE PROJECT SCALE

Outreach materials describe the central aims of the OASIS initiative
as "integrat[ing] nature-based solutions to the schoolyard design as a

mitigation measure for the Urban Heat Island effects and storm water flooding."[14] Most Parisian schoolyards are dominated by low maintenance and durable surfaces, such as concrete and asphalt, which absorb heat and are impervious to rain and stormwater. By replacing impervious surfaces with sand, wood chips, pavers, grass, and planted earth, the OASIS initiative simultaneously reduces local heat absorption and encourages on-site stormwater infiltration, reducing pressure on the city's drainage systems. The renovations also incorporated strategies to create shade, including preserving existing trees, planting new trees, and building vegetated trellises and other structures. OASIS schoolyards also incorporate vegetated swales, rain gardens, and rainwater harvesting to reduce runoff and capture water for irrigation (figure 3.5).[15] While these green infrastructures provide benefits in any weather, designers hoped to "create islands of freshness" in which "vulnerable people could be welcomed during heat waves."[16] Evaluations of pre- and post-intervention environmental conditions on the OASIS sites have not been released at the time of writing, but project planners anticipate that the schoolyard renovations will reduce daytime summer air temperatures between 1°C and 3°C (roughly 2°F–5.5°F).[17]

OASIS interventions primarily focused on excessive heat and stormwater flooding, but renovations at some pilot sites also revealed soil contamination. Remediation of contaminants increased project costs and delayed openings, but also reduced the exposure of children and other users to toxins.

SCHOOLYARDS FOR CLIMATE AND ENVIRONMENTAL LITERACY: INDIVIDUAL-SCALE ENVIRONMENTAL RESILIENCE

Building off of existing research and a study tour to renovated schoolyards in Brussels, Antwerp, and Barcelona, the OASIS team focused on transforming schoolyards into sites for experiential environmental education. OASIS schoolyards replaced barren asphalt lots with learning landscapes featuring topographic relief, climbable structures, quiet nooks, and diverse vegetation, all designed to encourage exploration and "controlled risk-taking."[18] A project leader spoke to the central place of environmental education, saying, "The first thing is to reconnect children with nature, and let them be able to touch vegetables, to see how from a little seed you can grow a plant, to see how seasonal changes can change the schoolyards, and to

a

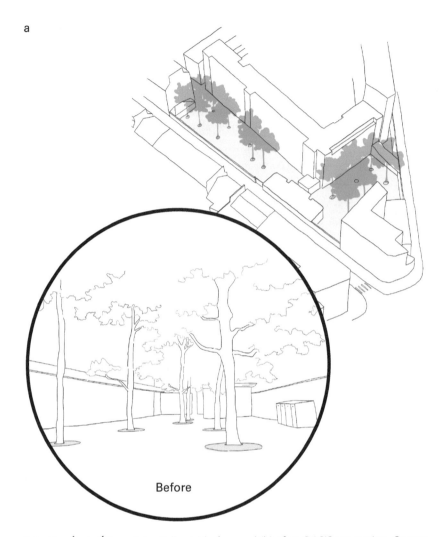

Before

3.4 The École Élémentaire Keller (a) before and (b) after OASIS renovation. *Source:* Mora Orensanz.

see that they are part of this system. In pre-transformation schoolyards, nothing changes because there are no living beings, only concrete."[19] The adapted schoolyards encourage students to interact with their environment through such features as vegetable gardens and composting. Flexible elements within the schoolyards encourage children to "remodel their environment."[20] Renovated schoolyards invite children to make contact with other species, including chickens, pollinator insects, and wild bats

b

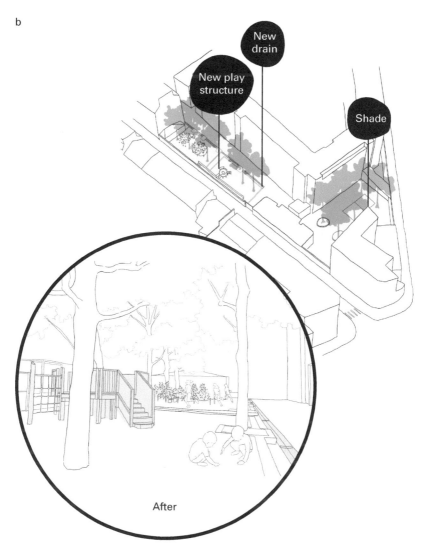

New drain

New play structure

Shade

After

3.4 Continued

and birds.[21] Designers also sought to make environmental processes legible to students. For instance, one project leader remarked, "It is very important that the cycle of water is visible, that the children see it and understand that water is a resource and not garbage."[22]

In addition to creating stimulating spaces for experiential education, proponents used the schoolyard conversion process itself to teach students about climate change and adaptation. During the co-design processes at

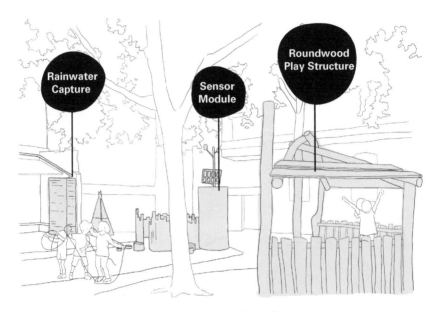

3.5 Pilot OASIS renovations, such as those at École Élémentaire Quatre Fils, included new planting, play structures, green infrastructure, and monitoring equipment both inside the schoolyard and in surrounding areas to assess environmental conditions. *Source*: Mora Orensanz.

each school, Météo France led Weather School sessions to teach students about the basics of weather, climate, and climate change. The Paris Ligue de l'Enseignement created educational toolkits to accompany OASIS projects, including a unit called "Adapt My Schoolyard to Climate Change," which explains climate change and nature-based adaptation. A second unit, called "Live in My Redeveloped Schoolyard," includes games and activities that teachers can lead in renovated schoolyards. These curricula are intended to be used in schools across Paris as the model from the OASIS pilot projects is scaled up.

Preliminary analysis of the social and educational impacts of the OASIS interventions suggests that they are improving conditions on several fronts. In nursery and primary schools in particular, evaluations by researchers at Sciences Po show that OASIS renovations are associated with reductions in student conflict, more mixed interactions between students of different genders and ages, and more positive interactions with nature.[23] In post-intervention surveys, teachers and students express appreciation for the diversity of spaces and experiences available in the renovated schoolyards.[24]

ENVIRONMENTAL RESILIENCE BEYOND THE SITE

The OASIS renovations promise environmental benefits beyond local climate adaptation and experiential education. Cooling and on-site stormwater infiltration benefit neighborhoods beyond the schoolyards. To reduce negative environmental impacts from construction, designers selected sustainable and low-carbon materials, including wood, stone, and recycled materials. By preserving existing trees and planting a diverse mix of local species suitable for projected climate change conditions, designers aimed to support biodiversity within the city. While the schoolyards are not large enough to provide extensive habitat, they can serve as "back-up," supporting populations from larger nearby green spaces.[25]

CO-DESIGN AND NEIGHBORHOOD SOLIDARITY: GOVERNANCE FOR EQUITABLE RESILIENCE

One way that the OASIS projects departed from the City of Paris's earlier technically focused schoolyard renovations was in their participatory design and inclusive programming. Two core aims of the projects reflect this emphasis on the self-governance dimension of equitable resilience. First, the team pledged to "adopt a bottom-up design approach by co-designing the schoolyards with their everyday users; the children, as well as by engaging the broader neighborhood in the process." Second, the project was to "establish a shared understanding of the co-use and co-ownership of the schoolyards by introducing an innovative governance scheme based on the principles of participatory democracy."[26]

The CAUE guided an intensive seven-week co-design process for each of the ten OASIS pilot projects, aiming to familiarize pupils and adults with the issues of resilience and the objectives of the project, inventory uses of the spaces before transformation, identify the needs of the pupils and adults for the new schoolyard, and reimagine the use and management of the new sites by the school and the local community.[27] In the words of one researcher involved in the project, "the philosophy of the ten [pilot projects] is pretty much the same; they are all nature-based and low tech," but "no two are alike, because each one is tailored and co-designed by different communities."[28] Following an inventory of physical conditions, the teams engaged intensively with one class of students in each school and gathered input from the rest of the school community through surveys (figure 3.6).

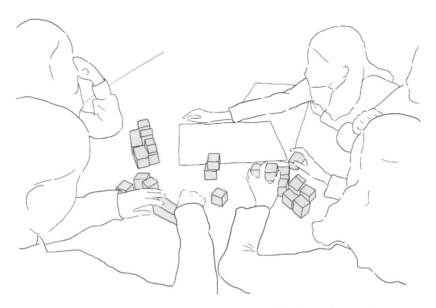

3.6 OASIS co-design process with students in a pilot school. *Source*: Mora Orensanz.

In addition to the CAUE workshops, the co-design process also included Météo France's Weather School and workshops organized by La Ligue, including sessions on soft skills such as conflict resolution and artist-led sessions on sustainability issues.

Following the iterative co-design process with students, teachers, and administrators, CAUE designers developed construction plans with the city departments responsible for construction, project management, and landscape work. In some instances, students planted trees and gardens, integrating lessons about plant life cycles and other topics into the renovation process. Beyond in-person engagement of students, the OASIS team also created a mobile app to gather ideas for schoolyard renovations.

The CAUE-led co-design process with students and other members of the school communities shaped the physical elements of the OASIS schoolyard renovations, but the OASIS initiative also included outreach to the wider communities around each school to inform public programs and uses for renovated schoolyards. La Ligue used participatory democracy strategies for a series of three citizens' assemblies in four different neighborhoods in 2019 and 2020. In the first meeting, a photographer and a local theater group participated in workshops, inviting residents to question uses of neighborhood schoolyards. In the second assembly, a comic-book artist visualized ideas for schoolyard uses in real time. In the third assembly, participants advocated for and voted on new public uses of schoolyards. In addition to informing the public about the projects and gathering ideas for schoolyard programs, the citizens' assemblies were meant to help form OASIS Collectives—groups of volunteers from the surrounding communities who would contribute to the safety and success of the newly public schoolyard spaces after construction was completed.

The OASIS initiative's goals for participatory design, planning, and governance ran into several challenges. COVID-19-related restrictions cancelled many in-person workshops or required that they be convened remotely. A transportation strike in Paris hampered outreach. On top of these exogenous challenges, project participants noted other problems with the design and implementation of the programs. For instance, participants in the citizens' assemblies expressed confusion and frustration because the process felt disconnected from the CAUE-led co-design process. Because the citizens' assemblies preceded most of the physical design choices made in the co-design process, assembly planners could not adequately answer some questions. Questions about funding and staffing for off-hours security, cleaning, and maintenance contributed to a delay of several months in opening the schoolyards to the public.

Perhaps most concerning with respect to the pursuit of equitable resilience, the outreach process was uneven both within each school community and across participating pilot schools. Participation was more robust in wealthier neighborhoods and among affluent residents of mixed-income neighborhoods. An evaluation of the program reported that "the working classes were relatively absent from these citizens' assemblies (including in working-class neighborhoods) and the public mainly came from more affluent socio-economic classes."[29] This uneven participation raises concerns that the programming and design of newly public schoolyards might be skewed toward the interests of affluent residents and might not reflect the priorities of less privileged residents.

The selection process for the OASIS pilot schools presents further equity concerns. The OASIS schoolyard renovations were designed as pilot projects that would inform similar renovations in all of Paris's 761 schoolyards. Early in the process, the city government engaged Bloomberg Associates to analyze UHI conditions across the city so that intervention could be targeted to those sites most in need of cooling refuges. Dissatisfied with the opaque methodology of the Bloomberg analysis, however, academic partners on the OASIS project analyzed a suite of factors for selecting the pilot sites, including minimum size requirements to provide significant cooling, stormwater, and biodiversity improvements. Other selection criteria included the age of the students, geographic dispersal across the city, and the public accessibility of schoolyards. Sites were scored on a scale from 0 to 100. Early analyses also included the prevalence of vulnerable populations, including people living in poverty and elderly residents. However, it appears that officials largely ignored this analysis in the interest of political and functional expedience. One project participant, who wished to remain anonymous, discussed the disconnect between the equity-weighted, data-driven analysis and the pilot site selection, viewed as driven largely by the need to please each arrondissement's mayor. They said, "Some sites should have not been chosen for different reasons. There was a big plan for choosing the sites with different objectives and criteria. And there was a big Excel file and a map. At the end, it was mostly political. The mayor of each district said, 'I want this school to be on the list and not this one.'"[30]

Other project participants reported that the primary criteria used for selecting the pilot sites were whether they were "already going to do renovation work" and whether school leaders "were going to be for the project, or were they going to be against this new approach."[31] While all three of these factors—political support, preexisting renovation plans, and internal support within schools—are understandable for an experimental project requiring buy-in across a range of stakeholders, relying on such unofficial criteria could lead to the selection of already well-resourced schools as pilot sites. If the selection process had truly been focused on advancing equitable resilience, the needs and vulnerability of the school communities and neighborhoods would have been central criteria and the project might have ended up selecting a different group of schoolyards for OASIS renovations.

LIVELIHOODS

The OASIS initiative did not emphasize support for residents' livelihoods, but the program did make some contributions in this realm. The projects

included training for school staff and city workers, introducing new potential uses and functions of schoolyards to advance climate adaptation and
experiential education. Further, the central focus of the initiative on building robust connections between neighborhoods and schools can be seen as
an intervention that supports livelihoods because it enables these essential
community spaces to fulfill new functions, both for students who benefit
from experiential education and for surrounding community members.

SECURITY

Similarly, although security was not central to the conceptualization of
OASIS, the initiative did raise some important questions in this domain.
One OASIS project leader described "the true resilience value" of the interventions as lying "beyond the environmental challenges." They went on
to say that "the OASIS schoolyards aim to strengthen the neighborhoods'
social cohesion, by becoming the neighborhood's meeting place, as well
as to promote civic participation, by fostering the development of citizen initiatives, (including) the envisioned 'OASIS Collectives.'"[32] Another
project leader reported that the initiative's goals included "reinforcement
of social cohesion" in response to Paris's increasing "social inequality."[33]
In Paris, as in many other cities, inequality is shaped by the structural
disadvantages faced by racial minorities and immigrants. Project leaders
hoped that OASIS would "engage parents who are not familiar with how
the French educational system works, so that they can better understand
the cooling strategy adopted by the school and feel the school as a welcoming place."[34]

Even though OASIS proponents prioritized building social cohesion
by encouraging residents to take broader ownership and find new neighborhood value, the community design processes struggled to engage
lower-income residents. The formation of OASIS Collectives, volunteer
community stewardship committees envisioned as an important mechanism for connecting schools to their neighborhood communities, has been
slow, initially hampered by COVID-19 restrictions.[35] Some project participants raised concerns that schoolyard renovations could actually reduce
some residents' security by contributing to green gentrification and climate
gentrification. They worried that while the Collectives were envisioned as

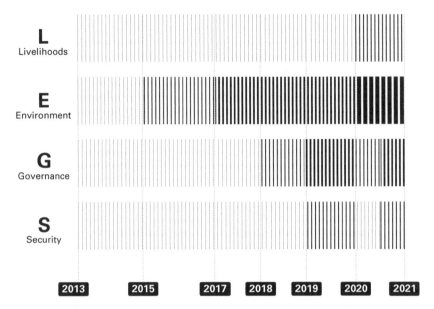

3.7 Paris OASIS pilot projects focused primarily on environmental improvements and co-design governance process. *Source*: Smriti Bhaya. 2013: Heat wave in Paris; 2015: Paris Resilience Strategy; 2017: Initial schoolyard renovations completed; 2018: Urban Innovative Actions grant awarded for OASIS initiative; 2019: Curriculum development and co-design processes; 2020: Schoolyard design and renovation proceed despite COVID-19-hampered outreach; 2022: Continued development of public uses and expansion to one hundred additional schoolyard sites.

a means to affirm belonging, without further progress in these domains, renovations could accelerate displacement in surrounding neighborhoods.

CONCLUSION: A PROMISING PILOT

In the face of daunting challenges presented by the pandemic, a transit strike, and other impediments, the OASIS initiative successfully used co-design and participatory democracy processes to guide ten pilot climate-adaptive schoolyard renovation projects (figure 3.7). The initiative's layering of physical adaptation and experiential environmental education with participatory design and planning clearly reflects the complex socio-ecological nature of climate resilience. However, the initiative may also have fallen short of some aims because of the unsystematic selection process for pilot sites, struggles to engage disadvantaged groups adequately, and limited attention to concerns about how OASIS renovations might contribute to green gentrification.

Although the OASIS pilot projects were rightfully regarded as successful, the team faced challenges in building off of these demonstration projects. As with the externally funded Gentilly Resilience District in New Orleans, it was initially unclear how the EU-funded OASIS pilot projects could be scaled up beyond the pilot phase. While conceiving of such adaptation interventions as discrete *projects* enables them to access funding and makes the interventions imageable to outside observers, a project orientation can cause problems when project funding runs out. One challenge to scaling the OASIS projects came from the very site-specific, community-engaged process that made the pilots so compelling. Project participants recognized that "there is no one size fits all solution" and that every schoolyard renovation requires site-specific and community-specific engagement because "you cannot do the same thing everywhere."[36] This deep engagement and site specificity allow for schoolyard projects that fit the needs and circumstances of local communities, but also make it challenging to scale up adaptive renovations to a larger group of schools.

As the city government has expanded OASIS across Paris, they have maintained many of the processes and principles from the pilot phase while adapting some aspects for efficiency. By the end of 2022, one hundred schoolyards had been renovated, using co-design processes with students to create climate-adaptive pedagogical landscapes. Due to limited staff resources within the city government, the scaled-up projects no longer deploy citizens' assemblies during co-design. However, the renovations are still intended to benefit their wider neighborhoods. The city maintains an open call for proposals soliciting ideas for how community groups might use the spaces during non-school hours. Many OASIS schoolyards are open on Saturdays as part of Paris's 15-Minute City initiative, a model for planning and urban design that promises to reduce the need for car travel to access basic needs.

After some initial coordination challenges, these scaled-up efforts signal promising linkages between OASIS and related projects in Paris, including a massive expansion of bicycle infrastructure, greening and pedestrianization of many formerly car-choked streets, and the renovation of 168 *rues aux écoles* or school streets, reducing vehicle traffic and introducing green infrastructure in streets near city schools.[37] These initiatives share the OASIS initiative's vision for spatially distributed climate action and resilience investments. Initially, OASIS participants reported little synchronization with these projects, but subsequent efforts suggest that robust coordination could ensure that renovations of schoolyards are sequenced and designed to advance climate equity, efficient design and construction, and effective functional adaptation.

The successes and shortcomings of the OASIS schoolyard renovation initiative offer especially valuable insights because such a program is theoretically replicable in many cities around the world facing extreme heat and other climate change impacts. In every city, school facilities present a ready-made network of geographically distributed, socially embedded institutions that could be redesigned and reprogrammed to advance equitable resilience. The OASIS initiative initially struggled to engage equitable participation and faced difficulties coordinating with other climate action and resilience initiatives. As the projects expand, however, their allied curriculum development, data gathering, and documentation of lessons learned have laid the groundwork for future schoolyard renovation projects in Paris and beyond.

LEARNING FROM THREE STRUGGLES FOR EQUITABLY RESILIENT ENVIRONMENTS

Although the three cases presented in this section are all government-led initiatives to reduce vulnerability to climate change–related hazards, they differ dramatically in both their contexts and the foci of their interventions. The Gentilly Resilience District promises a new model of distributed stormwater infrastructure for an economically and racially diverse suburb-like neighborhood in New Orleans. By contrast, the Paraisópolis Condomínios were built to rehouse thousands of residents safely from nearby hyper-dense favela areas exposed to landslides and flooding. Finally, as in New Orleans, the Paris OASIS initiative deploys a distributed green infrastructure model. The pilot OASIS schoolyard renovations demonstrate the potential of this strategy to create a network of neighborhood heat-mitigation sites across the dense French capital.

The cases demonstrate strategies focused on city centers and suburbs. They aim to transform infrastructural systems, low-income housing, and schoolyards. They intervene in celebrated centuries-old urban fabric, in informal settlements, and in car-oriented postwar suburbs. Despite this diversity, the cases do share several elements. Each recognizes that vulnerability to climate hazards arises not from geophysical risk alone but also from a confluence of social and environmental determinants. As such, each intervention aims to address climate threats in communities with significant socioeconomic diversity and disadvantage. The projects in each case attend,

Table I.2

Multiscalar analysis of environment cases

	Gentilly Resilience District	Paraisópolis Condomínios	Paris OASIS
Individual/ Household Landscape Literacy	Community events, teaching landscapes	Little engagement on environmental education	Curriculum development, experiential education
Project Risk Mitigation	Green infrastructure for urban heat and stormwater management	Rehousing from landslide and flood-prone areas	Green infrastructure for urban heat and stormwater management
City/Region Ecological Systems Health	Water quality, biodiversity, heat mitigation	Energy efficiency and green building	Water quality and efficiency, biodiversity, materials selection

at least partially, to all three scales of intervention that we outline (table I.2). Each of these first three cases also goes well beyond a narrow focus on conventional environment and hazard concerns, incorporating elements of the other core dimensions of equitable resilience to varying degrees.

The three projects profiled in the "environment" section also share some common struggles constraining them from advancing equitable resilience more fully. To progress, they must move beyond project-based interventions, improve conditions for disadvantaged populations without inadvertently threatening their security or livelihoods, and overcome the well-earned distrust that many disadvantaged groups feel toward formal planning, design, and city government institutions.

Organizing interventions into discrete projects is often viewed as a practical necessity of contracting and management, but projectification can run counter to the demands of equitable resilience.[1] A project is necessarily defined in both space and time. Projects have beginnings, endings, and territorial constraints. While a project may be pronounced to be complete once its stated goals have been achieved, these cases and others profiled in subsequent sections make clear that achieving transformative systems change in pursuit of equitable resilience cannot be cleanly segmented into discrete projects. This tension is most evident in the Gentilly Resilience

District, where federal contracting requirements and municipal manage-ment practices have led to a project-by-project approach that may under-mine the overall vision of distributed, integrated, and adaptable systems. Focusing on the green infrastructure projects that make up the GRD may also obscure the limitations of this approach in addressing the persistent underlying vulnerability of Gentilly to future gray infrastructure failures. Similarly, as impressive as the scale of the Paraisópolis Condomínio proj-ects may be, they do not represent a systemic change capable of addressing the underlying vulnerability of the favela's residents more broadly. More auspiciously, as of 2023, the OASIS program has been able to expand from the ten initial pilot sites to renovate more than one hundred schoolyards throughout Paris.

Although proponents of each of these three cases envisioned them as providing broad-based benefits to diverse urban populations, participants have acknowledged that each of these interventions could actually threaten the security of low-income residents seeking to remain in their neighbor-hoods. Project leaders in both New Orleans and Paris expressed concerns that, without further attention to housing affordability, their projects could drive green gentrification. For many new residents of the Paraisópo-lis Condomínios, the shift from informal to formal housing has strained their household finances and social structures, causing some to sell their units to more well-off people and move back to informal neighborhoods.

Finally, in each of these three cases, well-intentioned project leaders encountered entrenched distrust from residents in disadvantaged commu-nities, challenging their ambitions for building participatory, community-engaged interventions. GRD officials remarked on the unrealistic expectation that Gentilly residents would volunteer their time to maintain city-built green infrastructure, particularly in the wake of the green dot map, a green infrastructure mega-plan that resonated with past episodes of racialized planning violence and clearance in New Orleans. Paraisópo-lis Condomínio planners encountered intense suspicion from favela resi-dents, weary and wary from past and current clearance efforts. Resident mistrust, coupled with COVID-19 restrictions, meant that Paris OASIS planners struggled to create their OASIS Collectives of volunteer neighbor-hood stewards and struggled to engage low-income and immigrant com-munities. Solidarities emerge only slowly and can be derailed by delays.

Mitigating environmental risk is necessary but not sufficient to pursue equitable resilience. Overcoming persistent distrust will require equitable resilience interventions that go beyond environmental goals to support dignified livelihoods and enable self-governance for disadvantaged communities. Perhaps most fundamentally, environmental improvements such as reducing climate risk, improving hazard literacy, and strengthening ecosystem health will not provide benefits to disadvantaged groups if those same people lack the security to resist displacement. Our exploration of the equitably resilient city turns next to the pursuit of security.

II.
SECURITY

EQUITABLY RESILIENT SECURITY: STABILITY, COHESION, AND RECOGNITION

Reducing the impacts of climate-related hazards is of little use to poor people if resilience efforts undermine their security—especially security against displacement. This section explores how the pursuit of climate resilience can threaten or reinforce the security of vulnerable residents.

The term "security" carries varied connotations in the study of planning and urbanization. "Urban resilience" in research and public policy after the 9/11 attacks meant security from terrorist violence.[1] Many governments invoke security in pursuit of "law and order" to counter social movements demanding justice, from the widening of Paris's streets to forestall regime-threatening barricades in the nineteenth-century to violent police responses to the Black Lives Matter protests of 2020. Privileged groups from British colonial India to Apartheid South Africa have justified the segregation of cities by race and class in the name of security.

Appeals to security are frequently associated with reactionary and regressive urban politics, but the term is also used by insurgent actors and grassroots organizations fighting for humane conditions for disadvantaged groups. We embrace this latter progressive framing of security as one of the four "LEGS" of equitable resilience. More precisely, our framing focuses on security from displacement—an especially salient concern as critics raise alarms that adaptation and resilience efforts can directly and indirectly push disadvantaged residents from their homes and communities.

Climate change and resilience interventions can be associated with various forms of insecurity, including both displacement and threats of conflict,[2] violence,[3] and crime.[4] While we briefly discuss how planners and designers have confronted other forms of insecurity, the primary focus of this chapter and the subsequent case studies is security from displacement because planning and design interventions have a long track record of displacing disadvantaged people from their homes and communities and because people who are forcibly displaced from their communities and social networks often experience a form of violent "root shock."[5]

This chapter explores how planning and design disciplines have embraced the concept of security in the past, discusses the relationship between climate change and (in)security with particular attention to displacement, and presents a multiscalar conceptualization of how security contributes to equitable resilience. The case studies that follow illustrate partial successes and ongoing struggles associated with pursuing security in the name of equitable resilience.

PLANNING, DESIGN, AND SECURITY

The disciplines of urban planning and design have a multifaceted relationship with security. In exploring how security from displacement contributes to equitable resilience, we take stock of some of the ways that planning and design disciplines have dealt with security in the past.

There is a long track record of urban design interventions being justified in the name of improving security in the face of crime, violence, and social disorder, extending to late twentieth-century invocations of "crime prevention through urban design" and the "broken windows" theory of policing.[6] Some researchers frame urban resilience as centrally focusing on the impacts of armed conflict, crime, and state violence.[7]

While urban inventions to increase security from violence and crime present vital areas for research, our focus is more centrally on security from displacement. Around the globe, grand urban projects have displaced whole communities, frequently placing the heaviest burdens on the poor. The mid-nineteenth-century "renovations" of Paris under Georges-Eugène Haussmann—Napoleon III's civil servant responsible for Paris's new infrastructures, grand boulevards, and parks—demolished an estimated 27,500 buildings and displaced some 350,000 people. These projects pushed many poor Parisians into shantytowns at the urban periphery to make way for glittering new urban spaces largely meant for the growing bourgeoisie.[8] Subsequently, in city after city, similar Haussmannization projects have displaced poor people to benefit more affluent residents, ranging from urban renewal projects in mid-twentieth-century US cities to the clearance of "urban villages" in favor of high-end shopping malls in fast-growing Chinese cities such as Shenzhen.

Interventions that could today be touted as advancing urban resilience have often dislocated disfavored people, both within cities and in their hinterlands. Dams intended to provide flood protection and water supplies have displaced tens of millions of people around the world.[9] In the name of improving urban resilience, city leaders often evict poor urban residents from hazard-exposed landscapes such as steep slopes and floodplains, from New Orleans to Dhaka to São Paulo. While leaders justify these evictions in the name of reducing hazard vulnerability, the projects often also include new urban green spaces that align with the desires of "bourgeois environmentalism."[10]

Even as government interventions have led to mass displacements in cities around the globe, planners and allied professionals are increasingly attentive to ways that security of housing and land tenure enable urban poor residents to improve their lives. Peruvian economist Hernando de Soto's influential book, *The Mystery of Capital*, advocated mass titling for residents of informal settlements in fast-growing cities of the Global South. The World Bank and other development institutions widely adopted these recommendations, arguing that formalizing land and housing tenure in informal settlements could "unlock" vast stores of wealth.[11] Subsequent research found that while security of tenure did lead to improved conditions in many dimensions of life for the urban poor, including increased investments in household adaptation to environmental risk,[12] obtaining formal legal title to land and housing was less important than community-sanctioned security and property rights, which often does not require legal recognition.[13] Further, evidence from around the world has shown that promoting individual freehold property regimes in formerly informal settlements can lead to speculative pressure and gentrification, displacing the very residents who were supposed to benefit from formalization.[14] More recent research has argued for going beyond simplistic binaries of "formal–informal" or "strong–weak" tenure to recognize the wide range of tenure arrangements that govern how people live, how they adapt to environmental risk, and how they maintain their access to the advantages of city life.[15]

Disruptive change in built environments can undermine security by directly displacing residents or threatening their right to stay by accelerating gentrification. Yet, as we argue in this chapter and in the case studies that follow, design and planning interventions can also improve security.

CLIMATE CRISES AND SECURITY

Climate change threatens the security of urban residents unevenly, both through direct hazard impacts and by motivating heavy-handed interventions with disastrous consequences.

Households lacking secure tenure, including those in informal settlements, are especially vulnerable to climate change and other environmental hazards for several reasons.[16] Hazard-prone areas, including steep hillsides and floodplains, are often home to people without formal housing and land tenure because poverty and other forms of social marginalization exclude disadvantaged groups from safer formally urbanized areas.[17] Informal settlements are also often more sensitive to hazards because they frequently do not have adequate public infrastructure, such as slope-stabilizing retaining walls and adequate drainage. People threatened with displacement are less likely to make investments such as structurally reinforcing homes against high winds or earthquakes or elevating living areas above floodplains. People without strong property rights claims can also struggle to recover from disruptions because of limited access to government information and assistance.[18]

Like residents of informal settlements, people who formally rent housing also confront disproportionate disadvantage during times of environmental upheaval. Renters face greater vulnerability than those who own their homes for several reasons, including more hazard-prone locations, poorer physical conditions, less control over buildings, lower social connectivity, and reduced access to resources after disruptions.[19]

Disaster events can also lead to displacement by upsetting private housing markets, reducing housing supplies in impacted areas, inflating rents in remaining homes, and inspiring evictions,[20] especially in areas with weak tenant protections.[21] Rents frequently increase after hazard events.[22] Even when disasters render large portions of a city's rental housing uninhabitable, renters often receive less government support than homeowners,[23] and rental housing is often slower to be rebuilt than for-sale housing.[24]

Researchers have raised alarms about emerging patterns of climate gentrification, in which low-income residents of relatively safe areas face increased housing insecurity as more affluent people move in.[25] In other cases, investments in green infrastructure can create new amenities that contribute to green gentrification.[26] Even in cases such as New York City's post–Hurricane Sandy East Side Coastal Resiliency Project, where community input from low-income residents and anti-displacement efforts have been central to planning new flood resilience infrastructure and parkland, slow progress can allow existing gentrification pressures to overwhelm well-intentioned efforts.[27] Recognizing these threats, low-income residents of relatively flood-safe neighborhoods in Miami have called for the city's climate adaptation–focused Miami Forever bond to fund affordable housing as part of increasing resilience.[28]

Disasters can also threaten tenure security through other means, including the destruction of physical documents proving ownership, post-disaster mass evictions, and privatization drives.[29] Disruptions from hazards can precipitate what Naomi Klein has called "disaster capitalism," enabling opportunistic privatization of formerly public spaces, goods, and services that are essential to the lives of poor people.[30] For instance, policymakers and development interests in the island nation of Barbuda used the devastation wrought by Hurricane Irma in 2017 to justify privatizing lands long held in common by island residents.[31]

Security threats from climate change often become visible in sudden violent episodes. Yet, disproportionate impacts on disadvantaged groups are rooted in generations of structural oppression, often embedded in racialized and colonial property regimes that enable marginalization and exploitation.[32] As one example, residents displaced by the post-Katrina demolition of New Orleans's public housing were overwhelmingly Black. Black New Orleanians were also much more likely to be displaced by earlier urban interventions, from slum clearance to redlining to urban highway construction.[33] Often, climate disasters create new justifications for "purging the poorest" in "twice-cleared communities."[34] In a similar pattern of repeated displacement and climate vulnerability, North American Indigenous populations that had been forcibly displaced by settler colonialism are now more exposed to climate threats.[35] Elsewhere, colonial regimes imported and imposed systems of privatized property that enabled settlement, extraction, and accumulation.[36] These individualizing and inflexible forms of property are now coming into conflict with changing climates and landscapes—a mismatch that is especially problematic in dynamic landscapes such as those along the Mississippi River[37] and in the Bengal Delta.[38] Such patterns demonstrate how racialized property regimes entrench inequality by widening disparities in vulnerability to climate change.[39]

Governments around the world increasingly propose to relocate people away from the landscapes most exposed to climate change. Whether framed as "managed retreat," "relocation," or "buyouts," many planners and residents regard some degree of climate change–induced settlement restructuring as inevitable, especially along flood-prone shorelines.[40] Some researchers argue that managed retreat can be a form of equity-enhancing "transformative adaptation," creating new opportunities for otherwise disadvantaged people.[41] Others point out that, too often, relocation has been a driver of "adaptation privilege" and "adaptation oppression," rewarding groups who are already invested in dominant property regimes and harming groups, including Indigenous peoples, whose relationships with land and property follow other non-privatized logics.[42] Post-disaster relocation can also harm disempowered groups by disrupting social and economic networks, depriving people of community solidarities and livelihoods.[43] Successful managed retreat also requires that residents of receiving communities not be excessively disadvantaged by new arrivals.[44]

Official judgments about which places, people, and ways of life will be rendered nonviable by climate change can also lead to stark inequities, threatening the security of already disempowered groups. One researcher has identified what they call "anticipatory ruination" in which government and development actors treat the climate-induced destruction of farming communities in Southern Bangladesh as inevitable. Labeling these places as doomed to climate destruction, officials have argued for the radical transformation of landscapes to accommodate export-oriented shrimp farming in the name of climate-adaptive development.[45] Such practices underscore ways that evocations of "adaptation" and "resilience" can advance agendas that are not *equitably* resilient.

Whether through direct displacement due to climate impacts and adaptation-related evictions or indirect displacement through uneven patterns of climate change devaluation and gentrification, climate crises can exacerbate uneven development and insecurity. As we noted in the Introduction, critical scholars have charged that policies focused on urban ecological security can lead to conditions of "eco-apartheid" or "climate apartheid" in which a privileged few enjoy life in "premium ecological enclaves," while most people suffer tremendous insecurity and vulnerability.[46] Even as we later document cases of more equitably resilient efforts to enhance security, we cannot lose sight of such dangers.

EQUITABLE RESILIENCE THROUGH SECURITY ACROSS SCALES

Improved security—from displacement and other forms of violence—for disadvantaged people is an integral part of equitable resilience. An intervention advances the security dimension of equitable resilience if it improves conditions for disadvantaged people across three scales by advancing (1) *stability* (individual or household scale), supporting or maintaining residents' sense of safety, belonging, and place attachment; (2) *cohesion* (community or project scale), supporting or maintaining solidarity among residents as a means of protecting their right to remain in place; and (3) *recognition* (city or regional scale), supporting residents in gaining or maintaining formal security and acknowledgment from outside authorities in order to resist displacement and other forms of violence.

STABILITY: RESILIENCE THROUGH INDIVIDUAL- AND HOUSEHOLD-LEVEL SECURITY

Security depends on social relationships, both within a community (cohesion) and between people and larger institutions, including governments (recognition). However, the sense of connection that people have to a particular place, often referred to as "place attachment," can also significantly impact security and resilience in the face of displacement and violence. Place attachment, which Dolores Hayden describes as being made up of the "material, social, and imaginative" connections between people and places,[47] is intimately intertwined with how people experience and respond to environmental changes. Climate change and adaptation interventions can strengthen, weaken, and otherwise alter place attachment. Communities in areas that are highly vulnerable to climate change, such as the South Pacific and the Arctic, have been confronted with dramatic alterations to their climate, landscapes, resources, and livelihoods. In some cases, these changes have upset long-held traditional ecological knowledge and place attachments.[48] While this area needs further research, place attachment shapes how and whether people adopt significant climate actions, both adapting to vulnerability and adopting behavioral and technological changes for decarbonization. Research has shown that stronger place attachment may impact climate adaptation by making people less accepting of protective infrastructure and less willing to relocate or change their livelihoods in response to climate threats. In some cases, resisting disruptive change from adaptation may actually help people to maintain their sense of well-being and security.[49]

While environmental psychologists analyze and measure place attachment, planners and designers have sought to strengthen place identity through deliberate intervention. Echoing Kevin Lynch's conceptualization of "imageability" in city design,[50] researchers argue that particular qualities of places, including defined edges and prominent or personally meaningful landmarks, can strengthen place attachment.[51] Design professionals and others emphasize "placemaking" and, more recently, "placekeeping" as ways to foster or maintain a sense of belonging and engagement with the public realm.[52] Placemaking can take many forms, use diverse strategies, and advance divergent agendas, including real estate–driven place marketing,[53] "tactical" experiments to gather data about potential interventions,[54] and resistance against the erosion of urban commons,[55] as in the case of community-driven "Black placemaking" in American cities.[56]

Despite substantial research showing the important role of place attachment in shaping climate adaptation and mitigation choices,[57] it is still rare for adaptation interventions to address place attachment and placemaking explicitly. Nonetheless, several of the projects that we feature include efforts to enhance the public realm in support of place attachment. The Paris OASIS pilot projects (case 3) open schoolyards to surrounding communities to foster greater cohesion while incorporating physical retrofits to improve resilience to extreme heat and other climate threats. The Kounkuey Design Initiative's Kibera Public Space Project (case 11) seeks to reduce environmental threats such as flooding in one of Africa's largest informal settlements while creating much needed public spaces that accommodate markets, community meetings, and informal gathering. The Gentilly Resilience District in New Orleans (case 1) includes arts-based community outreach, linking public art to water and climate resilience.[58] Thunder Valley Community Development Corporation (CDC)'s new community on the Pine Ridge Reservation (case 10) explicitly uses Indigenous design elements when siting homes and public space. Finally, Living Cully (case 7) integrates art and placemaking in new spaces designed with and for residents of a historically disadvantaged Portland, Oregon, neighborhood.

For some households, neighborhood stability also depends centrally on perceptions of safety from physical harm, including domestic violence, crime, and gang activity. Several of the cases discussed in this volume— notably the Paraisópolis Condomínios project in São Paulo (case 2) and the KDI interventions in Kibera (case 10)—make clear that efforts to uphold security from violence can both guide personal struggles and propel community solidarities.

COHESION: RESILIENCE THROUGH SECURITY AT THE COMMUNITY SCALE

Community solidarity can enhance equitable resilience by increasing security in the face of displacement. In many cases, informal norms and expectations can be more important to tenure security than formal titling and recognition, especially because—as seen in Paraisópolis—formalization of tenure often comes with increased speculative pressure and household financial burdens. While many development agencies and governments embraced de Soto's view that formalizing land and housing titles is essential to "unlocking" the development potential of the urban poor, research suggests that formalization and tenure security are not the same thing and that informal tenure security can actually allow for greater levels of flexibility in adaptation in some circumstances.[59]

RECOGNITION: RESILIENCE THROUGH STATE RECOGNITION

State recognition is not always necessary to resist displacement, but formal legitimacy bestowed by legal recognition can provide added security. The upgrading of the Quebrada Juan Bobo community in Medellín, Colombia, is a powerful example of how state recognition, aligned with physical upgrading of homes and infrastructure, can build equitable resilience. The 1,200 residents of the settlement on the steep hillsides along a creek faced enormous social and environmental challenges due to their lack of formal tenure security and inadequate housing and infrastructure. In the late 2000s, the municipal government undertook an ambitious upgrading program in Juan Bobo, including new retaining walls, pedestrian infrastructure, and public spaces along the creek; improvements to some private homes; relocation of households from precarious areas along the creek to new nearby multifamily buildings; and new sewer and water infrastructure to reduce flooding and water pollution. In addition to physical upgrades, municipal authorities vowed that the project would not displace any residents and formalized the housing tenure of many households.[60]

While tenure formalization, coupled with physical improvements, can contribute to equitable resilience, there is also ample evidence that individual formal tenure can actually undermine community security by driving speculation and gentrification.[61] Fortunately, there are many alternatives beyond precarious informal arrangements or formal individual private tenure. By one account, more than three billion people on more than half of the territory on Earth hold land through various forms of collective, customary, and community-based tenure.[62] There are an enormous variety of alternative and collective property regimes that aim to deliver security and stability without privatization, including Indigenous property regimes, public ownership, nonprofit ownership, cooperatives, and community land trusts. Many of our case studies feature such structures.

Public and municipal ownership is one common pathway to provide secure and affordable housing for low-income urban residents. Cities from Hong Kong and Singapore to Vienna and Zürich have a long track record of successful public housing. In many other areas, including US cities facing affordable housing crises, advocates and policymakers are increasingly calling for reinvestment in publicly supported "social housing"[63] after generations of disinvestment and privatization.[64]

In China's urban villages, including Dafen village (see case 9), a model of collective land ownership and self-governance has enabled many villages to resist displacement in the face of urbanization pressure while also making upgrades benefiting both villagers and migrant laborers.[65] Urban village councils, like the governing boards in resident-owned manufactured home parks in the US (case 4) and other cooperative property regimes, can provide a means for low-income residents to safeguard their collective housing security. Unfortunately, decision making in these settings can also be cumbersome and prone to distortion by internal conflicts and power differentials.[66]

After early experiments in rural Black farming communities in the southern US, community land trusts have gained traction as a form of collective ownership in many rural and urban areas around the world.[67] CLTs can provide security of tenure for residents, and their inclusive board structure typically links them to the well-being of the wider community. There is growing interest among advocates, policymakers, and researchers in using the CLT model to support community resilience in climate-vulnerable communities from the Florida Keys and New Orleans to San Juan, Puerto Rico.[68] However, CLTs frequently struggle to attain significant scale,[69] and few CLTs, aside from Fideicomiso de la Tierra del Caño Martín Peña in San Juan (case 12), have placed climate resilience as a central goal.

Drawing together the range of strategies discussed in this overview, table II.1 sets out the challenges of equitable security at various scales.

PARTIAL SUCCESSES IN EQUITABLE SECURITY RESILIENCE

Climate change threatens to undermine the security of poor people in urban settlements around the world by increasing their vulnerability to displacement. Interventions undertaken to promote climate resilience can strengthen or weaken the security of disadvantaged urban residents. As with struggles of the urban poor more generally, unequal insecurity in the face of climate threats is intertwined with the ownership and control of property and land. Privatized urban property regimes enable the accumulation of wealth by favored groups while exposing historically marginalized groups to heightened threats from environmental hazards, displacement, and violence.

Each of the case studies that follow demonstrates the transformative potentials and the challenges of resisting both privatized property regimes and the precarity of insecure tenure. In the Pasadena Trails (case 4) community near Houston, Texas, Latin American immigrant residents of an

Table II.1

Three scales of equitable resilience through improved security against displacement

Scale	Equitable Resilience Principle	Question	Examples
Individual/ Household	Stability	Does the project support or maintain residents' sense of safety, stability, belonging, and attachment to their place?	Climate resilience through placemaking/ placekeeping: KDI Kibera Public Space Project, Gentilly Resilience District Public Art Project, Native Gathering Garden at Cully Park
Project/ Community	Cohesion	Does the project support or maintain residents' informal security through collective solidarity to resist threats, including displacement?	Comunidad María Auxiliadora
City/Region	Recognition	Does the project help residents gain formal recognition to resist threats, including displacement?	Juan Bobo upgrading, Medellín, Colombia; ENLACE Caño Martín Peña

MHP joined together to form a cooperative to buy and manage the land and infrastructure on which their homes rely. The community is one of a growing number of resident-owned communities (ROCs), an alternative to conventional MHP tenure which often burdens residents with dual vulnerability to displacement and environmental hazards. At Pasadena Trails, the Latina-led cooperative prioritized investments in improved drainage to address chronic flooding and energy-efficient street lighting and traffic calming for resident safety.

While the residents of Pasadena Trails faced excess water, residents of the Comunidad María Auxiliadora (case 5) struggled to secure enough water in the high mountain desert of the fast-urbanizing periphery of Cochabamba, Bolivia. Like Pasadena Trails, leadership from women residents proved essential to CMA's creation as a collective alternative to privatized property. While CMA residents succeeded in establishing a common property regime that enabled them to build homes and collective water infrastructure, the lack of official recognition of their collective tenure rights made the community vulnerable. After years of success, internal divisions and speculative pressure undermined the community's solidarity, and the collective property regime collapsed.

The final case study in this section is also the most ambitious in scale. For more than two decades, the Baan Mankong program (case 6) has included collective land tenure as part of regularization and upgrading informal settlements across Thailand. We focus on a subset of Baan Mankong communities along the canals of the capital region of Bangkok. Although the program's collective ownership and networked governance model has proven successful at improving security for formerly precarious communities and upgrading their physical infrastructure and housing to reduce flood vulnerability, recent attempts to scale up and speed up the program threaten to undermine its place-based and community-driven legitimacy.

After the three case-study chapters, we conclude the section with a brief synthesis of common themes across the three cases and highlight areas for future research and action to promote equitable resilience through enhanced security.

Case 4

PASADENA TRAILS: RESIDENT-OWNED RESILIENCE IN MANUFACTURED HOME PARKS

OVERVIEW

Residents of manufactured home parks (MHPs) face a dual burden of vulnerability to displacement and environmental hazards. The model of resident-owned communities (ROCs) championed by ROC USA, a national advocacy organization and network of communities, represents a powerful example of low- to moderate-income people building secure communities with the resources, agency, and self-governance structures to address these vulnerabilities. Drawing on site visits and interviews conducted at a ROC

4.1 A streetscape in the Pasadena Trails resident-owned community. *Source*: Jesús Contreras.

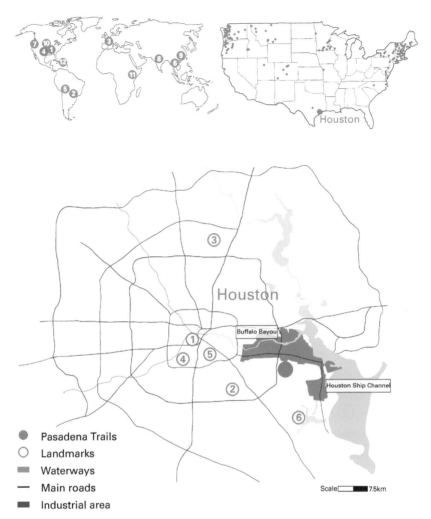

- ● Pasadena Trails
- ○ Landmarks
- ▬ Waterways
- — Main roads
- ▬ Industrial area

4.2 Pasadena Trails location within the ROC USA network (above right) and within the Houston metropolitan area (below). *Source*: Mora Orensanz. 1. Downtown Houston; 2. Hobby Airport; 3. George Bush International Airport; 4. Texas Medical Center; 5. University of Houston; 6. Space Center Houston.

USA community near Houston, Texas, and other interviews from across the country, we show that while ROC residents still face elevated vulnerability to climate change, the ROC USA model provides the tools and resources to build equitable resilience (figure 4.2).

INTRODUCTION: A MANUFACTURED SOLUTION FOR HOUSING

More than twenty million people across the US live in manufactured housing, a form of factory-built homes. Manufactured housing became a major part of the US affordable housing sector following housing shortages after World War II, when millions of families took up permanent residence in what were then called "mobile homes." [1] While many people place mobile homes on privately owned land, about forty percent are situated in specifically equipped neighborhoods called "mobile home parks" or "manufactured home parks."[2]

MHPs commonly operate according to a land-lease arrangement in which residents own their homes but pay rent to owners who control the land, common facilities, and infrastructure. This land-lease arrangement, along with the lower construction costs of factory-built housing (approximately half the per-square-foot expense of site-built housing),[3] makes living in an MHP a rare affordable option in many places. In fact, manufactured housing is the largest source of unsubsidized affordable housing in the US, serving more households than public housing and federally subsidized rental housing combined.[4]

Although MHPs provide affordable housing for millions of people, residents are vulnerable to displacement and financial exploitation. Despite the "mobile home" label, it is often expensive or impossible to move these housing units, leaving residents with little recourse if MHP owners allow park conditions to deteriorate, raise rents or fees, or close a park to redevelop the land for other uses.[5] The precarious condition of MHP residents is growing worse in many parks as large investors, from private equity and pension funds to sovereign wealth funds, buy MHPs. For investors, the fact that residents have few alternative housing options and little agency when faced with rising costs makes these communities ideal assets, ensuring that they can ratchet up returns year after year.[6]

On top of this housing insecurity, MHP residents face disproportionate vulnerability to environmental hazards, including climate threats. MHPs are often located in hazard-prone sites because such land tends to be inexpensive and because biased zoning regulations often bar MHPs from residentially zoned areas. An analysis of MHPs in nine states estimated that 22 percent are located within one hundred-year flood zones.[7] MHPs are also disproportionately located in areas that are poorly served by existing infrastructure and services and in areas exposed to extreme heat and wildfire.[8] When hazards strike, MHP residents often suffer more because social and physical factors make them more sensitive. Many homes predate the adoption of federal building standards in 1976, and many are not installed on secure foundations, making them vulnerable to earthquakes, high winds, and flood waters. Because many parks were built with inexpensive water, sewer, power, and road infrastructure to save costs and maximize profits, the infrastructure in MHPs is also often vulnerable. Socially, MHP residents are more sensitive too because they are more likely to have low incomes, to be elderly, to have disabilities, and to have low educational attainment, as well as having other demographic factors associated with elevated "social vulnerability."[9] Finally, MHP residents have less capacity to adapt and recover from disruptions not only because of socioeconomic disadvantages but also because land-lease arrangements substantially limit their agency to make adaptive changes in their built environments.[10]

While MHP residents typically face heightened vulnerability to housing insecurity and environmental hazards, the ROC model presents an alternative that can help residents improve both their housing security and their resilience to climate change and other threats. There are an estimated one thousand ROCs nationwide.[11] While many ROCs in Florida and California and elsewhere are home to relatively well-off retirees, ROC USA, an organization based in New Hampshire, has pioneered a limited equity cooperative ROC model that makes resident ownership available to lower-income residents. Since beginning as a project of the New Hampshire Community Loan Fund in the 1980s, ROC USA has built a national network of more than 280 MHPs that are home to more than eighteen thousand households. ROC USA offers subsidized financing for resident cooperatives to buy their communities and to make capital improvements. ROC USA financing comes with ten years of technical assistance from a network of regional experts who help residents implement a model of democratic self-governance to confront whatever challenges may arise. The ROC USA model has proven effective in giving residents the resources and capacity to own and manage their own communities. To date, no ROC USA community has ever sold or reverted to private ownership.[12]

THE ROC USA MODEL AND EQUITABLE RESILIENCE

The ROC USA model was primarily created for community development and affordable housing preservation, but it can also help residents build resilience to climate change and other threats. With their newfound housing security, residents of ROCs across the country have made investments in reducing their hazard vulnerability at both the household and community scale. ROC residents govern themselves through democratically elected cooperative boards empowered to make critical decisions about community finances, capital investments, and rules. With their secure community tenure and self-governance mechanism, ROC residents have the agency to act upon their own priorities and to mobilize their own experiential knowledge in addressing community challenges, including climate hazards. ROCs within the ROC USA network also benefit from access to public and philanthropic resources as well as insights from peer-to-peer networks connecting ROCs across the country.

This chapter discusses how the ROC USA model of cooperatively owned MHPs can enable equitable resilience. We draw from the experiences of several communities in the ROC USA network that have taken action to reduce their vulnerability to climate change and other hazards. The chapter focuses particular attention on Pasadena Trails, a ROC near Houston, Texas, where the predominantly Latinx[13] residents have forged a strong collective identity and made major investments, including upgrading their drainage infrastructure and energy-efficient lighting (figures 4.3 and 4.4). Resident ownership and upgrades have enabled Pasadena Trails residents to withstand extreme events such as 2017's Hurricane Harvey and a prolonged cold snap in the winter of 2021 that left much of Texas without power and water for several days.

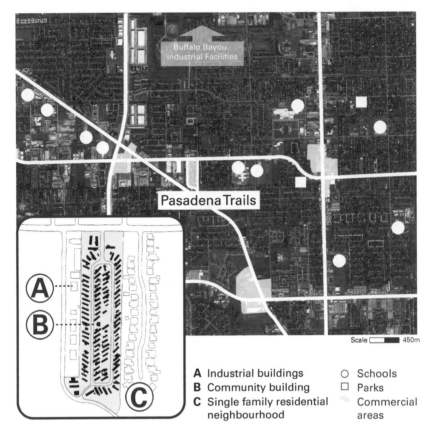

4.3 Site plan and context map of Pasadena Trails. A. Light industrial facilities; B. Small community building; C. Single family residential. *Source*: Mora Orensanz.

SECURITY: RESILIENCE THROUGH CONVIVENCIA

Security from displacement and financial exploitation is central to the ROC USA model. Many MHP residents nationally suffer enormously because of the split-tenure land-lease arrangements that give them control over their homes but not the land and infrastructure on which they depend. As in many privately owned MHPs, before Pasadena Trails became a ROC, the absentee owners prioritized profit over community and safety. A ROC USA official said that the previous Arizona-based owners "had never actually ever been to the community" and had "done nothing" to improve conditions.[14] The previous owners hired a resident as property manager, who,

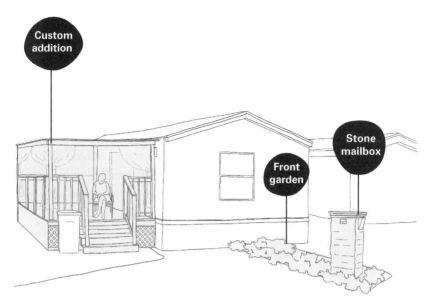

4.4 Without the threat of displacement, residents of Pasadena Trails invest in improving their homes with custom additions, landscaping, and other features. *Source*: Mora Orensanz.

according to residents, ignored complaints and took the proceeds from arbitrary fines.

Interviews with residents, ROC USA staff, and technical assistance providers make clear that Pasadena Trails is not alone in suffering when investor owners maximize profits by deferring maintenance, neglecting necessary upgrades, and ignoring community facilities. One official described "famous" seminars that instruct MHP investors "to invest as little as possible in your park," suggesting that "the first thing you want to do is shut down the community building . . . put a sign on the door that says closed for repairs, and then just never reopen it."[15] When residents take ownership of a park after investor owners have neglected repairs, the cooperative is confronted with a range of challenges, including failing infrastructure and distrust between residents and management. Once residents take ownership and build a sense of internal trust and solidarity, they no longer fear forced displacement. With their newfound security, residents can confidently make investments to improve livelihoods, safety, and health.

RECOGNITION: "WE HAVE THE SECURITY THAT THIS WILL NEVER BE SOLD"

In 2009, residents of Pasadena Trails formed a cooperative and purchased their community from its absentee owners. For Troy Thomason, resident ownership represented a rare opportunity to secure his family's housing future. Before moving to Pasadena Trails, he and his wife had been evicted and forced to pay US$2,300 to move their home when their previous MHP was sold.[16] Estela Garza, a longtime Pasadena Trails co-op board member, spoke about the security imparted by the ROC's legally recognized cooperative ownership, saying, "We have the security that this will never be sold, that no one is going to tell you 'you have to vacate the property.'"[17] The transition to resident ownership has been overwhelmingly positive for residents, as seen in the growing demand for lots at Pasadena Trails. When the ROC was founded, an estimated 15 of the community's 113 lots stood vacant. Under resident ownership, there have been few vacancies, and there is typically a waiting list to fill any opening.

The security that ROC residents gain through resident ownership is a major selling point. The marketing materials for Pasadena Trails describe the community as offering "a path toward financial security and stability" because "unlike most mobile home communities, WE the homeowners of the Pasadena Trails Resident Owned Community own the land beneath our neighborhood." In turn, this means "WE are secure—there is no commercial-owner who can decide to close the community."[18]

CONVIVENCIA: THE POWER OF SOLIDARITY

While the legal recognition of ROC ownership forms the bedrock of residents' security, other forms of security—including internal solidarity among residents and individual place attachment—are also critical to the success of the model in advancing equitable resilience.

Estela Garza, who has lived in Pasadena Trails for eighteen years, notes, "When the ROC purchase went through, everything changed. People were happier because there were more opportunities for *convivencia*" or community building. She goes on to say, "It is very beautiful . . . We all know each other" and "knowing each other allows us to protect each other." Community connectivity is especially important for families, Garza says:

"As fathers and mothers we have the confidence that our neighbors will keep an eye out for our kids around the park and contact us for any reason, and that is very beautiful." With an estimated 90 percent of Pasadena Trails households identifying as Latino and many first-generation Latin American immigrants, co-op leaders liken their tight-knit community to their home countries, saying it is "like how we were raised in Mexico" and "it is a blessing to live in a community like this."[19]

The solidarity forged by ROC residents through shared ownership and governance is especially helpful in facilitating mutual aid during challenging times. Blanca Vega, a co-op leader who has lived at Pasadena Trails for thirty years, reported that if, for instance, "something with the water breaks, I know that if I call a neighbor, they will come and help me fix it and vice versa."[20] When a house in the community had a fire late at night several years ago, neighbors had each other's phone numbers and were able to call nearby neighbors to make sure they could evacuate.

Pasadena Trails, like many other MHPs, has few shared spaces. The laundromat at the center of the park serves as the de facto community hall, a meeting space for people to connect and reinforce their mutual ties. Estela Garza reports that before the residents took ownership, "Some of us mothers would bring our kids to play behind the laundromat . . . and the property manager would come and say we couldn't be there." As a ROC, residents can use their limited community spaces as they like: "We pass out notices that we're going to hold an event for the kids with games, prizes, and food here behind the laundromat . . . we buy pizza and cookies. We have a raffle and people never want to leave. We stay till dark, just talking."[21]

At Pasadena Trails and in other ROCs, leaders report that social events are important to building solidarity, encouraging people to get involved in volunteer committees, reducing the need for outside property managers. Garza reports that the co-op board "communicates to people that keeping the rents low" requires that residents serve on the "half dozen committees that are staffed by volunteers."[22] A ROC leader from New Hampshire reports a similar relationship between social events, volunteerism, and reduced costs for residents, stating, "If you have a great social atmosphere—which we do—you'll have people who want to be involved. If you have people who want to be involved, it takes the burden off of hiring other outside people to do things."[23]

Solidarity within ROCs also enables security from other threats, including crime, violence, and unsafe drivers. Garza reports that, at Pasadena Trails, "When cars or people pass by that we don't recognize, we communicate to the residents to be aware."[24] To improve the safety of the community while reducing energy costs, the board installed LED streetlights.

The legally recognized collective tenure security and informal solidarity enabled by resident ownership has been transformative for residents of Pasadena Trails and other ROCs. Garza says succinctly, "Before [becoming a co-op], this was not a community; it was just a trailer park . . . There was no community like the one we have now with ROC."[25] Residents of ROCs can leverage their hard-fought security to improve environmental conditions and reduce vulnerability to climate hazards and other threats.

ENVIRONMENT: MOBILIZING EXPERIENTIAL KNOWLEDGE TO REDUCE VULNERABILITY

The ROC model is primarily focused on preserving community stability and affordability. Yet, residents of Pasadena Trails and other ROCs have also used their newfound security to build resilience to climate change and other environmental harms. Absentee owners have little incentive to invest in community resilience, but resident owners can take action to reduce their vulnerability to threats, including flooding, storms, extreme temperatures, and drought.

As part of ROC USA's resident ownership conversion process, co-ops work with technical assistance providers and consultants to assess the conditions of community infrastructure and to plan for capital investments. This due diligence process includes not only expert evaluations but also a survey for residents to report their own experience with infrastructure challenges, from street flooding to unreliable electricity and burst pipes. ROC USA's subsidized loans for resident purchases include funds to address deferred maintenance and capital improvements neglected by private owners.

At Pasadena Trails, residents had long complained of street flooding during rainstorms. The previous owners did little to address the problem. After taking ownership, residents prioritized improving drainage. With guidance from ROC USA, the co-op board set their rents and established budgets that allowed them pay down their community mortgage and to make capital improvements, including more than US$200,000 in drainage upgrades. Hurricane Harvey brought historic rains to the area in 2017, causing floods in approximately two hundred thousand homes and US$16 billion in damage across the region. Pasadena Trails saw no substantial flooding.

In addition to upgrading their "gray" drainage infrastructure, residents of Pasadena Trails also prioritized protecting their sheltering canopy of mature shade trees. In recognition of the importance of the trees, Garza said, "Look, this here is life—the tree. The shade from the tree is life." The trees at Pasadena Trails provide shade and evaporative cooling, crucial in the Texas heat. Residents of Pasadena Trails also credit the trees with improving local air quality. Garza explains, "We take care of [the trees]. Some people say they want to cut them down because they drop too much debris, but no, the tree is life. We are surrounded by many chemical plants. So, I think the oxygen these trees provide us is very valuable."[26]

Other ROCs have also invested in hazard mitigation and climate adaptation, including construction of a new tornado shelter and community hall at the Park Plaza cooperative in Minnesota and elevation of homes above the floodplain at Missouri Meadows in Montana. ROCs are also making investments in renewable energy. One ROC in upstate New York recently developed a large solar-energy project on twenty acres of adjacent land.[27] A ROC in Minnesota partnered with a community solar cooperative to access renewable energy from an off-site solar facility.[28] Renewable energy not only enables residents to improve their energy affordability and resilience but also reduces the greenhouse gas emissions that drive climate change.

In addition to building community infrastructure for energy and hazard resilience, ROC residents also take advantage of their tenure security to invest in their own homes, including adapting to hazard threats. Residents of investor-owned MHPs may not make substantial upgrades to their homes out of fear that they will lose those improvements if they are forced to move. ROC residents, by contrast, frequently make upgrades to their homes and lots that reflect their sense of security. They build stone pillars for their mailboxes and substantial garden beds. They plant trees and perennial shrubs. They construct covered porches and other structures to supplement the space in their homes. At Pasadena Trails, multiple households have built canopies over their homes to provide another layer of protection from heat and storms (figure 4.5).

ROC residents also enjoy improved adaptive capacity because of their social cohesion and collective efficacy. In the winter of 2021, when extreme cold weather struck Texas and hundreds of thousands of people were left without heat or electricity, the residents of Pasadena Trails rallied to each other's aid. Garza reported that "during the big freeze, there were fifty-five water pipes that burst [in the community] and our husbands were out there trying to fix them because there was no one else providing help." Because the residents are also the owners and managers of their community, they prepare for hazards. Garza said, "We were used to keeping back-up supplies—PVC pipes—which helped us address some of the water issues during the big freeze."[29]

Just as resident ownership and the collective efficacy that it engenders can help residents to cope with storms and extreme cold, ROC residents had additional resources to draw on in response to COVID-19. Early in the pandemic, residents reported a range of community responses, including sewing masks and gathering supplies for immunocompromised neighbors. ROC USA staff contacted leaders in every co-op and offered special loans and grants to reduce financial stresses on communities and households.[30] Estela Garza at Pasadena Trails reports that, during the pandemic, "Many people lost their jobs. Others worked a lot less, but through the [co-op board's] office, we helped people who lost their jobs by allowing them to pay their rents in smaller payments."[31]

GOVERNANCE: THE POWER TO ACT

Many of the resilience benefits enjoyed by ROC residents are rooted in their shared ownership. Yet, it is not collective ownership alone that is responsible. Rather, the adaptive capacity of ROCs comes from how communities leverage their ownership to make decisions together about how to improve their lives.

Under the previous ownership, Pasadena Trails residents were subject to the whims of managers they regarded as unfair and untrustworthy. Garza reports, "The previous manager constantly abused the residents. If you wanted something, you had to give her something."[32] A ROC USA technical assistance provider reports that residents "felt taken advantage of and oppressed" by the previous "toxic" ownership situation. This treatment made residents eager to become a ROC, so that they could "have their own rules."[33]

Garza describes the transition to resident ownership from the previous regime: "It's better now . . . We have meetings and people share their opinions or complaints. When you are a renter in any other place, you usually don't have that opportunity." She describes the annual residents' meeting at Pasadena Trails, saying, "Once a year we have a meeting with all the residents to review our laws and make any big decisions for the park."[34] Pasadena Trails marketing materials lay out the benefits of the ROC model, combining democratic self-governance with access to outside assistance, asserting: "WE approve community rules and park improvements . . . While the ROC's Members control the budget and elect the Board, we do not 'manage' the cooperative by ourselves." Instead, "the ROC board hires management and repair services and is a part of a national network of resident-owned parks for access to training, support and resources to help our community and its Members."[35]

With the ROC self-governance model, residents use their own experiential knowledge to address predicaments. If someone has a problem in or around their own home, Pasadena Trails management often does not call in outside repair services. Instead, longtime board member Blanca Vega reports, "I would say 'What you need here is the bucket with all the tools. I'm going to leave it and tomorrow I will come by to pick it up, but you are going to fix it.'"[36] Having the agency and capability to solve their own problems is critical during extreme events, enabling ROC residents to cope when outside help may be slow in coming. ROC USA reported that during Hurricane Harvey at Pasadena Trails, "Two homes reported leaky roofs . . . and Board members dispatched volunteers to cover them with tarps."[37] A Minnesota ROC board member describes a similar experience when an overnight windstorm knocked down a tree, blocking a resident's home. After noticing the damage, she contacted another resident. "Within like a fifteen-minute period," a crew of residents had already assembled with chainsaws to remove the tree. She proudly reports, "I don't think that would have happened if we weren't a co-op."[38]

Regional and national ROC organizations enable peer-to-peer learning and training for board members. Speaking about a national gathering of ROC leaders, Natividad Seefeld, a co-op leader from Minnesota, says, "[it] was great to be able to be with so many people that are on the same level as you, having the same problems as you, trying to figure out how do we . . . make things work."[39] Thomas Choate of the Cooperative Development Institute, a New England ROC USA technical assistance provider, described another network function, saying, "We have a kind of informal group of ROC leaders called ROC Strong that meets once a month. Just to skill share and talk about issues that are coming up."[40] The multilevel structure of the ROC USA network provides opportunities for co-op leaders to develop their leadership capacity. One national ROC leader describes such a progression, saying that she encourages co-op leaders to "think of [co-op leadership] as a stepping stone to represent nationally, to help things really rock!"[41]

One critical element of ROC leadership development is the capacity and confidence to secure outside resources, including from government agencies and philanthropies. Minnesota ROC leader Seefeld details helping leaders in a New York co-op overcome their hesitancy to seek support from local political leaders: "They asked me . . . how did you ask a legislator to coffee?" With the confidence of a seasoned leader who knows that she and her community are worthy of respect, she told them, "You just invite them."[42] The ROC structure improves the collective efficacy of communities by helping them to secure benefits of various kinds. When Pasadena Trails residents were concerned about unsafe conditions on a nearby public road, Garza reports, "Residents of the park gathered signatures to get the city to install a speed bump. They collected ninety signatures even though they only needed fifteen, and the city eventually installed the speed bumps."[43] Elsewhere, ROCs have successfully advocated for outside resources to help them adapt to changing environmental conditions. The ROC in Minnesota that built the tornado shelter and community hall could do so because it sought and received state funds. An Oregon ROC and their technical assistance provider secured a grant from the Ford Foundation to improve their drinking-water infrastructure after drought conditions rendered their water unsafe.[44]

Although ROC USA co-ops benefit from the model's self-governance structure, self-governance comes with challenges. Cooperative governance takes work and time. In many communities, residents are hesitant to serve on boards. Garza reports that in Pasadena Trails, "it is difficult to get people to join [the board] because they say they don't have time."[45] Co-op board members often serve very long terms, with a small number of residents taking on a large share of the self-governance burden. In Pasadena Trails, as in some other co-ops, the governance work is overwhelmingly done by women. One researcher labeled the unequal burden of co-op governance as "differential commoning."[46] The uneven distribution of self-governance work in ROCs can lead to two distinct problems: burnout among residents who serve on boards and concerns about biased leadership voiced by those residents who do not serve.

Social complexity always shapes democratic self-governance. When people make decisions about how to live together, those processes can be riven with factional differences. Thus, sustaining equitable resilience through collective ownership and self-governance requires continuous care and support. The ROC USA model provides this.

LIVELIHOODS: SUSTAINING ACCESS TO JOBS AND SERVICES

The ROC model also contributes to residents' resilience by supporting their livelihoods, both by keeping their cost of living low and by maintaining their access to jobs and services. The monthly lot fees at Pasadena Trails have only increased by US$2 since the co-op was formed in 2009, making monthly costs there at least US$100 less than those at other nearby parks. Pasadena Trails marketing celebrates the agency that residents have in keeping their costs low, saying, "WE vote on the budget and set the rent. There is no profit margin in your rent."[47]

Residents at Pasadena Trails are especially grateful to have stable housing in a location with access to work and community services. Garza reports, "All the schools we have around here are very good, the stores are very accessible. We have a good location."[48] Many residents work in nearby industrial plants. Others do construction and landscaping work—trades that require space to store work vehicles, tools, and materials, which is possible at Pasadena Trails (see figure 4.5). Co-op leaders report that, along with a new bus shelter for children and a new community space, future

4.5 Residents have space to store vehicles, materials, and tools—a critical feature for livelihood support. *Source*: Mora Orensanz.

priorities include installing a community-wide wireless internet network to reduce costs and enable better connectivity for school and work.[49]

CONCLUSION: RESIDENT-OWNED RESILIENCE

The ROC USA model of limited-equity resident-owned MHPs has proven to be a robust and scalable model for collective ownership and self-governance (figure 4.6). More than eighteen thousand households across the US enjoy stable affordable community-controlled housing through the model. The security of tenure and well-supported self-governance that ROC USA affords can also be a powerful mechanism, enabling resident-driven equitable resilience.

Even with the many benefits of the ROC USA model, residents still face many of the same hazard challenges as residents of privately owned MHPs.

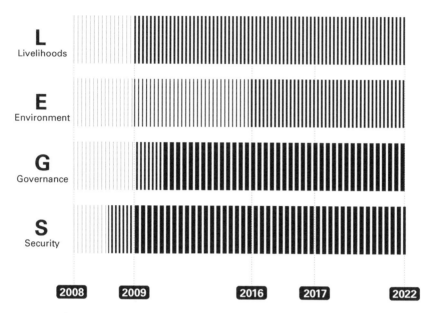

4.6 Pasadena Trails ROC, like other ROC USA communities, was founded to advance security and self-governance. The livelihood and environment enhancements follow from enhanced security and governance. *Source:* Smriti Bhaya. 2008: ROC USA founded; 2009: Pasadena Trails becomes a ROC; 2016: Pasadena Trails upgrades drainage and other infrastructure; 2017: During Hurricane Harvey, community sustains little flood damage, continues incremental adaptation.

ROCs are often located in hazard-exposed areas. Their infrastructures and homes are still sensitive to extreme environmental conditions. However, ROC residents' secure shared tenure encourages them to make adaptive investments in both community infrastructures and individual homes. ROC residents enjoy the benefits of access to nearby jobs and services without the threat of displacement. Crucially, the democratic self-governance model that ROC USA has developed enables residents to leverage their own experiential knowledge to take action to reduce their vulnerability while also helping ROCs access outside resources. In short, ROC residents have "the power to act" in the face of climate change and other stresses.[50]

Case 5

COMUNIDAD MARÍA AUXILIADORA: SEEKING WATER AND SECURITY IN COCHABAMBA, BOLIVIA

OVERVIEW

The fast-growing city of Cochabamba, Bolivia, faces significant water-supply challenges coupled with dwindling affordable land and limited services for disadvantaged residents. In 1999, to combat these conjoined problems, a group of women launched a cooperative and bought land for a new community: Comunidad María Auxiliadora (CMA) (figure 5.1). They pursued affordable land and housing through collective land ownership and mutual aid construction and savings, targeting the needs of low-income women and children. By constructing its own water supply and sewage system, the organization promoted an autonomous political movement of families, contributing to a common struggle for urban services. The community became home to more than seven hundred households. However, they faced factional infighting and formally dissolved in 2019, revealing how difficult it can be to sustain innovative tenure arrangements, especially without state recognition and support. Drawing on more than a dozen interviews with both community leaders and dissidents as well as a site visit and consultation of a range of documents, it becomes clear that the story of CMA reveals both substantial achievement and disheartening limitations.

5.1 Comunidad María Auxiliadora, 2019. *Source*: Fundación Pro Hábitat.

INTRODUCTION: WATER WAR AND PEACE

The rise of CMA in peri-urban Cochabamba, Bolivia's third-largest city, has been both a resolutely local endeavor and a far-reaching confluence of three global processes: inequitable urbanization, an ongoing climate crisis, and a surging anti-globalization movement (figure 5.2).

COCHABAMBA'S WATER WAR

For three millennia, agriculturalists relied on elaborate irrigation works to bring water from the Andes to grow crops in Cochabamba's semi-arid valleys, persisting through successive rule by Incan, Spanish, and independent Bolivian regimes. Recently, the region's droughts have deepened; glaciers in the Tunari Mountains have disappeared, and the climate has become hotter and more erratic. In combination with what one assessment calls "inept local and state policy," climatic changes "produced a long-term drought that has turned the region into a dustbowl." Migration to Cochabamba from rural areas accelerated in the 1950s and dramatically increased in the late 1980s, following closure of outlying tin mines. By 2020, migration had transformed a small city into a sprawling metropolis of nearly 1.3

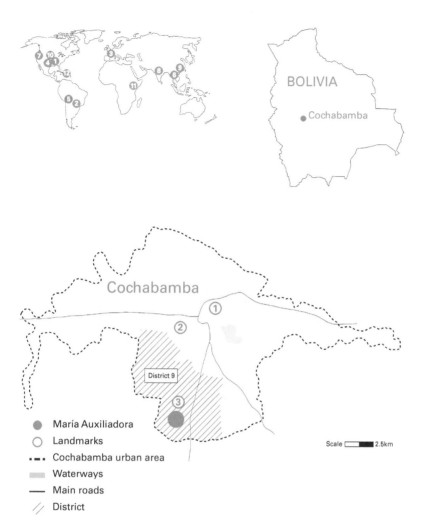

5.2 Comunidad María Auxiliadora's location, in the far south of Cochabamba, Bolivia. *Source*: Mora Orensanz. 1. Cochabamba Downtown; 2. Jorge Wilstermann International Airport; 3. Antonio Diez School.

million people, with urban services lagging badly behind. The influx of people prompted city officials to pave over now-dry lakes and waterways, but they did nothing to forestall the escalating water crisis.[1]

Urbanization and environmental challenges came to a head in 1999 when the Bolivian government privatized water provision in the Cochabamba Valley. Seeking to comply with the market-friendly Washington Consensus requirements of a US$138 million IMF loan and other aid, Bolivian leaders embraced privatization in the face of growing popular frustration. With a US$2.5 billion forty-year contract awarded to a consortium led by a subsidiary of US-based construction company Bechtel (the sole bidder), mass resistance intensified, led by Coordinadora de Defensa del Agua y de la Vida (Coalition for the Defense of Water and Life). The new private company, Aguas del Tunari, doubled or tripled some water rates and aimed to take over water systems in some peripheral areas where local communities had constructed and paid for their own infrastructure. Making matters worse, Aguas del Tunari showed little interest in expanding services to the impoverished southern part of the city, seeing no opportunity for profit. Across Cochabamba, residents resisted through strikes and roadblocks that shut down the city for four days. In January 2000, the government imposed martial law. In February, riot police responded with tear gas. The violence escalated in April. With opposition spreading across the country, the government backed down. The protesters formally won what became widely known as the Cochabamba Water War when the Bolivian Congress rescinded the privatization law. The return of SEMAPA, the former local water utility, now with greater community control, lowered prices but still left half the city without water service.[2]

Environmental historian Sarah Hines views the Cochabamba Water War as a key battle in a longer struggle to democratize water access. Hines argues that the struggle was about reasserting "democratization and popular control of the region's water sources and infrastructure." Those who had constructed and maintained these systems in a region with a long history of communal governance had long been exploited by others. In *Water for All*, Hines traces 140 years of community-based efforts to transform Cochabamba's "water tenure regime" through "their labor, planning, protest, purchases, and seizures of previously hoarded water sources." The 2000 water revolt also profoundly shaped Cochabamba's future. It ignited five years of "rallies, strikes, blockades, and marches" that upended two presidencies, culminating in December 2005 with the election of Evo Morales Ayma, Bolivia's first Indigenous leader. As Morales put it in his inauguration speech, "Water is a natural resource that we cannot live without and so cannot be a private business." The local celebration of Cochabamba's victory over neoliberal privatization gained global visibility, sparking years of protests against World Bank and IMF policies.[3]

COMMUNAL WATER PROVISION FOR MARÍA AUXILIADORA

In parallel to the high-profile Water War, impoverished residents of Cochabamba's South Zone organized their own efforts to secure water and land to live. The privatization battles largely focused on areas that already had municipal water service. Areas that lack piped water rely on water trucks with high prices, irregular service, and dubious quality. This situation prompted Rose Mary Irusta to envision a new community, premised on collective land tenure rather than individualized property, that would be self-reliant for its water. Irusta's husband was active in the Cochabamba water protest movement, and so she understood the high stakes. In 1999, she joined with five other women to plan a new community, Comunidad María Auxiliadora, intended as a *Hábitat para la Mujer* (habitat for women). With a name chosen to honor the Catholic figure Mary, Help of Christians, the group set out to assist low-income women, especially single mothers impacted by domestic violence.[4] Irusta obtained a loan to purchase ten hectares of land with minimal government services on Cochabamba's southern periphery. She sold her own house to afford the payments on the land. She then divided the large parcel into two-hundred- and three-hundred-square-meter plots, where households could build homes on collectively owned land. Irusta later extended the community in two further phases (figure 5.3), with additional land acquired in 2001 (eleven hectares) and 2008 (eighteen hectares).[5] The community charter specified that both the president and vice president of the residents' council be women. All but fifteen of the first 450 households in the community were female headed.[6] By 2001, the first phase of the community obtained electricity. By 2003, residents established formal organizational guidelines for various services and activities, ranging from loan funds and mutual aid to festivals and social events. In

5.3 Site plan of Comunidad María Auxiliadora, as of 2019. *Source*: Mora Orensanz. 1. Playground; 2. Soccer field; 3. Multipurpose field; 4. Library; 5. Nursery school and after school support space; 6. Community center; 7. Playground; 8. Multipurpose field; 9. Soccer field.

2004, the community installed a ninety-three-meter-deep well, with half of the cost covered by a Swiss NGO, PAMS-Suiza.[7] The community portion of the cost of the well construction came from a loan from the Fundación Pro Hábitat, a Cochabamba-based NGO that is part of the Habitat International coalition.[8] Because the well was recognized by the Bolivian state as the source of water, CMA gained a guarantee that no other wells would be dug in the immediate vicinity. Leaders such as longtime resident Juana Aguilar Zeballos did not want CMA to connect to SEMAPA's water system, preferring to have community control rather than rely on the mistrusted municipality or private water trucks. "When we built [our system], we were proud," she recalls. We were the only neighborhood in our district to have water and sewerage."[9] While construction of water and sanitation systems for the 373 households in Phase 1 was completed in 2005, Phase 2's infrastructure was never completed, and Phase 3 gained a well only in 2018.[10]

MARÍA AUXILIADORA: FROM COMUNIDAD TO INDIVIDUAL PROPERTY

Conflicts about both infrastructure and collective land tenure in Comunidad María Auxiliadora came to a head in 2011. Some members complained about unequal treatment; only about a third of households had functional sewerage.[11] Confrontations mounted in 2012 with the passage of Bill #247 in the national legislature, promoting "Regularización de los Derechos de Propiedad Individual" (Regularization of Individual Property Rights). The bill undermined CMA's core commitment to collective property. Stress over infrastructure and collective tenure intensified, sparking lawsuits and violence. Eventually, these conjoined pressures overturned Irusta's vision of the Comunidad, dismantling CMA as a collective and imposing a new system of individual property ownership.

COMUNIDAD MARÍA AUXILIADORA AND EQUITABLE RESILIENCE

At its core, CMA was an attempt to foster security through shared ownership and communal governance, while also supporting livelihoods and improving environmental conditions. Although some aspects of CMA's achievements have been undermined by dissenters, the effort points to both the promises and the perils of pursuing holistic forms of equitable resilience.

The saga of CMA entwines the rise and fall of an innovative collective tenure model with improvements to safety and a women-empowering model of shared governance in a community with low incomes and high environmental stress. As such, this account begins with CMA's pursuit of equitable forms of security and governance before considering efforts at community water and sanitation provision and livelihood enhancement.

5.4 Founder Rose Mary Irusta sought to build a secure community for women and boldly announced this with signage. *Source:* Mora Orensanz.

SECURITY: BUILDING SOLIDARITY THROUGH COLLECTIVE OWNERSHIP

Rose Mary Irusta and her partners pursued an alternative future for low-income women centered on collective land tenure and community-supported infrastructure (figure 5.4). Irusta and other leaders evaluated families seeking to join the community, developed a registration system, and assigned plots. Before joining the Comunidad, prospective residents attended meetings introducing community rules and principles, including ideas about collective ownership, empowerment of women, youth protection, mutual aid (*ayni*), and a rotating community savings and loan system (*pasanaku*). Determined to build a safe community where women would no longer be threatened by abusive partners, CMA's leadership prohibited the sale of alcohol and instituted a system of whistles to alert community members whenever a person was in distress.[12] Each household developed an installment plan to pay its share of the land costs. They required household representatives to attend workshops about house self-construction. Juana, one of the earliest arrivals, describes a triumphant first decade: "During these ten years, the community was so amazing; we progressed so much—we were like a huge family."[13] Another resident, María Serrano, concurs, praising the sense of freedom that came from sharing space with other families and bypassing landlords who refused to rent to her household with five children.[14] Ema, who arrived in 2004 and built her house the following year, found the transition from being a lifelong renter to a homeowner transformative. When first shown her future home site, she remembers, "I got down om my knees on the lot and was so grateful to have this. All the women that came here felt so supported."[15]

While 115 families partnered with Hábitat para la Humanidad (Habitat for Humanity), which provided a standard housing type, most CMA households constructed their homes themselves over time when they had the necessary resources. Fundación Pro Hábitat, where Rose Mary Irusta's son Manolo Bellott now works as a lawyer, provided two different schematic designs for homes, enabling hundreds of households to build their own homes.[16]

COMMUNITY FACTIONS AND FRACTURES

True security of tenure, however, proved elusive. In part, insecurity persisted because the Cochabamba municipality resisted CMA leaders' efforts to formalize their communal ownership. Even without local recognition, their model gained national and international acclaim, inspiring a national bill supporting Comunidades Solidarias (Solidarity Communities) and influencing language about the "right to housing" in a 2008 revision of Bolivia's constitution. CMA earned global plaudits as a World Habitat Award Finalist in 2008[17] and received recognition from the American Planning Association in 2012.[18] Even so, the internal dissent that percolated through 2011 and 2012 boiled over to undermine the system of collective deeds.

One of the chief dissidents, Osvaldo Rocha, had arrived as an early CMA resident. He says he came because he expected to get a formal individual title for a portion of the land and was told, "Be patient, Osvaldo, the documentation is in the municipality." In addition to feeling misled about titling, he came to resent CMA's *anyi* mutual aid requirements, claiming, "If you didn't go to one meeting, your lot was given to another person."

He also bristled at the notion that CMA promoted itself as a habitat for women: "I started hearing that it was a community that prioritized women and I didn't think that was right—that was discrimination. . . . Respect [between men and women] is what needs to be prioritized, not the women. Men weren't allowed to be presidents or be in charge! This was not a democratic process if men were left aside."[19] In 2011, he and ten others founded a Grassroots Territorial Organization or OTB (Organización Territorial de Base). Bluntly, he observes, "the 'community' divided in two: us—and those who supported Rose Mary." At first, he complains, "Every time our group gathered, the police would come and we had to stop our meeting." But, if OTB members enlisted the police, "they wouldn't come." Believing that Rose Mary would not, or could not, provide valid land titles and upset that the land remained listed as under her personal ownership, the OTB went on the offensive.[20]

For her part, Irusta describes the conflict as a matter of gendered power relations: "A group of men . . . broke the community in two: those who advocated for collective tenure and those who wanted individual property," leading to "much violence between these groups."[21] According to one independent outside account, "The dissenting group withheld payments and contributions, destroyed the water pump that left the community without water, and sabotaged community work in an 80-meter [infrastructure] trench (which caused the project funders to pull out)."[22] Then, about twenty people on the anti-collective side sued Irusta, the first of three groups to charge her with *estafa y estelionato* (scam and fraud), claiming that she had sold the same parcels to multiple buyers. She was jailed for four months while awaiting trial.[23] Human rights advocates objected, claiming that her arrest was illegal. Ana Sugranyes, the secretary general of Habitat for Humanity expressed frustration, praising Irusta as "a Bolivian leader who dedicated her life to fighting for the right to housing."[24]

Eventually, the court dismissed the charges,[25] but the damage was done. Operationally, the lawsuits and the additional legislative push to dissolve CMA, coupled with municipal recognition of the OTB's alternative governance system, forced the community to develop individual titles in 2018. The government supported the OTB alternative, claiming that collective land tenure could exist only in rural areas (a matter that remained constitutionally ambiguous). Faced with opposition from the courts, the legislators, and the majority of the community that she had built, Irusta mourns the loss: "We cried a lot; it was such a disappointment." Her distress marked a recognition that "we couldn't keep CMA and allow such a friction between groups in the community because, after all, they would all be neighbors."[26]

In 2019, she formally eliminated the Comunidad model and changed the name of the settlement. It is now merely María Auxiliadora. Today, residents in all three phases of the erstwhile CMA have individual titles, and each phase has a different name—OTB María Auxiliadora, El Pedregal (stony field), and Juana Azurduy (named after a female Bolivian independence fighter).[27]

SECURITY FOR WOMEN

The broader aims of security, mutual aid, and support for women in the community seem to have risen and fallen along with the communal tenure model. At Irusta's insistence, CMA membership documents stipulated that "in case of divorce, the person taking care of the children would keep the property," a provision viewed as "a way of mainly protecting women."[28] At CMA's peak, Irusta attests, "people knew we intervened if there were any violent situations . . . Many times, kids would run to us in the night crying because their dad was beating their mom. Our committee would gather and rush to the house to intervene and mediate the situation." Aided by an ongoing prohibition against drugs and alcohol (supported by city inspections), Irusta credits the Comunidad for having "stopped terrible situations from happening."[29] Ema, an early CMA resident, recounts interventions to stop domestic violence and describes how literal whistle blowing led to the capture of a burglar. Since the fall of CMA, however, the whistles have gone largely silent, the small women's support network has disbanded, and "now people solve their issues by hitting each other."[30] With the rise of the OTB, moreover, liquor stores are now allowed (including one organized by Osvaldo Rocha himself), since, he says, "people need to survive and have their commercial spaces."[31]

Interviewed in June 2022, Irusta describes the neighborhood as "now dangerous," with "young people doing drugs, robberies in homes, alcohol and who knows what else." "I used to know everyone," she recalls, "and now I don't know anyone." Community facilities, including a kindergarten and library, were shut down because, she alleges, the new OTB administration "does not prioritize the social and cultural activities that are so important for community building."[32] Jhenny, a resident for nearly two decades, concurs: "It was very sad that the CMA dissolved." She misses the community's mutual aid construction and mourns the loss of the committee formed to "support family issues and avoid femicides." Now, because of the community schism, there is more violence and "so many beautiful things have been destroyed." She blames both "a lot of mean people" and lack of municipal support for "the concept or benefits of collective tenure."[33]

With the vision of collective land ownership upended, families have a certificate attesting to individual property ownership. Even so, their tenure is not fully approved and registered by the Cochabamba municipality. Irusta and several other residents we interviewed in 2022 blame this reluctance to formalize registration on "land mafia leaders" engaged in corrupt speculation and informal sales. Nonetheless, the neighborhoods remain stable, and many families continue to invest in expanding their houses.[34]

GOVERNANCE: A STRUGGLE FOR CONTROL AND COMMUNAL AGENCY

At its core, CMA's battle over property regimes is a contest over governance and enduring questions of gendered power relations. Male-led groups undermined the women-led collective, leaving profound scars. This is also a fight rooted in economic desperation, a search for ways to leverage land and homes for personal profit. All of this is made more urgent by roiling national politics, equally unsettled climate conditions, and protracted uncertainties triggered by the COVID-19 pandemic (figure 5.5).

Despite such challenges, CMA's original communal impulse has not been wholly lost. The OTB system of neighborhood councils is still meant to facilitate grassroots democracy. However, the OTB model is centered on individual, rather than collective, aims. Rocha became OTB's third president in the spring of 2022. He leads the council in his area, where about half of the residents are affiliated with the organization. Those who are, he insists, benefit from "having the chance to participate and shape the budgeting for improvements," including an additional bus line and

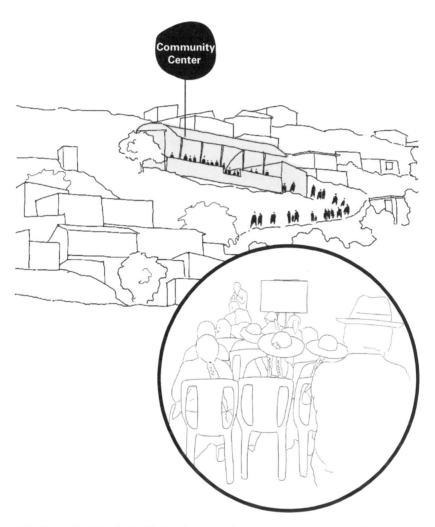

5.5 Comunidad María Auxiliadora began with a variety of community facilities and women-led collectives but later split into factions. *Source*: Mora Orensanz.

expanded street paving. By contrast, he argues that the former "community" model benefitted only Rose Mary Irusta and her allies.[35]

Looking back on the collapse of the communal governance structure at CMA, Manolo Bellott, Pro Hábitat lawyer and son of Rose Mary Irusta, wonders whether his mother took on too much personal responsibility. He notes that she was "mostly alone in the process of interviewing and selecting families from 1999 to 2009." It would have been much better, he now thinks, if they had had "an interdisciplinary team from the beginning" and had implemented "more strict rules and filters."[36] Irusta herself thinks she should have kept the settlement smaller, to enable "a tighter community."[37]

Despite ongoing animosities, some long-term residents have navigated the conflicts with equanimity. Margarita, currently the vice president of the OTB directorate that upended the former Comunidad, nonetheless remains grateful to Rose Mary Irusta: "Many people are against Rose Mary but thanks to her I have a house. Thanks to her everybody here has what they have."[38] Leonor, who serves as OTB's records manager, supported the ouster of its previous president (Sandro Castro), and insists that "we have included OTB directors from both sides." This includes "those like us who have been opposed to Rose Mary Irusta" as well as "her supporters." Leonor values the "beautiful" ideal of living in a mutual aid community but regrets that selfish "interests got mixed up." In the end, she argues, "it wasn't clear that a 'community' meant having collective tenure over the land." Pragmatically, she maintains that the OTB offers "more opportunities to get economic help from the mayor to the neighborhood."[39] This is a reminder of just what Sarah Hines demonstrates in *Water for All*: governance is about more than elite power and control; it is "also the work of collectivities who manage resources like water with varying degrees of autonomy from the state."[40]

ENVIRONMENT: INCOMPLETE PROTECTION

As the Water War made abundantly clear, many discussions of governance in Bolivia quickly turn to issues of water and sanitation resources (figure 5.6). Since María Auxiliadora arose beyond the reach of SEMAPA, its infrastructure has typically been built by local residents using funds from external NGOs. The sewerage system, a pioneering effort in the southern part of Cochabamba when it was initiated in 2004, never reached all phases of the settlement. Worse, even the Phase 1 sewerage system has struggled. In August 2022, OTB records manager Leonor characterized the system as "completely collapsed," adding that the community lacks "enough money to improve the water connections and the sewage."[41] Meanwhile, houses in Phase 3 have implemented "eco-toilets," which operate without water, and may well be a more appropriate technology.[42]

Environmental battles in the community extend beyond water and sanitation shortages. In the early days of CMA, occasional flooding from

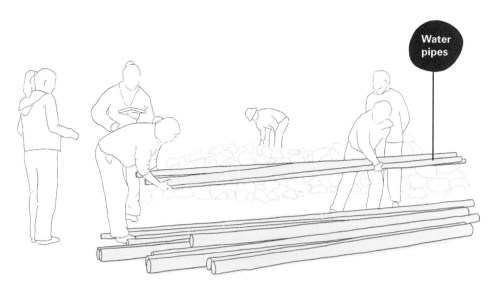

5.6 Many of the CMA's environmental improvements were built by the residents. *Source:* Mora Orensanz.

an adjacent river brought both problems and opportunities. Floods carried water and rocks, supplying raw material for collectively built gabions used for retaining walls and homes. As one longtime resident recalls, "When it rained, we all went down to collect rocks."[43] As another positive development, Rose Mary Irusta's early plan retains ample open space. CMA added its first park in 2009, and Irusta sought to reserve "a lot of empty land" for other parks and future community facilities such as a market, a health center, and another school. Since the dissolution of CMA, however, residents rely on the municipal government for these services. Irusta worries that residents will sell off community land rather than use it for community needs.[44] Although many of CMA's early hopes for modern water, sanitation, and green development faltered after the first phases, the settlement has attained more amenities than most nearby communities and has done a great deal to support the livelihoods of residents.

ACCOMMODATING LIVELIHOODS AND BUILDING CAPACITIES

During its first decade, CMA supported the livelihoods of its community members in many ways. The Comunidad helped residents to develop skills and income-generation opportunities. The shared governance structure also encompassed a range of community services and extended into the *pasanaku* and *ayni* mechanisms for community savings and mutual aid. CMA operated a "Tool Bank" for shared equipment[45] and supported "specialized training in building techniques."[46] They sponsored workshops for nearly three hundred women "focused on leadership, community building, human rights advocacy and the right to housing, and technical entrepreneur skills." CMA partnered with the University of San Simón in Cochabamba to provide a two-month training course in plumbing and electricity, awarding certificates that enabled residents to be "paid workers in the construction of future houses."[47] Fundación Pro Hábitat organized women to support food production. Other groups held weekly sewing and handcrafting workshops[48] as well as instruction on savings and financial management.[49] More recently, the mayor's office has offered free workshops on topics ranging from electricity to baking.[50]

At its peak, the community included a daycare facility, youth center, and library, which not only provided children with food and education but also supported mothers in engaging in other productive activities. Kermess events, held at least three times a year, raised funds from food sales to support loans for new families to build their homes.[51] Long-term resident Giovana points out that even more than this "objective of gathering money and loaning," a kermess became an occasion for new arrivals to "present themselves." So, it became a way "to get to know each other and have a more secure community."[52]

Pro Hábitat's design guidelines for CMA houses required each to have a "working space"—either "commercial or productive like a workshop." They implemented parallel workshops to train residents and develop capacity. Although most residents work outside the community in informal commerce, construction, or transportation, some use their homes to enhance livelihoods, including production of everything from jams, ice cream, and baked goods to metalwork and homemade cosmetic creams (figure 5.7).[53] Giovana observes, "CMA was thought to be a place where women could have their commercial activities in their houses so we could care for our children."[54]

In the early years, CMA remained quite isolated, requiring a long walk to reach the nearest bus line. Over time, however, the community obtained its own Trufis (small buses carrying up to fifteen people), eventually adding

5.7   Some houses accommodate space for livelihoods. *Source*: Mora Orensanz.

multiple new bus lines.[55] For residents such as Jhenny, who arrived with her family as a six-year-old in 2003, "growing up in the CMA was a great experience." She praises both the solidarity and the practicality of "so much group and community work," and credits a good local school for launching her educational journey, which culminated with an economics degree from the University of San Simón. Now twenty-five years old, she lives with her siblings in María Auxiliadora, teaching classes about garbage disposal, recycling, and theater and providing after-school support for kids she hopes will follow her path to college. Disappointed that the Comunidad dissolved, she has not joined the OTB.[56]

CONCLUSION: THE PARTIAL SUCCESSES OF AN AMBITIOUS COMUNIDAD

Despite the removal of "Comunidad" from the CMA name after 2019, nearly all the community members have stayed. Pro Hábitat's Manolo Bellott, whose organization worked with CMA until 2018, observes that "very few of the original families have abandoned the community." Out of 315

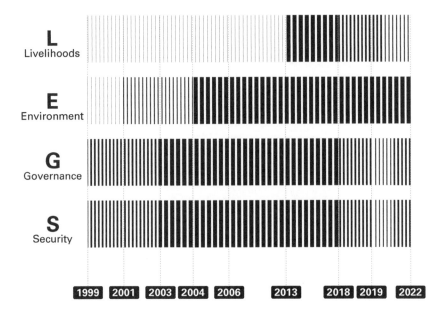

5.8 For its first decade, CMA made strong progress in the dimensions of governance, security, and environment, but its plan for collective tenure faltered after 2018. *Source:* Smriti Bhaya. 1999: CMA Founded; 2001: First phase electricity; 2003: First organizational guidelines; 2004: Potable water and sanitation; 2008 Constitutional "right to housing"; 2013: Community center; 2018: Internal conflict; 2019: CMA dissolved.

families that arrived in Phase 1, 80 in Phase 2, and 170 in Phase 3, he estimates that "only about twenty-five left."[57] Rose Mary Irusta points out that there are still CMA supporters in each phase of María Auxiliadora, but realistically believes that only about 10–20 percent of each group still fully embrace the original CMA vision.

Despite setbacks and what Bellott calls insufficient external "political will" to "operationalize this [communal] model,"[58] many of the community's accomplishments are substantial and lasting (figure 5.8). Low-income residents of southern Cochabamba asserted themselves as the kind of "insurgent citizens" extolled by James Holston.[59] Sarah Hines does not discuss CMA in *Water for All*, but the collective water management practices of these women resonate with her broader notion of "vernacular environmental governance"—community management of natural resources. CMA created an entire inhabited community, not just a collaborative approach to infrastructure provision.[60] Even after the faltering of the Comunidad, the settlement persists, and the struggle to build and sustain it is worth analyzing and celebrating.

Looking back across her own rocky experience of community building, Rose Mary Irusta is still optimistic. She points to the ways that the CMA school, established so many years earlier, has launched many children on successful paths to university: "Thanks to that, we have many professionals here. For me," she reflects, "this means it was all worth it."[61]

Case 6

BAAN MANKONG: "PEOPLE-DRIVEN" SECURE HOUSING IN BANGKOK

OVERVIEW

Over two decades, the Baan Mankong program has developed a "people-driven" approach to upgrading low-income communities across Thailand, achieving transformative results across the four dimensions of equitable resilience. While "Baan Mankong" literally translates as "secure housing," the program treats housing and land security not as an ultimate goal but rather as a foundation for improving livelihoods, reducing environmental vulnerability, and increasing low-income residents' capacity and power for self-governance. Drawing on site visits and interviews with community leaders and other project participants, this chapter discusses two distinct generations of Baan Mankong projects along a single canal (*khlong*) in Bangkok to highlight the importance of investing in community solidarity to sustain the often slow and complex work of participatory design and planning.

INTRODUCTION: INFORMAL URBANIZATION AND UPGRADING IN THAILAND

Thailand rapidly urbanized in the late twentieth century. In 1950, only 20 percent of Thailand's population lived in cities. By 2000, that proportion had increased to more than 50 percent. Bangkok, the capital and

6.1 A path in Bang Bua, an area with several Baan Mankong upgrading projects along Bangkok's Khlong Lat Phrao. *Source*: "Upgrading" by Flickr User Cak-cak, CC BY-NC-ND 2.0.

largest city, grew from around seven hundred thousand residents in 1950 to an estimated eight million in 2000.[1] Urbanization led to housing crises and the growth of informal settlements in Bangkok and elsewhere. After widespread criticism of government mass housing efforts, in 2003, the Community Organization Development Institute (CODI), a government agency founded in 2000, started an experimental program called Baan Mankong. The program was informed by radical housing initiatives elsewhere in the world, including the projects of the Society for the Promotion of Area Resource Centers (SPARC) and Mahila Milan in India (see case 8). Whereas previous top-down efforts relocated poor people away from centrally located informal settlements to housing projects at the urban periphery, CODI designed Baan Mankong as a "people-driven" and "demand-driven" alternative, providing subsidized financing and secure land tenure to allow residents themselves to make decisions about how their communities should be built or rebuilt.[2]

Between 2003 and 2018, CODI initiated over one thousand Baan Mankong projects across Thailand, serving approximately 105,000 households. The model is flexible enough to meet local needs, but Baan Mankong upgrading generally includes some combination of the following steps: (1) initial meetings between community members and program representatives, often precipitated by threats, including eviction; (2) establishment of community savings groups in which participating households collectively generate savings in lieu of collateral for subsidized community upgrading loans; (3) community surveys to understand existing conditions; (4) participatory planning and design for new settlement and housing configurations; and (5) implementation of upgrading or redevelopment plans. Following upgrading, Baan Mankong communities repay their loans as they adjust built environments to suit their needs and access a range of other services.

CODI has implemented Baan Mankong projects in some 405 localities across seventy-six of Thailand's seventy-seven provinces. The rapidly urbanizing Bangkok region has been a primary area of focus (figure 6.2). Unlike large informal settlements elsewhere such as Dharavi in Mumbai or Kibera in Nairobi (see case 11), Bangkok's informal settlements tend to be small clusters of self-built structures distributed across the urban region. With late twentieth century urbanization, an estimated 23,500 households took up residence along the banks of the canals that once irrigated and drained the deltaic region's farmland. The ambiguous public jurisdiction for the canal areas allowed poor residents to establish settlements, many of which are well located with respect to transportation and livelihood opportunities.

By the late 1990s, the Bangkok Metropolitan Administration (BMA) was threatening residents of canal-side settlements with mass eviction, blaming them for water pollution and increased flooding.[3] In response to mounting environmental complaints and expulsion threats, residents of Bang Bua, an area along Khlong Lat Phrao, began organizing in the early 2000s, becoming among the first settlements to benefit from Baan Mankong. When Bangkok suffered widespread flooding in 2011, government officials again blamed canal-side settlements for obstructing drainage and exacerbating floods. Following a military coup in 2014, officials in the new government prioritized flood mitigation in Bangkok to demonstrate their problem-solving capacity. The coup government adopted the Baan Mankong approach to upgrading as part of a 2016 masterplan for Bangkok's

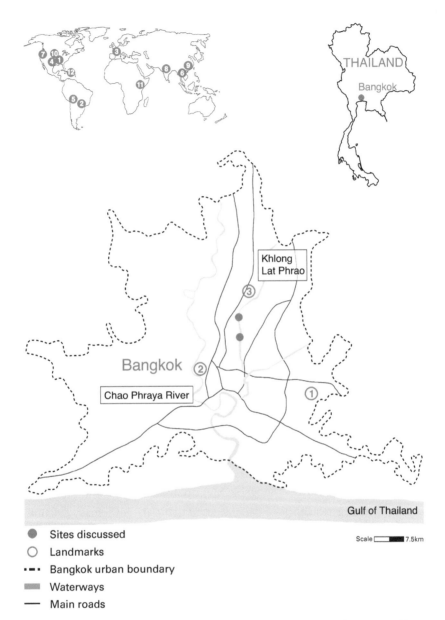

● Sites discussed

○ Landmarks

▪▬▪ Bangkok urban boundary

▬ Waterways

— Main roads

Scale ⬛ 7.5km

6.2 Location of selected canal-side Baan Mankong sites within Bangkok. 1. Suvarnabhumi Airport; 2. Grand Palace; 3. Don Mueang International Airport. *Source*: Mora Orensanz.

canals, spurring the creation of a new subprogram, focusing exclusively on canal-side settlements. This initiative has radically expanded and accelerated upgrading along the Khlong Lat Phrao and a second major Bangkok canal, Khlong Prem Prachakorn. As of May 2019, the Canal-Side Baan Mankong Program had upgraded nearly three thousand units of housing and approved projects for another five thousand households in thirty-eight communities. The twenty-year vision for the Canal-Side Baan Mankong Program includes upgrading in seventy-four communities benefiting some eleven thousand households, with a possible expansion beyond Bangkok to other Thai cities.[4]

The Canal Master Plan and canal-side Baan Mankong projects have accelerated upgrading, but these projects have shifted the focus of Baan Mankong away from people-driven community development and toward improving drainage to reduce flooding throughout the capital region. The plan calls for standardizing the flood capacity of major canals to meet the drainage demands of one hundred-year flood events and designates the BMA's Department of Drainage and Sewerage as the primary agency responsible for implementation.

Given the long track record and variety of Baan Mankong projects across Thailand, numerous case studies could offer insights for the pursuit of equitable resilience. For the remainder of this chapter, we focus on Baan Mankong efforts in Bangkok's canal settlements. Specifically, we compare early upgrading efforts in the Bang Bua area of Khlong Lat Phrao to more recent, post-coup and post–master plan, upgrading in the nearby Lang Wor-Kor Chandrakasem (LWKC) community.

BANG BUA

In 2004, a coalition of residents from twelve eviction-threatened settlements lining Khlong Lat Phrao in the Bang Bua area of north central Bangkok began working with Baan Mankong representatives to secure their rights to stay and to upgrade their homes and infrastructure (figure 6.3). The group, which eventually expanded to include more than three thousand households, collectively negotiated a renewable long-term lease to remain in their communities. They then worked with CODI to secure subsidized loans and grants for community upgrading. Operating in small

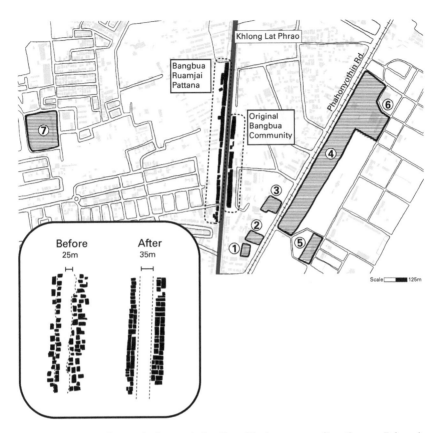

6.3 Bang Bua settlement before and after Baan Mankong upgrading. *Source*: Eakapob Huangthanapan and Mora Orensanz.

groups, residents collaborated with designers and planners to generate flexible upgrading plans to meet their constraints and needs.

LANG WOR-KOR CHANDRAKASEM

The 2016 Bangkok Canal Master Plan called for upgrading fifty-one canal-side communities, home to some seven thousand households. The expanded canal-side Baan Mankong effort included upgrading the LWKC community, named for its position behind Chandrakasem Rajabhat University. The LWKC community, which has recently been relabeled as Siam Venice by some local leaders, stands a few kilometers south of Bang Bua, also on the banks of Khlong Lat Phrao (figure 6.4). As in Bang Bua, LWKC's

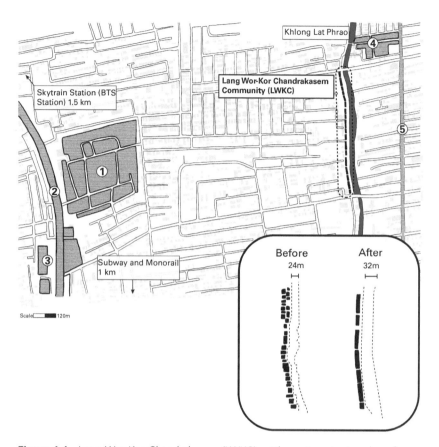

Figure 6.4 Lang Wor-Kor Chandrakasem (LWKC) settlement context and configuration before and after Baan Mankong upgrading. 1. Chandrakasem Rajabhat University; 2. Ratchadaphisek Road (Bus Lines); 3. Courts and Administration Complex; 4. Temple: Wat Siri Kamalawat; 5. Lat Phrao Wanghin Road (Commercial Corridor). *Source*: Eakapob Huangthanapan and Mora Orensanz.

residents (approximately 500 people in 123 households) were already well organized to fight an eviction threat before they joined the Baan Mankong program. While many of the steps in LWKC's upgrading parallel those in the earlier Bang Bua efforts, this latter project manifests some significant differences, reflecting shifts by the military government.

Upgrading in LWKC moved with greater speed and prioritized flood-control improvements. These changes constrained the upgrading process, both in terms of its participatory design and in the form of its new buildings and community spaces. Where earlier Baan Mankong projects clearly reflect residents' values and priorities, reports from project participants and evidence from the newly upgraded settlements, including LWKC, suggest that self-governance and participation are being sacrificed in the name of efficiency and speed. Researcher Hayden Shelby characterizes this transition in Baan Mankong as a transformation "from a grassroots movement to a 'best practice' program."[5]

BAAN MANKONG AND EQUITABLE RESILIENCE

The following sections briefly discuss how the Baan Mankong model, as implemented in the Bang Bua and LWKC communities, addresses the four dimensions of equitable resilience in our LEGS framework: livelihoods, environmental safety, governance, and security. These projects hold insights related to all four dimensions, but we focus particular attention on how Baan Mankong's model of collective ownership and "people-driven" upgrading strengthens the security of residents, providing a foundation for pursuing other community and household priorities.

ENVIRONMENT: "WE STARTED WITH THE ISSUE OF DIRTY WATER"

Environmental concerns, from water pollution to flood vulnerability, are often central to Baan Mankong projects in Bangkok's canal-side communities. Residents of these settlements, like people in similarly situated communities elsewhere, are often blamed for water pollution and for exacerbating flooding by encroaching on drainage canals and reducing their capacity. As Somsook Boonyabancha, a founder of CODI and Baan Mankong says, "Governments always accuse the poor rather than accuse themselves."[6]

Baan Mankong project participants have fought to change this dynamic, arguing that they too suffer from environmental harms and that they are on the front lines of addressing these challenges. Khun Gomeen, a leader in Bang Bua, reported that the community's organizing efforts "started with the issue of dirty water."[7] During upgrading, designers incorporated inexpensive homemade kitchen sink grease traps, invented by a resident.[8] Through these and other efforts, including periodic canal cleanups and negotiations with upstream polluters, CODI argues that residents are "demonstrating that these canal-side communities are not polluters but are an important asset to the city in its efforts to maintain its canal system."[9]

Instead of self-built homes directly over the *khlong*, upgraded communities include wide paths along the canals to allow access for fire trucks, waste collection, and maintenance. Upgrades have also included infrastructure to intercept untreated sewage from both canal-side settlements and surrounding areas, reducing pollution in the waterways.[10] With respect to reducing dumping in the canals, Bang Bua leaders reported that, after upgrading, "everything is much tidier and more clear, people can see and point at you if you carelessly dump waste in the river. So now, no one dares to publicly dump waste in the canal."[11]

In addition to pollution reduction, upgrading in both Bang Bua and LWKC aimed to reduce residents' flood vulnerability. The projects incorporated new canal walls, and homes are generally built on elevated plinths to avoid flooding. The Canal Master Plan and recent upgrading projects have included widening and dredging of the canals to increase drainage capacity.

Although the Baan Mankong upgrades along Bangkok's canals have taken steps to reduce pollution and mitigate flood risk, problems remain. By converting the soft and irregular canal edges to hardened walls, Baan Mankong upgrades may actually cause a range of negative social and ecological impacts, including harms to ecological health and water quality.[12] Some residents and professionals involved with Baan Mankong expressed concern that the new emphasis on city-wide flood mitigation has made later canal-side upgrading less sensitive to resident needs. One said that placing the Department of Drainage and Sewerage and not the Social Development Division of Bangkok in charge "shows that the first priority of the project is the canal, not the social development."[13]

SECURITY: LAND FOR HOUSING THE POOR

Environmental concerns have played an important part in motivating canal-side upgrading, but secure housing and land tenure for poor people has been the central focus of Baan Mankong since the program's earliest days, as underscored by the program's very name. Government agencies around the world often respond to informal settlements with mass evictions, labeling residents as "encroachers" and "invaders." However, as Boonyabancha explains, Baan Mankong emphasizes secure housing and land because "evicting slums just leads people to move"; it does not get rid of the slum.[14] Threats of eviction motivated organizing in Bang Bua, LWKC, and many other Baan Mankong communities. Many aspects of the Baan Mankong approach to security have been preserved in recent canalside upgrading efforts. Yet, there is evidence that, as in the environmental domain, some security-related efforts have become more rigid and less effective in later projects.

Most Baan Mankong residents are rehoused on the same site. In 2018, CODI reported that 61 percent of Baan Mankong households had been rehoused in situ—as was the case in Bang Bua and LWKC. When Baan Mankong cannot secure arrangements for residents to stay in place, households are usually relocated within five kilometers.[15] Baan Mankong residents secure access to housing and land through a variety of legal mechanisms. As of 2011, 44 percent of Baan Mankong households had collective long-term leases, 35 percent held land through cooperative ownership with titles, 8 percent had short-term leases, and 13 percent had other forms of official permission.[16]

BUILDING PRIDE OF PLACE

Research in Baan Mankong settlements suggests that the organizing and upgrading processes often increase residents' pride in their communities. A CODI-affiliated designer reported that before Baan Mankong upgrading, residents' "sense of belonging or attachment to place are very low."[17] A resident in an early Baan Mankong community told a Thai researcher, "In the past when we went back [to our home villages], we were afraid to say [where we live]. People might look down [or] have antipathy. However, recently we speak proudly that we live [here]. We speak louder than before."[18] This increased sense of pride and belonging contributes to the community cohesion and solidarity necessary to succeed in the hard work of upgrading a settlement through Baan Mankong.

BUILDING SECURITY THROUGH COMMUNITY COHESION

Much as the Baan Mankong process can build community pride, the collective work of saving, reimagining, and rebuilding a settlement can both require and reinforce solidarity. Residents of Baan Mankong communities reported to researcher Hayden Shelby that prior to upgrading, they lived as *dtang kon dtang yuu*—"to each their own."[19]

Surveys of residents in early Baan Mankong communities, including Bang Bua, suggest that the program significantly increases perceptions of security of tenure and willingness to invest in housing.[20] In some Baan Mankong settlements, savings groups help to build and reinforce internal solidarity. CODI reports that, after upgrading, twelve communities included in the early Bang Bua projects expanded their savings group to provide low-cost loans for other needs beyond housing, including births, funerals, school fees, and marriages.[21] A leader in the LWKC community, Prapassom Chuthong, has also emphasized the importance of community cohesion, noting, "It's the internal solidarity and unity that get things done. Usually, these external organizations [like CODI] . . . would not know what's going on internally. We have to manage ourselves internally." Chuthong went on to remark on the ways that life in LWKC is distinct from surrounding formal urban areas: "We live very close like other non-urban communities. Comparing to other apartments or condominiums located behind us, they live differently, and I don't think they are as happy as we are."[22]

Not all Baan Mankong communities achieve such high levels of cohesion. Holdouts can obstruct savings groups and collective tenure security for various reasons. Informal landlords and opposing political groups may seek to advance their own interests by encouraging resisters.[23] Some residents may be hesitant to join the collective for fear of increased costs that can come with upgrading and formalized tenure.[24] A CODI official indicated that building unity and solidarity can be especially difficult in linear canal-side communities "that are very long . . . where competitions for leadership" can persist between factions from different sections of the community.[25] To help build and maintain internal solidarity, CODI officials report that they "constantly arrange different festivals throughout the year . . . to always keep the communities active and going."[26] The formal structure of Baan Mankong's collective tenure regimes also helps to encourage internal solidarity. Residents who want to move elsewhere are required to offer their homes for sale to the community cooperative first before selling to outsiders. The collective ownership of the land limits the profits available to sellers in Baan Mankong communities, encouraging residents to treat their homes not as speculative investments but rather as a secure foundation for building their lives.

The structure of Baan Mankong encourages residents to build solidarity, not only within their own community but also with residents of other Baan Mankong communities with whom they can share lessons learned and build collective power.[27] CODI and Baan Mankong founder Boonyabancha refers to this larger sense of collective solidarity: "As a single community," she observes, "you cannot negotiate much, but as a network you can always get what you want. Networks have a strong political negotiation power!"[28] One early indication of the collective power of residents across the Baan Mankong network came when the twelve Bang Bua communities collectively negotiated inexpensive rent and a renewable thirty-year lease with the Department of the Treasury, the agency that owns the land under their homes.[29]

THE SECURITY AND COMPLEXITY OF OFFICIAL RECOGNITION

Along with informal solidarity, the Baan Mankong approach to securing land and housing for residents crucially also entails attaining officially recognized tenure. Boonyabancha explains, "When a community of poor people moves from being squatters to owning their own land, that's a very big change in their lives, and it represents a very dramatic swing from informality and illegality to full citizenship with legitimacy and formal status" (figure 6.5).[30] To achieve official recognition, residents who have initiated a savings group must also establish a legally recognized cooperative that can negotiate with and enter into contracts with government entities to secure land rights, loans, and other services. While establishing a registered cooperative is required for official recognition, some residents point out that the accounting and legal requirements mandated by these arrangements add complexity and challenges. They recognized the value of being a legally recognized group but gaining acceptance from the government authorities means "everything must be surveyed . . . with fixed

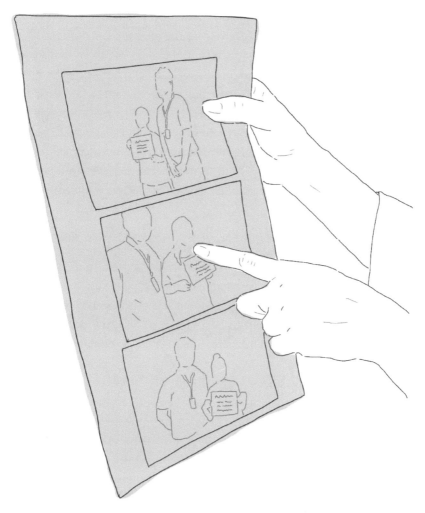

6.5 Community leaders in Bang Bua documented each household's official notice of secure tenure rights. *Source:* Mora Orensanz.

rules and regulation." By contrast, "the saving group is somewhat more informal and sometimes doesn't get recognized by the government and only gets approved and recognized by CODI."[31]

According to Baan Mankong participants, as the canal-side upgrading program scaled up, regulations grew more rigid. One CODI official commented that with the post-2016 policies, "the whole process must be 100 percent legal with all those construction permits, environmental assessments etc." Such official processes can slow upgrading, "and with this time lost, the people's confidence toward the program has decreased."[32]

Attaining official tenure security is central to Baan Mankong, but achieving recognition also comes with increased financial costs. Research in Baan Mankong communities noted that while residents "appreciate their new houses and improved living environment, they are wary of the resulting debt burden."[33] Researcher Shelby refers to the political and social control enacted through Baan Mankong communities' collective debt as "responsibilisation," saying that debt "can lead to personal stress, interpersonal conflict, and tensions between individual and collective well-being."[34]

LIVELIHOODS: PEOPLE-DRIVEN WORK, INCOME, AND SERVICES

In Baan Mankong projects, secure land and housing provides a foundation. However, housing alone is not enough to build a better life. According to one CODI official, "If there is a house but the people can't look after it, then the project is considered to be a failure."[35] Securing collective land and upgrading housing and infrastructure must be accompanied by ongoing livelihood support.

Baan Mankong's livelihood contributions take several forms. Reflecting the "people-driven" spirit of the program, residents in the early Bang Bua projects had considerable flexibility in deciding how to spend CODI loan funds. Some residents chose to do building work themselves to reduce costs. One group with construction skills started a cooperative firm called Chang Chumchon (Community Builders) to provide building services to other Baan Mankong communities.[36] Other residents received small loans to launch a fisheries business in the neighboring canal. In some later projects, livelihood capacity building became more institutionalized. For instance, a post-coup canal-side Baan Mankong project included a partnership with Phranakhon Rajabhat University to train and equip residents to do catering work for the university.[37]

The intensive participatory design and planning process of early Baan Mankong projects (described below) allowed residents to shape the spatial configuration of their communities to accommodate existing and new livelihood activities. Canal-side walkways often host vending carts for

6.6 Canal-side upgrading incorporates commercial storefronts and other livelihood-supporting spaces. *Source*: Mora Orensanz.

residents to sell food and other products (figure 6.6). Community centers and open spaces are used for workshops and small enterprises, including catering, laundry services, artificial flower making, and vegetable market gardening.[38]

In addition to supporting livelihoods through training programs, job creation, and accommodating small businesses in upgraded settlements, Baan Mankong also sustains residents' livelihoods by preserving access to jobs and services in and near their communities. Bang Bua, LWKC, and the other Baan Mankong communities along Khlong Lat Phrao enjoy central locations with access to jobs and critical transportation infrastructure, including a new Bang Bua Skytrain station (see figure 6.2). Early Baan Mankong interventions in Bang Bua also incorporated space for critical services, from public libraries to nurseries and kindergartens. Some Bang

Bua projects also included *baan klang* (welfare houses) for impoverished residents, sometimes subsidized by rental income from market-rate apartments in the same structures.[39]

Baan Mankong projects support livelihoods in several ways, although upgrading and formalization also carry some livelihood costs, including the increased debt burdens discussed above. Some residents report that their lives are less *sabai* (relaxed) after Baan Mankong. They work more and have reduced leisure.[40] Leaders in Bang Bua note, "In the past, only the husband would work in a typical household." After Baan Mankong, "many residents want to improve their living, like buying TV or air conditioning." So, "now both the husband and wife work to get more income."[41]

GOVERNANCE: COMMONING AND PEOPLE-DRIVEN DESIGN

Given Baan Mankong's emphasis on "people-driven" upgrading, self-governance is a central theme throughout the upgrading process and afterward. Shelby describes Baan Mankong's founding principles and early efforts as focusing on "commoning," or "the collective production of shared resources and institutions of democratic self-governance by commoners themselves."[42]

Commoning in Baan Mankong proceeds through a hybrid governance strategy that leverages the resources of the state while encouraging self-governance in poor communities. Self-governance helped inspire Baan Mankong's founding and design, an alternative to top-down slum clearance and urban renewal. As one CODI official describes it, in Baan Mankong, "every individual participates in making decisions for his or her own life."[43]

Baan Mankong empowers all residents through participatory savings, design, and decision making, but capable community leaders are still essential. Bang Bua residents stress that "one key to success is the leader," someone sufficiently respected in the community to make decisions that will be deemed legitimate.[44] A core group of community leaders typically forms in each participating settlement. A CODI representative describes the importance of this dedicated team, saying, "Once these ten to fifteen representatives are formed, then the project usually progresses. It is impossible for CODI or other agencies to go and talk to more than two hundred community members and sustain the progress." In early Baan Mankong projects, once community leaders were established, the program gave "complete autonomy to a community to make decisions and manage their own money."[45]

One way that Baan Mankong prioritizes community autonomy is through participatory planning and design, often led by professional architects and other designers. CODI founder Boonyabancha, herself trained as an architect, describes the role of architects in Baan Mankong, saying, "Architects can use their knowledge in the people-driven process to see the overall picture and interpret the strengths of each element."[46] In this vision, the architect's role is not to arrive at their own solutions. Rather, she states that professionals should "go and see how [people in informal settlements] are doing things, . . . make the form of change fit with them, . . [and] make people the active actors, . . . [since] poor people are the experts" in their own communities.[47]

Adopting a process in which professional designers act as humble translators of community knowledge may sound straightforward, but the process can be painstaking, slow, and rife with conflict. In some Baan Mankong projects, CODI employs designers directly. Other cases draw designers from allied universities or other institutions. Whoever is doing professional design and planning work, the process typically follows similar steps. Following initial meetings, commitments from community members, and formation of savings groups, Baan Mankong communities work with planners and designers to conduct surveys to document existing physical and social conditions. In many cases, larger communities will subdivide into smaller clusters—often on the basis of family relations, friendships, home villages, religious affiliations, or existing spatial relationships—to negotiate sensitive re-blocking decisions, reallocating space for public facilities, housing, and infrastructure. Over the course of several meetings with community members, architects develop and present iterations of housing types, often using physical models and even full-scale mock-ups to demonstrate spatial layouts. In some cases, Baan Mankong upgrades are phased, first building houses for leaders and other committed community members to demonstrate options and to convince hesitant residents to join the program (figure 6.7).[48]

This in-depth participation process enabled early Baan Mankong projects, such as Bang Bua, to reflect the specific needs and constraints of

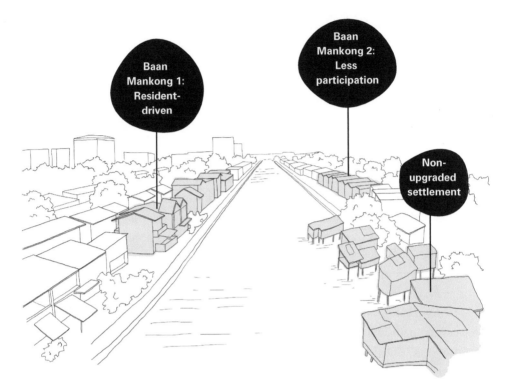

6.7 Early Baan Mankong projects improved conditions in canal-side communities using participatory planning and design, yielding idiosyncratic environments. Later projects are more standardized. *Source*: Mora Orensanz.

communities, a process that Boonyabancha says "makes more efficient use of state resources for the poor and promotes variation rather than standard solutions."[49] Houses vary according to family size and resources. Typically, Baan Mankong houses provided only basic enclosures, leaving families to install finishes according to their resources. In early projects, households could reduce their debt by using recycled materials—for example, incorporating salvaged windows and doors from their old homes.[50] Bang Bua upgrading projects employed local contractors and laborers, saving money, supporting local livelihoods, and ensuring access to ongoing support for future changes or repairs to their homes after construction.[51]

Participatory design also allowed residents of Bang Bua and other early Baan Mankong projects to incorporate desired community spaces

6.8 Robust participatory design in early Baan Mankong projects allowed for preservation of community assets such as shrines and venerated trees. *Source*: Mora Orensanz.

into upgraded settlements, in many cases carving out common spaces by shrinking the private domain of homes.[52] Indoor and outdoor community spaces in these settlements fulfill many functions, including accommodating meetings and workshops, enabling livelihood generation, and protecting revered trees (figure 6.8).[53]

In Baan Mankong communities, following participatory design and upgrading, secure collective land tenure and community self-governance practices provide structures for ongoing decision making. Savings groups often continue to make loans for household needs, and cooperative organizations link communities to outside resources long after upgrading is complete. Researcher Diane Archer describes CODI as "bridging the gap between the state and poor communities,"[54] and CODI officials describe the Baan Mankong cooperatives as "like a power plug socket, in a sense that they are prepared to receive external help and aid and be able to autonomously communicate and coordinate with external agencies."[55]

Baan Mankong also forges bridging connections between poor people, providing linkage not only within a community but also between cooperatives across a city, between groups in other Thai cities, and with networks of poor people around the world.[56]

WEAKENED COMMITMENT TO PARTICIPATION AND SELF-GOVERNANCE?

The coup government's efforts to scale up and accelerate canal-side Baan Mankong upgrading has threatened the vigorous self-governance and participatory planning practices that characterized early interventions in communities such as Bang Bua. While the military government's canal-side program delivers more housing at a faster pace and offers more per household subsidy to ease recruitment, some project participants and researchers worry that Baan Mankong is being transformed into a top-down housing program. Shelby refers to the shift from the early days of Baan Mankong to the program's later instantiations as a transition from "commoning to being commoned."[57]

One study of Baan Mankong remarked that, in its early iterations, "the very strength of the programme—collaboration between community and other actors—limits the speed and scale of change."[58] In recent canal-side Baan Mankong upgrading projects, participants see those strengths being sacrificed in the name of rapid expansion. One senior Thai academic who has participated in past Baan Mankong projects describes these changes in the canal-side projects: "In terms of the constructions, since they were rushing the projects, they would just try to find the fastest way to build within the unit costs. This is in contrast to the programs before the 2011 flood where the process tried to find solutions where they utilized local labor in order to absorb and lower construction costs."[59] A well-informed project advisor describes a shift to more coercive tactics by the coup government, saying that more recent projects operated under "conditions set by the military government with a clear goal and fixed timeline. The community participation process was not fully inclusive—almost like a participation process at gunpoint—the military was involved in the process as the facilitator where the communities could either join the program or just be evicted."[60]

Residents of an early Baan Mankong upgraded community expressed concern that more recent projects have seen a "lack of community participation in the process from the residents." They went on to say, "There is currently a lack of communication of the residents' needs . . . There are differences between being told to do or change [something] and being a part of the conversation and saying your needs, which subsequently turns into a genuine ownership."[61]

The reduced scope for self-governance and participation in recent Baan Mankong projects is reflected in their spatial and material form. Because the new canal-side projects are being implemented under the Canal Master Plan, administered by the Department of Drainage and Sewerage, upgraded communities must fit into a narrow twelve-meter right-of-way between the widened and regularized canal and adjacent private lands. The LWKC community, with its long line of standardized blocks, reflects these spatial constraints. Unlike earlier projects such as Bang Bua, the design of LWKC does not reflect idiosyncratic household and community conditions.

CONCLUSION

For two decades, Baan Mankong has facilitated upgrading projects that deliver transformative results across the environment, security, livelihoods, and governance dimensions of equitable resilience (figure 6.9). However, the contrast between early Baan Mankong projects along Bangkok's canals and more recent efforts carried out under the coup government's Canal Master Plan suggests serious challenges to the program's "people-driven" model. While earlier projects invested the necessary time and resources to ensure that upgrading efforts reflected the priorities and values of residents, later efforts have sacrificed some of that commitment in pursuit of standardization and efficiency.

Early Baan Mankong projects were not perfect. The model relies heavily on the commitment and labor of already overstretched poor people, often women residents. Baan Mankong has long been criticized for placing heavy debt burdens on poor households. The solidarity required for collective land ownership and self-governance also makes the program vulnerable to

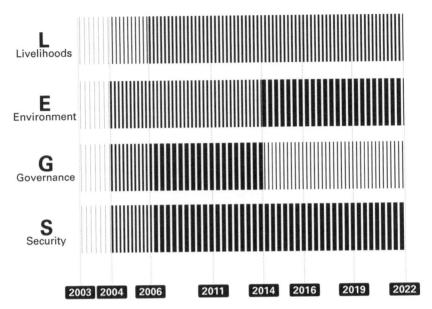

6.9 Canal-side Baan Mankong upgrading in Bangkok significantly improves conditions across all four dimensions, although the recent acceleration and expansion of the program threatens the robust self-governance model of earlier projects. *Source*: Smriti Bhaya. 2003: Baan Mankong founded; 2004: Bang Bua residents begin Baan Mankong process; 2006: First phase of Bang Bua upgrading completed; 2011: Bangkok floods; 2014: Military coup; 2016: Canal Master Plan issued; 2019: Canal-Side Baan Mankong Program completes upgrading for five thousand households, including in LWKC.

holdouts, political opposition, and entrenched interests that benefit from existing arrangements (e.g., informal landlords). More recent projects such as LWKC deliver in situ upgrading to a huge number of canal-side residents while improving urban drainage, overcoming some of the challenges of earlier projects through increased subsidies, reduced debt burdens, and streamlined planning and design. Nonetheless, reflections from residents and other project participants suggest that pursuing lasting equitable resilience that is embraced by residents may require slower, sometimes-messier, people-driven processes that sacrifice some measure of efficiency for greater popular legitimacy.

LEARNING FROM THREE STRUGGLES FOR EQUITABLY RESILIENT SECURITY

The three chapters in this section profile community-scale efforts to secure land and housing, enabling low-income people to build resilience in the face of climate change and other threats. All three of these cases include neighborhood-scale collective property regimes. Yet, there are many other mechanisms for securing stable and affordable housing, including public ownership (case 2: Paraisópolis Condomínios; and case 9: Dafen village), subsidized housing built by independent nonprofit developers (case 7: Living Cully; and case 10: Thunder Valley CDC), and community land trusts (case 12: Caño Martín Peña/ENLACE). In this section, we have focused on three different forms of community property to highlight the diversity, promise, and complexity of such models applied in radically different contexts facing different climate challenges, from flooding in suburban Houston and central Bangkok to water scarcity in the peri-urban desert environment of Cochabamba.

Environmental challenges play very different roles in these three communities. Concerns over pollution and flooding motivated the upgrading of canal-side Baan Mankong communities, who were often unfairly blamed for these problems. Founders envisioned CMA as a place where low-income women could overcome inadequate public and privatized services and obtain secure access to water, along with housing, land, and physical safety. The ROC USA model of cooperative manufactured home parks focuses primarily on community development and affordable housing preservation. Yet, residents of Pasadena Trails and other resident-owned communities have also used their newfound security to invest in hazard mitigation and environmental improvements.

One critical issue emerges from the three cases profiled here: the struggle for security at different scales, ranging from individual place attachment to informal solidarity to formal recognition from outside institutions (see table II.2). For Pasadena Trails and other ROCs, formal recognition of their cooperative ownership and ROC USA's model of democratic self-governance are essential to forging collective solidarity and place attachment, transforming residents from precarious tenants to cohesive and empowered stewards. For many Baan Mankong communities, savings groups and participatory planning processes deepen preexisting solidarities, but the formal cooperative structures required by the government can strain their financial and administrative capacity. Finally, the rise and struggles of CMA reflects the challenges of maintaining solidarity in the face of internal division and speculative pressures without the external legitimacy and structures imposed by formal recognition (table II.2).

While Pasadena Trails, CMA, and Baan Mankong vary dramatically in their contexts and trajectories, there are common themes that emerge from these stories, including (1) threats to community security from factionalism and social division, (2) tensions related to scaling up and speeding up people-driven deliberative and participatory processes, and (3) the value of embedding community tenure institutions within broader networks of support and knowledge sharing.

Social divisions and the uneven burdens of commoning can impede communities from realizing their collective goals, including strengthening security of tenure and improving physical conditions. Each of the three collective property regimes profiled here rely on sustained work from a small group of dedicated individuals, often women residents. This "differential commoning" can lead to burnout among leaders[1] and to distrust from others who feel shut out of decision making. The perils of internal division are most apparent in CMA's reversion to individualized property, but similar challenges emerged in the other two cases.

Table II.2

Multiscalar analysis of security cases

	Pasadena Trails (ROC USA)	Comunidad María Auxiliadora	Baan Mankong
Individual/ Household Place Attachment	Community gatherings reinforce place attachment	Early workshops reinforced shared principles	Early projects featured participatory design and preservation of key sites (e.g., shrines and trees)
Project Cohesion	Mutual aid during COVID-19 and extreme events	Informal collective tenure and mutual aid in home and infrastructure construction	Savings groups, participatory planning, and mutual aid
City/Region Recognition	Cooperative owner-ship and self-governance of land and infrastructure	Collective tenure persistently questioned by government	Recognition of community cooperatives required for lending and services

CMA leaders saw the collapse of their collective property regime as rooted in their attempt to scale up. As the settlement grew from one phase to another, it became difficult to ensure that new entrants shared the vision for communal tenure security, mutual aid, and women-led governance. Similarly, many Baan Mankong project participants mourned the shift from "people-centered" participatory upgrading in early projects (e.g., Bang Bua) to fast-paced mass production in later post-coup projects (e.g., Lang Wor-Kor Chandrakasem). The question of scale manifests differently in Pasadena Trails because residents inherited an existing MHP and have made only rela-tively modest alterations to the built environment.

While growth and change in individual ROCs such as Pasadena Trails is constrained by their inherited territorial and infrastructural constraints, ROC USA has grown their national network quickly in terms of geographic distribution, number of parks, and number of households. Despite rapid expansion, ROC USA has preserved local collective security and democratic self-governance by building networks of regional technical assistance pro-viders and ROC resident leaders to facilitate peer-to-peer knowledge sharing and leadership training. Baan Mankong similarly deploys their proven model for building collective security through both expert networks and community-to-community networks across scales, from local alliances to global coalitions with groups such as Slum Dwellers International, the Asian Coalition for Housing Rights, and SPARC. While CMA's early efforts enjoyed support from international groups such as Habitat for Humanity, leaders operated informally, building solidarity on the basis of mutual aid and relationships. This improvisational approach yielded transformative early results, but without formal transparent structures and external recognition, their gains proved tenuous.

These three cases demonstrate that a foundation of collective tenure security can enable people to build the power to act in the face of threats such as climate change and displacement. Nonetheless, just as a house is not finished after the foundation is built, equitable resilience requires more than security and environmental risk mitigation. For people to build lasting equitable resilience, they must have reliable and dignified livelihoods. It is to the matter of livelihoods that we turn next.

III.
LIVELIHOODS

EQUITABLY RESILIENT LIVELIHOODS: CAPACITY, ACCOMMODATION, AND ACCESS

Part III examines how stable and dignified livelihoods support equitable resilience. Many planning and design interventions undertaken in the name of urban resilience threaten the livelihoods of disadvantaged people. Even in cases where people can stay in an area with reduced environmental hazard risk, such security is worth little if residents cannot satisfy their daily needs.

Climate change and climate adaptation can both impact residents' livelihoods, displacing people and harming their ability to earn a living. Urgent calls to "build back better" after major disruptions too often weaken worker protections, undermining the well-being of people who build and sustain cities. Recent popular mobilizations around the world have recognized that addressing climate change should not harm the livelihoods of already disadvantaged people. Rather, movements such as the Green New Deal in the US place job training and worker protections at the center of "just transition" away from fossil fuels and "just adaptation" of threatened communities. This approach links job creation and training to climate justice, ensuring that climate action benefits disadvantaged and climate-vulnerable groups.[1]

This chapter outlines the relationship between livelihoods and resilience, presenting examples in which narrowly conceived resilience projects have threatened the livelihoods of disadvantaged residents. We also discuss cases where livelihood considerations are integral to resilience interventions across scales, including housing and communities designed to accommodate residents' livelihoods; settlement location decisions that ensure low-income people have access to jobs, resources, and transportation; and efforts to create "resilience jobs" and train disadvantaged people to benefit from climate action.

PLANNING, DESIGN, AND THREATS TO LIVELIHOODS

Planning and design interventions can shape the livelihoods of disadvantaged people, both positively and negatively, enabling access to jobs, resources, and services or leading to isolation that makes it difficult to satisfy everyday needs. Constructing, repairing, and maintaining urban environments also requires tremendous effort and labor, supporting both direct construction jobs and vast networks of services and suppliers. As such, making and remaking cities can also impact the stability and dignity of livelihoods by shaping the terms on which people are employed.

LIVELIHOODS IN INTERNATIONAL DEVELOPMENT

The concept of livelihoods as a central concern of social policy, planning, and design largely emerged from the field of international development. Development scholar Robert Chambers argued that questions of wealth and poverty are complex social, cultural, and political self-assessments, often measured quite differently by professionals than by the poor themselves. Chambers defines livelihood as "the means of gaining a living, including livelihood capabilities, tangible assets and intangible assets," encompassing not just employment but also other activities that yield "food, income and security."[2] Threats to livelihoods cannot always be captured in quantifiable metrics such as employment and income. Other factors, Chambers argues, are inextricably intertwined, including "social inferiority, isolation, physical weakness, vulnerability, seasonal deprivation, powerlessness and humiliation."[3] Differential access to livelihoods is inevitably tied to other axes of structural inequality, including gender, race, age, disability, and caste.[4] Socially favored groups enjoy preferential access to stable and dignified livelihoods, while work performed by marginalized groups is frequently devalued or rendered invisible.

Chambers's view that livelihoods depend on not just wealth but also well-being aligns with Amartya Sen's call for a *capabilities* approach to human development.[5] Livelihoods are a measure of individual and household well-being, rooted in larger social processes that enable people to meet a range of needs, from the technical to the sociocultural. Assessing livelihood therefore includes socio-spatial analysis of how and where people access work, resources, services, social networks, and cultural practices.[6]

Livelihood generation is more than an individual pursuit. It is the work of households and larger communities, supported (or undermined) by governments, private-sector employers, and civil society organizations. A widely utilized sustainable livelihoods framework (SLF) captures the complexity of this concept. The SLF was developed after the 1992 Earth Summit in Rio based on ideas from Chambers and Gordon Conway[7] and has been adopted by international development institutions. In its original form, sustainable livelihoods were envisioned as a five-sided "asset pentagon," including social capital (i.e., supportive bonds, bridges, and linkages to friends, family, and others), natural capital (i.e., access to common resources, such as fishable waters), financial capital (i.e., wages, savings, credit and debt, remittances, and pensions), human capital (i.e., personal factors such as health and skills that affect capacity to work), and physical capital (i.e., infrastructure, tools, and technology). Some assessments also include a sixth dimension—cultural capital—recognizing the importance of identity, spirituality, and attachment to place.[8]

LIVELIHOODS IN PLANNING AND DESIGN

Scholars have long debated how the configuration of settlements relates to economic activity and labor. From nineteenth-century industrial urbanization to late twentieth-century "suburban sprawl" and early twenty-first-century "suburbanization of poverty," from "urbanization without industrialization" and "informal settlements" in the Global South to Chinese "urban villages," many of the dynamics that have preoccupied urban research for generations relate to the spatial relationship between where people work and where they live.

Inexpensive housing that is well located with respect to livelihood opportunities and other urban amenities is perennially under threat by interventions promising greater profitability or alluring visions of modernity and prosperity.[9] Across the Global South, areas that provide well-located affordable housing for urban workers are often labeled "informal settlements." Such areas not only house more than a billion people globally but also enable residents' livelihoods, allowing access to jobs and in-home income generation. Beyond the private residential realm, public spaces in streets and plazas are essential to many forms of informal commerce that are essential to the urban poor, such as street vending. Accommodating informal livelihoods is central to the pursuit of equitable resilience.[10]

Too often, when authorities intervene to create new settlements for the urban poor, they disregard livelihood considerations. Throughout the mid-twentieth century, housing authorities across the US erected large public housing projects, deliberately bereft of nonresidential amenities and often sited in isolated locations lacking public transit and work opportunities. Across the globe, from Mexico to Brazil to Angola, new settlements funded by national governments and international organizations such as the World Bank have focused on maximizing production of housing units, often failing to design communities with access to livelihoods. In many cases, massive housing developments arise on urban peripheries where large parcels of land can be inexpensively assembled, enabling economies of scale.[11] When asked, recipients of such housing often acknowledge that although these projects have improved their housing situation, they have also brought social isolation and stagnant incomes.[12] Even programs that provide families with a "free house" can leave people poorer if new housing is disconnected from employment, services, and affordable transportation.[13] In the most disheartening cases, intended beneficiaries never move in, leaving abandoned "ghost cities."[14]

Despite the many examples of interventions that destroy residents' livelihoods, there are promising counterexamples. Places such as Singapore and Hong Kong have largely succeeded in creating public housing that serves sizable percentages of the population in developments that include commercial space, social services, and transportation infrastructure.[15] The extensive social housing created in interwar "Red Vienna" presents another promising historical example of integrating housing and livelihoods. Between 1918 and 1934, Vienna's socialist municipal government undertook an ambitious program of social housing construction to address crises in housing affordability and inequality. Through aggressive taxation of luxury goods and tight rent control regulation of private apartments, the administration effectively destroyed the speculative housing market and funded construction of more than sixty-four thousand apartments in four hundred state-owned complexes in well-located areas of Vienna. These *Gemeindebauten*—or municipal buildings—often included nonresidential uses and services, incorporating ground-floor commercial space, public schools, laundries, and clinics.[16] Later observers, from Walter Gropius to Dolores Hayden, followed the model of the *Gemeindebauten* in calling for centralized spaces for cooking, laundry, and other care work to free women from undervalued domestic work.[17]

Red Vienna's social housing program supported stable and dignified livelihoods, not just for residents but also for the workers who built the buildings. Rather than reducing labor costs through industrialization and prefabrication, the projects intentionally used labor-intensive methods of construction from stucco-covered load-bearing walls to elaborate metal, tile, and terra-cotta decoration to employ more workers and artisans.[18] While many elements of the design and planning professions have long been agnostic (if not hostile) regarding the rights and welfare of workers engaged in constructing and maintaining built environments, organizing efforts by groups such as The Architecture Lobby are building solidarity between design professionals and other workers, bringing attention to the relationship between climate change, the built environment, and the need for stable and dignified livelihoods.[19]

LIVELIHOODS, DISASTER, AND RECOVERY

Extreme weather and other hazards can threaten livelihoods, especially for already-vulnerable groups. Typhoons and hurricanes not only upend houses and communities but also ruin livelihoods. Beyond the direct impacts of hazards, demands to rebuild quickly frequently lead to the destruction of livelihoods for disadvantaged residents.

The understandable rush to replace damaged or destroyed housing regularly fails to consider the interrelationships of home and income generation. All too often, government directives to remove people from vulnerable areas following events such as tsunamis or storm surges are socially and economically destructive. In 2013, when Typhoon Yolanda (known internationally as Haiyan) wrecked the densely inhabited coastal area around Tacloban

City in the Philippines, the government proposed to relocate fourteen thousand households more than ten miles away. Despite government directives, many households chose to return to self-built housing rather than abandon their lives and communities.[20] In Sri Lanka, following the 2004 Indian Ocean tsunami, the government prohibited rebuilding coastal fishing villages but embraced proposals for high-end hotels on those same sites. Local groups fought to stop these land grabs, viewed as particularly egregious examples of "disaster capitalism." Policies that favored tourism and real estate development over existing community livelihoods revealed larger government priorities. On its "Bounce Back" website, the Sri Lanka Tourist Board breathlessly and callously touted the tsunami as having "presented Sri Lanka with a unique opportunity," predicting that "out of this great tragedy will come a world class tourism destination."[21]

Studies of disaster recovery around the world highlight the repeated pattern of governments, development agencies, and NGOs emphasizing physical rebuilding and paying insufficient attention to necessities beyond housing and other buildings.[22] Too often, building-centered reconstruction ignores cultural preferences, yielding buildings that may be more resistant to hazards but which are also ultimately resistant to humane inhabitation. Even during desperate times, hazard-impacted communities reject new facilities that are incompatible with their livelihoods.[23]

In China's Sichuan Province, following the devastation of the 2008 Wenchuan earthquake, the Chinese government marshalled enormous resources to build new housing and relocate villages, claiming that 92 percent of the 139,000 damaged businesses had reopened within four months. Even so, incomes for rural residents suffered, anticipated tourism failed to materialize, and new industrial parks struggled to attract companies. As elsewhere, mass resettlement of rural agriculturalists to urban apartments failed to accommodate livelihoods, including space for animals and access to fields.[24]

Even if people are not displaced, hazards can impact their livelihoods in other ways. In addition to damaging equipment, infrastructure, and ecosystems that people rely on, hazards can precipitate changes in policy that undermine livelihoods. For instance, after Hurricane Katrina, authorities undermined protections for worker safety, health, wages, and hiring in the name of rebuilding. Post-flood investments also accelerated a shift in the city's economy toward low wage and insecure tourism and service industry sectors.[25]

Disadvantaged groups bear lopsided burdens of hazard-related liveli-hood disruption. The IPCC reports that livelihood impacts are "felt dis-proportionately, with the most economically and socially marginalized being most affected," including along such divides as "gender, class, race, income, ethnic origin, age, level of ability, sexuality and nonconforming gender orientation."[26] Post-Katrina changes in the New Orleans labor land-scape particularly harmed Black workers.[27] Gender can also shape uneven vulnerability to hazards. A study in Bangladesh revealed that both flood-ing itself and flood protection infrastructure threatened gendered resources and livelihoods, including by destroying kitchen gardens and forcing households to sell jewelry and other forms of gendered property.[28] Lessons about uneven livelihood impacts from disaster and recovery are increas-ingly salient as the world's settlements confront climate change.

LIVELIHOODS RESILIENCE IN THE FACE OF A CHANGING CLIMATE

Rising sea levels, storms, extreme temperatures, drought, and wildfires all threaten livelihoods. Shifting patterns of temperature, rainfall, and other extreme events are devastating rural livelihoods by undermining ecosystem services and upsetting long-established agricultural, fisheries, and forestry practices.[29] In urban areas, perhaps the most direct climate change impacts on livelihoods come from increasing extreme heat, which exposes work-ers to risks from heatstroke and other health problems. Indoor work can be just as dangerous as outdoor work in extreme heat. From overcrowded workshops and factories to poorly ventilated home workspaces, excessive heat can increase fatigue and threaten well-being.[30] In 2019, the Interna-tional Labour Organization predicted that, by 2030, global warming will cause a drop in productivity equivalent to the loss of eighty million jobs.[31] Crucial from an equity perspective, the IPCC states that "adverse impacts on livelihoods will be concentrated in regions and populations that are already more vulnerable," especially in low-income communities across the Global South.[32]

Climate change can also undermine livelihoods by forcing people to relocate away from established social and economic networks. A 2022 IPCC report warns that "relocated people can experience significant finan-cial and emotional distress as cultural and spiritual bonds to place and livelihoods are disrupted."[33] Using a six-part SLF, a recent analysis of 142 studies documenting 203 cases concluded that "the resettlement process overwhelmingly resulted in negative outcomes for affected people across natural, social, financial, human and cultural assets."[34]

In many cases, people are forced to relocate not by direct climate change impacts but rather by infrastructural megaprojects undertaken in the name of adaptation, development, and modernization. Development-induced displacement has moved in tandem with the construction of dams in many countries, with recent large-scale examples in China and India.[35] Government initiatives to attract tourism as a source of revenue have forced residents out of long-treasured places. Those resettled often face a litany of negative consequences, including "landlessness, joblessness, homelessness, marginalization, food insecurity, increased morbidity, loss of access to common property resources, and community disarticulation."[36]

Research and organizing efforts focus increasingly on disruptions to livelihoods. Internationally, there is growing momentum behind movements for "climate reparations" to address the disproportionate climate change burdens facing poor nations that have historically contributed very little to the emissions causing climate change.[37]

RESILIENCE WITHOUT *EQUITABLE* RESILIENCE: THE CHALLENGE OF RECOVERING LIVELIHOODS

Many cities, even those far from coastal areas, have faced devastating floods exacerbated by climate change. In 1997, the Red River of the North overtopped levees and inundated Grand Forks, North Dakota. The flood displaced more than fifty thousand people, nearly all of the city's residents. Fires caused by the flooding and exacerbated by incapacitated response systems burned eleven buildings and sixty apartments. After the floods, the city used US$40 million from the Federal Emergency Management Agency, the state of North Dakota, and the US Department of Housing and Urban Development to buy more than eight hundred properties in high flood-risk areas, prohibited new reconstruction in the hundred-year floodplain, and planned a new subdivision for displaced residents a few miles away. Across the river, the smaller town of East Grand Forks, Minnesota, suffered damage to 99 percent of its homes. There, officials bought out four hundred residential properties, offering a premium above the assessed value to account for rebuilding costs.[38]

As recovery began, Grand Forks city officials embarked on a campaign with the slogan "Miracle in Progress."[39] A decade later, officials applauded the region's resilience, using the mantra "Rebuilt, Renewed, Reborn."[40] Celebratory events extolled post-flood projects, including a US$409 million system of levees, floodwalls, and pumping stations. A 2,200-acre greenway provided floodable "room for the river," along with trails, a golf course, campgrounds, athletic fields, and gardens. Downtown, Grand Forks embarked on a major effort to reimagine the city, including construction of "Noah's Ark," a large industrial building redeveloped as office space for displaced small businesses—and subsequently converted into an Amazon.com call center.[41] Marking two decades since the floods, the Federal Emergency Management Agency issued its own assessment, with the headline: "20 Years Later: A Resilient Recovery after Catastrophe."[42] Beneath the hype and beyond the reassuring refrain of "build back better," the saga of Grand Forks's recovery is certainly a tale of substantial accomplishment. However, the story is also an example of resilience that falls short of *equitable* resilience, especially in the domain of livelihoods.

A study of Grand Forks conducted by the Federal Reserve Bank of Minneapolis focused on "disaster recovery for low-income people," including the uneven livelihood impacts of the floods and reconstruction. The study found that "the community's affluent and moderate-income residents gradually returned to work and rebuilt homes," whereas flood recovery for poorer households was "more problematic," especially regarding housing affordability, employment, and childcare. Single mothers and low-income elderly residents fared worst. Many recovery programs targeted homeowners. Although many affordable and subsidized rental units were lost in the floods, "the new rental units [built in the recovery] have gone to middle income families." After suffering flood damage, many small in-home daycare facilities closed permanently, leaving low-income households unable to afford childcare.[43] City officials focused on reopening downtown businesses and financial institutions, but the floods also destroyed smaller institutions, including agencies focused on chronic challenges of homelessness, substance abuse, violence, mental illness, and unemployment.[44]

At the time of the 1997 floods, more than 90 percent of Grand Forks and East Grand Forks residents were white. Minoritized sections of the population, mostly Native American and Latinx farm workers, faced particular challenges in the recovery. Many Indigenous households relocated to nearby reservations to receive support from their tribal communities. Most migrant workers had not yet arrived for the growing season at the time of the floods. When they tried to return in 1998, many discovered a devastating catch-22: the manufactured homes that they inhabited during seasonal work were heavily damaged, but they were ineligible for compensation, since these were regarded as "second homes." As one study found, "Race, class, and gender inequalities—not simply proximity to hazard—set up some residents more than others to [suffer from] disaster long before floodwaters rise." Further, even as solidarity increased among white residents after the floods, other long-standing divisions hardened: "The solidarity of flood victims as a whole was undermined by an implicitly gendered, racialized, and classed vision of community recovery."[45]

The uneven livelihood impacts of the floods manifested not just in obvious harms to employment and housing but also in unequal mental and physical health impacts. A study of pre- and post-flood health conditions in Grand Forks found a 43 percent increase in domestic violence, a 45 percent increase in depression, a 129 percent jump in DUI arrests, and a 275 percent leap in drug/narcotics violations. Those without health insurance suffered most and faced the biggest obstacles to returning.[46] While officials invested in new infrastructure and buildings, issues such as surging domestic violence and insufficient human services received inadequate attention.[47]

In short, because officials pursued generalized resilience after the floods, they did not focus on the needs of those vulnerabilized residents likely to suffer most. One analysis of the uneven recovery pointedly concludes: "Some stories were silenced in order to present a public face of satisfaction and consensus rooted in cultural homogeneity."[48]

EQUITABLY RESILIENT LIVELIHOODS ACROSS SCALES

The aftermath of the Red River floods demonstrates the need to pursue equitable resilience across scales. Interventions ranged from the city-regional scale (riverfront green space and infrastructure in two states) to the project level (downtown revitalization and new housing) to household- and individual-scale needs for jobs and services. While the efforts yielded some great successes at each scale, these undertakings helped some residents more than others.

An intervention advances the livelihood dimension of equitable resilience if it improves conditions for disadvantaged people across three linked scales, by enhancing (1) access (city/regional scale), enabling residents' livelihoods through proximity to work opportunities and/or affordable transportation; (2) accommodation (community/project scale), creating spatial and material conditions that support residents' livelihoods, including through home-based and informal work; and/or (3) capacity (individual/household scale), supporting residents in acquiring or maintaining skills, tools, and capital necessary for stable and dignified livelihoods.

ACCESS: LIVELIHOODS RESILIENCE AT LARGER SCALES

Equitable resilience requires attention to the accessibility of livelihoods. Who has access to livelihoods and who does not? Is housing located near sources of employment? Who does transportation infrastructure serve, in terms of cost, routes, and service? Do households have access to affordable services, including childcare and healthcare?

Ensuring equitably resilient livelihoods at the scale of the city region has often proved difficult. In Indonesia's Aceh Province, site of the 2004 tsunami's most widespread devastation and greatest death toll, the Indonesian government quickly passed a land-use plan regulating coastal development within a two-kilometer coastal zone.[49] Resettlement away from coastal areas brought trade-offs between enhanced resilience in environmental terms and degraded livelihoods. The celebrated Indonesia–China Friendship Village, dedicated in 2007, illustrates the dilemma. The settlement, popularly known as "Jackie Chan Village" after the Hong Kong movie star who made a donation and paid a brief visit, is located three hundred meters above sea level and one and a half kilometers inland, beyond the reach of future waves. Unfortunately, it is also beyond the reach of most jobs. The Rehabilitation and Reconstruction Agency of Aceh-Nias gave 606 free houses to 2,400 displaced people. When assessed seven years after completion, the village was only half occupied. Located seventeen kilometers from the employment center of Banda Aceh, the hilltop village offered few livelihood options for the fishermen, drivers, traders, service workers, and small-scale entrepreneurs who resettled here. Moreover, farmers who gave up land for the village also lost their source of livelihood.[50]

In some cases, the challenge of securing safe living conditions with place-based livelihoods can be addressed by moving homes while retaining access to work in areas judged to be too risky for permanent settlement. In Southern Taiwan, following Typhoon Morakot in 2009, the government urged Indigenous Paiwan people in dozens of remote farming villages in the hills of Pingtung County to relocate to new communities in flatlands not exposed to landslides. In the case of Ulaljuc Village in Taiwu Township, this entailed replacement of an irregularly laid-out village with a gridded settlement of 157 modern houses located closer to the town. Some relocations after Typhoon Morakot proceeded in a top-down manner that ignored cultural practices and exacerbated long-term tensions.[51] The Ulaljuc plan, however, included attempts to connect with local tradition, incorporating traditional construction motifs in an elementary school and a version of the chief's traditional stone house and stone pillar, based on those in the original village.[52]

Rather than completely abandon the original village that was judged to be too dangerous for full-time occupation, the tribe maintained the village land (with its vital ancestral associations) but shifted to using it for growing coffee—a practice established before the typhoon. Paiwan tribal leaders chose their new village site from among six proffered locations, since it was the one closest to their mountain village. That choice enabled multiple generations of villagers to continue visiting and working the land on the mountain, providing a source of sustainable livelihoods to encourage tribal youth to remain. The new lowland village featured a tribe-owned cooperative coffee production factory with machinery obtained through government subsidies. Located near a scenic area with a well-known Catholic church, villagers marketed the new Ulaljuc village as a tourist destination, teaching visitors about organic coffee production and Indigenous cultural practices. More recently, the tribe established a biotechnology company devoted to research and development of new inventions from byproducts of coffee production.[53] Despite tensions over the decision to abandon residential life in the mountain village, the consultative aspect of the move coupled with the ardent commitment to supporting alternative livelihoods suggests avenues for achieving equitable resilience.

Although too-infrequently practiced, there are many instances of pro-active attention to livelihoods during in situ redevelopment or reloca-tion of housing for marginalized populations. Many of the cases already discussed—notably the Paraisópolis condominiums in São Paulo (case 2), the Pasadena Trails manufactured home park in Texas (case 4), and canal-side Baan Mankong upgrading in Bangkok (case 6)—emphasize main-taining access to place-based livelihoods. Similarly, the community land trust–driven upgrading of communities along Caño Martín Peña in San Juan, Puerto Rico (case 12), also preserves livelihood access.

ACCOMMODATION: LIVELIHOODS RESILIENCE AT THE PROJECT SCALE

At the project scale, when designing or retrofitting settlements, it is essen-tial to consider how built environments can accommodate livelihoods. Does an intervention include more than purely residential spaces? Does new infrastructure protect and preserve the livelihoods of poor people? In many cities around the world where most livelihood activities do not fit the narrow confines of formal work, interventions must accommodate informal livelihoods, including both in-home opportunities and work outside the home.

Although the track record of previous resettlement efforts does not bode well for climate-induced resettlement, there are encouraging exam-ples. One of the earliest documented climate change–induced relocation efforts—the relocation of the village of Vunidogoloa, Fiji—exemplifies the possibilities of livelihood-sensitive resettlement. Here, the entire commu-nity of 153 people moved two kilometers inland, away from the coast where residents had suffered from "slow-onset climate impacts including tidal inundation, coastal erosion and saltwater intrusion." Before moving, village leaders worked with other officials to plan for future livelihoods not just houses. The new village offered solar-powered electrification, modern plumbing, fishponds, pineapple plantations, and cattle grazing land, as well as access to a main road. Moreover, because the new vil-lage was within walking distance of the old settlement, villagers retained access to fishing grounds and culturally important plants, including coco-nuts and pandanus leaves for weaving. Researchers who interviewed villa-gers found "a strong sense of social cohesion and unity" and identified important gains across all six dimensions of the sustainable livelihoods framework.[54] Vunidogoloa, like Taiwan's Ulaljuc village, is a small place, but these kinds of settlements have large implications for the pursuit of equitable resilience.

In urban settings, city leaders have attempted in situ upgrading of low-income areas, aiming to increase environmental safety while retaining access to employment. The upgrading of the Quebrada Juan Bobo community in Medellín, Colombia (previously discussed in Part II), coupled housing and infrastructure improvements with attention to resilient livelihoods. The intervention included new multistory buildings with commercial space on the ground floor and a bridge across the Juan Bobo creek, providing access to a new cable-car system that connected residents to the rest of the city. Within the right-of-way below the elevated cable-car line, the project introduced a new pedestrian street, or *ramblas*, that has encouraged new economic activity. Architect Alejandro Echeverri estimates that an area that once had ten informal shops is now home to some four hundred businesses.[55]

In Buenos Aires, city officials have worked to preserve livelihoods when removing a multistory informal settlement with about 1,500 households built under an elevated expressway in Villa 31. In advance of relocations, they built new housing a short walk away, coupling this with a Labor and Entrepreneurship Center (CeDEL) that claims to have found jobs for more than a thousand people.[56] The new YPF Housing (named after the YPF company warehouses that it replaced), with more than a thousand apartments, includes space for ground-floor businesses, yielding new public-facing streetscapes while enabling some residents to recreate their former workspaces from beneath the highway. The YPF district is also served by a new bus route. Less auspiciously, interviews conducted by our team in early 2020 indicate that because the project houses many families who had not previously lived under the highway, it did not come close to rehousing all of the displaced residents and businesses. Rather than preserving existing social networks, as intended, the new development suffers from considerable internal conflict.[57] The ongoing Villa 31 redevelopment illustrates the difficulty of achieving equitable livelihood provision despite thoughtful policy and design engagement with these issues.

CAPACITY: LIVELIHOODS RESILIENCE FOR INDIVIDUALS
AND HOUSEHOLDS

Many positive and negative livelihood impacts are felt most deeply at the scale of individuals and households. CeDEL in the Villa 31 project exemplifies the role of job training and placement in pursuing individual- and household-level livelihood resilience. Similarly, the Juan Bobo upgrading project included job training programs in construction trades, food service, clothing, shoemaking, and entrepreneurship.[58] At the level of individuals or households, an intervention may enhance or harm people's capacity to secure their own livelihoods. Are members of the community supported, trained, and respected? If particular livelihoods are negatively impacted by an intervention, what provisions are made to ensure that people can transition to other sources of income and self-worth?

One especially successful example of livelihoods-centered recovery following the 2004 Indian Ocean tsunami came from the anti-poverty network Uplink Banda Aceh (UBA) in Indonesia. Their work bridged project-level attention to livelihood accommodation with emphasis on helping individuals and households. UBA took the stance that villagers should be able to rebuild where they previously lived rather than being relocated away from their homeplaces and livelihoods in the name of tsunami safety. Uplink coordinated efforts to build three thousand houses, re-creating village layouts and preserving access to mosques and other key sites. They raised some of the housing on stilts, creating covered space on the ground floor that could be used for business activity or to store equipment for fishing and farming. To build capacity to manage reconstruction, UBA helped form Jaringan Udeep Beusaree, a grassroots organization whose name means "a network for living together." UBA and its partners treated physical rebuilding as an entry point for building capacity, self-determination, and psychological healing. Residents managed construction of their own homes, as UBA sought to rebuild not just housing but trust. They organized art therapy programs and social events and helped restore income-generating opportunities in farming and other sectors.[59]

Elsewhere, innovative projects have sought to teach marginalized groups new skills to support their individual- and household-level livelihoods while enabling community adaptation to climate change. The Mahila Housing Trust, based in Ahmedabad, India, works with grassroots collectives to train and employ women to implement simple interventions for cooling homes in dense and often stiflingly hot informal settlements as part of their mission to advance "dignified home, dignified work, dignified life." Cooling strategies deployed by the Mahila Housing Trust "climate saathis" or "climate companions" include tree planting and retrofitting sheet metal roofs with insulation made from low-cost locally available materials. An improved roof and a cooler home are not just convenient amenities; they also allow residents to preserve and expand home-based livelihoods in the face of climate threats.[60]

EQUITABLY RESILIENT PATHS TO LIVELIHOODS

Pursuing equitable resilience with respect to livelihoods entails addressing vital questions at each of the three scales outlined above (table III.1).

Table III.1
Three scales of equitable livelihood resilience

Scale	Equitable Resilience Principle	Question	Examples
Individual/ Household	Capacity	Does the intervention help residents acquire or maintain skills, tools, or capital for stable and dignified livelihoods?	Uplink Banda Aceh, post-tsunami reconstruction; Mahila Housing Trust "climate saathis"; Villa 31 redevelopment, Buenos Aires
Project/ Community	Accommodation	Does the intervention accommodate residents' livelihoods, including home-based and informal work?	Vunidogoloa village relocation, Fiji; Juan Bobo, Medellín, Colombia; Villa 31 YPF Housing, Buenos Aires
City/Region	Access	Does the intervention enable residents' livelihoods through access to work opportunities and/or affordable transportation?	Ulaljuc Village relocation, Taiwan; Juan Bobo upgrading and transit interventions, Medellín, Colombia

In the case studies that follow, leaders have made low-income residents' livelihoods central to efforts to build resilience to climate change and other threats. These projects have also engaged other aspects of equitable resilience, but their successes in the realm of livelihoods stand out. The cases discussed in this section include Living Cully (case 7), a coalition that trains residents of a low-income neighborhood in Portland, Oregon, for jobs installing green stormwater infrastructure and retrofitting homes for energy efficiency; Yerwada (case 8), a collection of in situ upgrading projects of informal settlements in Pune, India, in which individual households shaped the design of their homes, maintaining their livelihoods both within and outside the community; and Dafen village (case 9), wherein drainage infrastructure retrofitting helped to protect a flood-prone urban village in Shenzhen, China, that is home to a thriving community of artisans. Following these three case studies, part III closes by assessing commonalities and lessons from these examples.

Case 7

LIVING CULLY: ENVIRONMENTAL WEALTH BUILDING IN A PORTLAND NEIGHBORHOOD

OVERVIEW

Living Cully is a coalition of community-based NGOs whose work links environmental improvement to anti-displacement and anti-poverty goals in a diverse low-income neighborhood in Portland, Oregon. The coalition's activities range widely. They include training and employing residents in energy efficiency and green infrastructure construction, building and preserving affordable housing, developing new parks and green infrastructure, and advocating for socially and environmentally progressive city policies. Across their activities, Living Cully aims to build dignified livelihoods for Cully residents as a means of building "environmental wealth." The chapter draws on a range of sources, including site visits; reviews of written plans, reports, and other documents; and interviews with coalition members, city staff, and other stakeholders.

INTRODUCTION

Low-income people are especially vulnerable to the impacts of climate change and spend a greater proportion of their income on energy. In theory, these disproportionate burdens should mean that low-income people would benefit most from climate action. In reality, climate action can hurt

7.1 Living Cully coalition members support livelihoods in the Cully neighborhood of Portland, Oregon, through training and employment in environmentally progressive fields such as green stormwater infrastructure installation and maintenance. *Source*: Living Cully.

the poor and even contribute to displacement. This dynamic has led some low-income residents to fight against improvements to parks and other public spaces in their neighborhoods out of fears that these investments will supercharge green gentrification and climate gentrification.[1] These contradictory conditions inspired the creation of Living Cully, a coalition of NGOs in Portland, Oregon, that promotes "sustainability as an anti-poverty strategy."[2]

Living Cully's work links environmental sustainability with struggles against poverty and displacement in Cully, a neighborhood in northeast Portland (figures 7.2 and 7.3). Oregon, like many other states, has a troubled history of racial exclusion and exploitation. When the state was established in 1859, it already held laws that explicitly barred Black people from working, living, or owning land—a restriction not formally repealed until 1926. As Portland's population grew more diverse, exclusionary practices fostered segregation and discouraged investment in neighborhoods of color. The

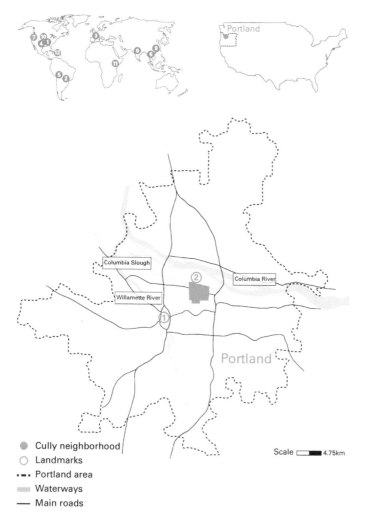

7.2 Living Cully sites within Portland, Oregon. 1. Downtown Portland; 2. Portland International Airport. *Source*: Mora Orensanz.

Cully neighborhood was largely urbanized between 1910 and 1960 and formally became a part of the City of Portland only in 1985, becoming one of the city's most racially diverse areas. Forty-five percent of Cully residents identify as people of color compared to 29 percent citywide.[3] The neighborhood's 13,300 residents have a median household income that is US$10,000 less than the city average.[4] Furthermore, 26 percent of Cully residents live in poverty (compared to 17 percent citywide), and 40 percent

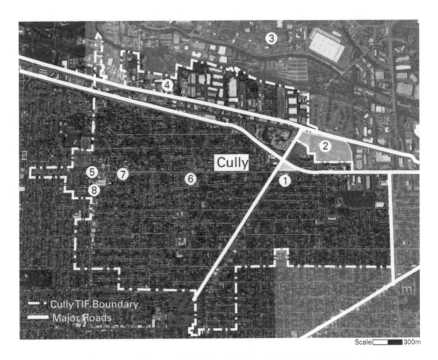

7.3 Key sites for Living Cully's work. 1. Las Adelitas; 2. Cully Park; 3. Portland International Airport; 4. Native American Youth and Family Center; 5. De La Salle North Catholic High School; 6. Khunamokwst Park; 7. Oak Leaf Mobile Home Park; 8. St. Charles Church Climate Hub. *Source*: Smriti Bhaya.

of Cully residents are renters, including some 10 percent of residents who live in the neighborhood's six manufactured home parks (MHPs).[5]

Cully residents are vulnerable to insecure housing and environmental harms from both pollution and mounting climate threats. The neighborhood, which includes a substantial industrial area along the Columbia Slough, has many brownfield sites and high levels of local pollution. Pollution comes from industrial sources such as a glass recycling facility and an asphalt grinding plant, as well as nearby highways, train lines, and Portland International Airport. According to one Living Cully staff member, neighborhood parents say things such as "I can smell my kids' clothes when they come in from outside; it smells like industry."[6] While industrial pollution has long threatened the health of Cully residents, the extreme temperatures that settled on the area during the summer 2021 "heat dome" event demonstrated that climate change is also a present danger.[7] A Living Cully staff member describes how the heat wave changed perceptions of climate risk in the area: "Portland . . . and low-income people and people of color really are being threatened. People are dying in Portland because of climate change."[8]

Cully residents, like residents of other low-income neighborhoods, face elevated danger from both climate change and displacement. The threat from extreme heat is exacerbated by the neighborhood's lack of green space and depleted tree canopy.[9] Although Cully is relatively low density and predominantly residential, only 24 percent of residents live within a quarter of a mile of a park compared to a regional average of 49 percent. Only 35 percent of Cully's streets have sidewalks, and residents have limited access to public transportation and commercial services.[10] Despite these challenges, Cully's relatively affordable housing has made the neighborhood attractive to buyers priced out of other neighborhoods. Between 2000 and 2022, home prices in the Portland metropolitan area increased more than 340 percent.[11] Between 2010 and 2015 alone, the median home price in Cully increased 57 percent, placing increasing pressure on its low-income population.[12]

In the context of increasing displacement pressure, four NGOs with established track records and community ties in the neighborhood came together to form Living Cully in 2010. The partners included Hacienda Community Development Corporation, a Latinx-led affordable housing provider; Verde, a social enterprise focused on job training in environmentally progressive industries; Habitat for Humanity Portland Region, the local chapter of the international affordable housing builder; and the Native American Youth and Family Center (NAYA), an organization serving area Native people with services and events grounded in the belief that "traditional cultural values are integral to regaining sovereignty and building self-esteem."[13]

Living Cully formed in response to an initiative from the Portland Sustainability Institute to create "ecodistricts." According to one early analysis, this coalition differed from other ecodistrict efforts "in its focus on equity and its participatory approach to organizing" to increase "the access of low-income communities, and communities of color, to environmental assets."[14]

The four partner organizations share a common vision articulated in a 2014 Memorandum of Understanding (MOU), but each pursues their own "signature projects." Verde serves as the "backbone organization," providing staffing and organizational support, but the coalition structure is intended to allow Living Cully to be more than the sum of its organizational parts.[15] In the words of Cameron Harrington, Living Cully's program manager, the coalition takes on "big projects that would be beyond what any of [the partner organizations] could do on their own and does them in a way that's better than any of them could do on their own."[16]

LIVING CULLY AND EQUITABLE RESILIENCE

Since Living Cully formed, the coalition has undertaken a range of projects that contribute to equitable resilience.[17] Through activities ranging from green job training and low-income home weatherization to developing new open space and affordable housing, Living Cully consistently links anti-displacement and poverty alleviation to sustainability and climate resilience. The organizations support residents in building sustained and dignified livelihoods through employment training, provision of living-wage employment in environmentally oriented social enterprises, pandemic relief, and language services—all part of efforts to safeguard the ability of residents to stay in a well-located neighborhood. While the focus of this chapter is on Living Cully's livelihood efforts, we also discuss the coalition's contributions to the environmental, security, and governance domains of equitable resilience, each of which is significant in its own right.

Living Cully has built an impressive track record. Yet, the coalition has also faced significant struggles. Some programs have been discontinued after short-term grant funding ceased. Neighborhood residents still face threats from displacement and from persistent violence. The sometimes-scattered and inconsistent nature of Living Cully's work is emblematic of broader challenges presented when NGOs are asked to deliver crucial services in underserved areas. Nonetheless, Living Cully is trying to address these structural challenges by establishing reliable sources of support for their work, forging strategies that could inform similar efforts to advance equitable resilience elsewhere.

LIVELIHOODS: TRAINING AND EMPLOYMENT FOR NEIGHBORHOOD RESILIENCE

Verde and other Living Cully coalition partners train Cully residents for jobs installing green infrastructure and energy efficiency retrofits. Living Cully's support for residents' livelihoods is primarily focused on two areas: (1) fighting displacement to ensure that low-income residents retain access to downtown Portland and other job centers, and (2) training and equipping low-income residents for meaningful employment, making their neighborhood and region more environmentally sustainable and resilient. We turn first to Living Cully's training and employment efforts and focus on anti-displacement efforts in the security section below.

The Living Cully coalition coalesced in the wake of the 2008 financial crisis, backed by federal American Recovery and Reinvestment Act (ARRA) stimulus funds supporting "green jobs" programs. This included Portland's Green Careers Training Program, which trained low-income residents and people of color in energy-efficient construction, insulation, HVAC technology, and related industries.[18] The ARRA-funded Clean Energy Works program supported Verde's equity-centered deep energy efficiency retrofit program. Andria Jacob, the Climate Policy and Program Manager at the City of Portland's Bureau of Planning and Sustainability, describes how these programs treated energy efficiency and renewable energy as strategies "for wealth creation, economic opportunity, rooting into a community, creating stability, preventing displacement and the onslaught of gentrification." Jacob says that because the "construction trades were the hardest hit" by the 2008 recession, the programs aimed to create "good, local construction" jobs with "apprenticeship pathways" into "union, licensed and . . . really good, livable wage jobs."[19]

Although these programs ended when the stimulus funding was exhausted, Verde and Living Cully have continued training and employing low-income residents in environmentally progressive industries. Verde Builds, which operates as a general contractor, has led worker training and installation of ductless heat pumps for a neighborhood buyers' cooperative along with continued weatherization and energy efficiency retrofit work.

Verde started as a spin-off of Hacienda CDC—another of the organizations that would later come together to form Living Cully—focused on training and employing local residents to do landscape work for properties owned by Hacienda. Verde has offered training and certification programs in several fields beyond energy, including irrigation installation, equipment maintenance, rain garden installation, and landscape maintenance. Verde also offers English as a second language classes as well as GED courses and programs in financial planning and management. While employed by Verde, trainees and employees are paid a living wage with benefits—a rarity among construction and landscaping contractors.[20] Employing local residents in landscaping and contracting provides broader benefits to the neighborhood. According to one analysis, "every dollar spent on a Verde Landscape project generates almost two dollars of economic activity in Greater Portland."[21]

In addition to their own training and social enterprises, Living Cully advocates for local employment for other projects in the area to ensure that investments in Cully benefit residents. In the construction of Cully Park, one of Living Cully's signature achievements discussed below, Verde Landscaping drew 18 percent of its labor force from among low-income neighborhood residents.[22] Fully three-quarters of design contracts and 60 percent of construction contracts for the park went to minority and women-owned businesses.

The COVID-19 pandemic disrupted many of Living Cully's activities. While Verde Builds and the other coalition partners continue their work, Verde Landscaping shut down during the pandemic. Living Cully partners suspended some programming and redirected their efforts to safeguarding the immediate well-being of Cully residents. Based on assessments of community needs, Living Cully adopted three priorities during the pandemic: food access, rental assistance, and technology access for remote school and work. With many residents out of work, Living Cully organized weekly food deliveries to low-income residents. To make sure residents were not displaced during the pandemic, Living Cully organized the Cully Neighborhood Renter Relief Fund, which raised and distributed around US$150,000. According to a Living Cully staff member, the Fund focused on "folks who did not receive federal relief payments, largely because of immigration status." In addition to raising funds themselves, Living Cully has enabled residents to access government relief funds to which they were entitled, totaling more than US$500,000. Finally, Living Cully also concentrated

COVID-19 relief efforts on "access to technology, especially for families with kids who are staying home from school and didn't necessarily have computers or internet access."[23]

While COVID-19 required significant shifts in Living Cully's activities, the group used these emergency relief projects to advance their organizing and community development goals. One Living Cully leader says the coalition is "using those new contacts we've made with folks that we wouldn't have had if we hadn't been doing direct service as a way to get people involved" in other Living Cully projects.[24] In sum, Living Cully supports low-income residents' livelihoods in multiple ways, combating threats ranging from COVID-19 to climate change through grassroots organizing that creates lasting community connections.

ENVIRONMENT: ENVIRONMENTAL WEALTH AND COMMUNITY WELL-BEING

Living Cully addresses residents' environmental and climate change burdens through two broad goals: (1) reducing energy demand via efficiency retrofits and renewable energy; and (2) green infrastructure to mitigate hazard risk from stormwater flooding and heat.

After Verde's ARRA-funded weatherization projects ran out of stimulus funding, Living Cully continued to train local residents in labor-intensive work such as air sealing, insulating, and installing efficient mechanical systems. Because their coalition includes Habitat for Humanity, Living Cully is also able to do basic home repairs to address health and safety issues as part of the weatherization process.[25]

The waxing and waning of ARRA-funded energy programs is emblematic of the unpredictable funding that plagues most NGO-driven community development. In response, Living Cully advocates for more reliable resources. In 2018, the City of Portland adopted the Portland Clean Energy Community Benefits Fund (PCEF), which taxes national and global retail businesses (e.g., "big box stores") to fund clean-energy projects focused in under-resourced areas such as Cully. After Living Cully drafted their own energy plan in 2018 calling for a reduction to neighborhood carbon dioxide emissions by 40 percent by 2030, the coalition successfully used PCEF funds for several projects, including a community energy education campaign,

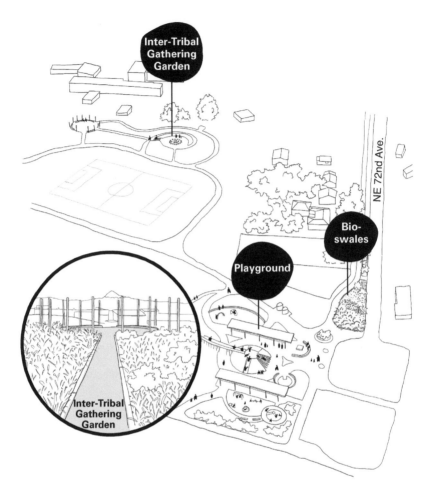

7.4 Cully Park converted a former landfill into much-needed green space and green infrastructure through a co-design process. *Source*: Mora Orensanz.

energy efficiency upgrades and solar installations at a nonprofit-owned MHP, a ductless heat pump buyers' cooperative to enable households to install efficient heating and cooling systems, and large solar energy generation and storage installations at a local church. The latter, one Living Cully staff member observes, will make the church "a climate hub in the case of disasters."[26] Building on the success of the PCEF, in 2017, Living Cully supported the successful passage of Portland's 100 percent renewable energy resolution (which will also prioritize community-based renewable energy infrastructure) and the city's Climate Emergency Declaration, adopted in 2020, which Andria Jacob states will "set a framework for a Climate Justice Initiative."[27]

In addition to their portfolio of energy projects, green infrastructure has been central to Living Cully's work. Living Cully's largest green infrastructure project to date is the creation of Cully Park, which converted a former quarry and landfill into twenty-five acres of public green space (figure 7.4). Living Cully helped instigate Let Us Build Cully Park!—a coalition of seventeen groups that began working with the City of Portland to create the park in 2010. When the park opened in 2018, it included community gardens, nearly two thousand square feet devoted to stormwater management green space, and an Inter-Tribal Gathering Garden—a space of some thirty-six-thousand square feet featuring traditional food, fiber, and ceremonial plants of the Northwest Indigenous Tribes that NAYA serves. One study of Cully Park's community health impacts described the park as a "place to celebrate life. A safe, outdoor space to connect with nature and community through multicultural and intergenerational learning, sharing, exercise, play, activities, well-being and balance."[28]

The same multidimensional, community-centered perspective on "environmental wealth" seen in Cully Park characterizes Living Cully's other green infrastructure initiatives. Working with the City, Verde Landscaping installed street-side bioswales along 72nd Avenue, one of the streets leading to Cully Park. One report describes the project as converting "an 860-ft crumbling asphalt street with no curbs [or] storm sewer" into bioswales "using native plants that create habitat for birds and wildlife" and "slow and filter stormwater, capturing contaminants." Further, the design of the project renders stormwater control legible to residents, enabling "community education about stormwater management and watershed health."[29] Living Cully is planning similar green infrastructure installations for the Oak Leaf Mobile Home Park in the neighborhood. A staff member describes the value of bringing "green assets" to low-income communities, saying they are "not just something that rich people can have; it's something that a mobile homeowner can have."[30]

Living Cully's approach to environmental improvement links resilience to heat and flood hazards with other crucial goals. In addition to supporting local livelihoods through job creation and training, their projects aim to support community cohesion for Indigenous residents and people of color, reduce the environmental health burdens of pollution, fight displacement by reducing energy costs, and facilitate experiential environmental education through both designed interventions and community programming.

SECURITY: "DISPLACEMENT MAKES OUR COMMUNITIES LESS RESILIENT"

With rapidly rising housing prices in Cully and across the region, fighting displacement has become a core component of Living Cully's work. A 2013 report by students at Portland State University found that 68 percent of Cully renters spent more than 30 percent of their income on housing compared to 50 percent citywide.[31] As Cully's property values rise, Living Cully has embraced anti-displacement initiatives, including building and acquiring affordable housing, promoting tenants' rights, and reducing living costs and energy burdens through weatherization and other efforts.[32] Living Cully also argues that Cully Park strengthens security for low-income residents by "fostering a sense of community" through creating spaces for "intergenerational activities and events" and "culturally-relevant programming."[33]

Bridging anti-displacement and environmental initiatives is crucial, given mounting concerns about green gentrification and climate gentrification.[34] Speaking about green infrastructure projects along Cully streets, Tony De Falco, Living Cully's former Executive Director, worries, "If you do a project like this enough times in a neighborhood like this, property values will increase without corresponding benefits to the community."[35] Living Cully's 2018 Energy Plan links decarbonization goals with security from displacement, noting, "While many sustainability advocates ignore gentrification and displacement or treat it as unsolvable, Living Cully directly addresses this issue as a core aspect of our work, advancing a model of environmental wealth alongside racial and economic diversity."[36] The plan explicitly argues that "displacement and climate change are deeply connected" because "displacement makes our communities less resilient, less able to survive and rebuild after a climate event."[37]

In addition to their energy and livelihood programs, Living Cully partners have also been instrumental in securing affordable housing in the neighborhood. Hacienda CDC owns and operates several income-restricted

7.5 Participatory design methods informed the design of the Las Adelitas affordable housing project that replaced a strip club. *Source:* Mora Orensanz.

apartment buildings in the area. In late 2022, Hacienda completed the replacement of the former Sugar Shack strip club with Las Adelitas, a development with 142 units of affordable housing and space for community programs for Latinx youth (figure 7.5).

Living Cully partners help drive the Cully Housing Action Team (CHAT), which organizes for tenant rights. As with ROC USA's nationwide efforts (see case 4), much of CHATs organizing work has focused on improving housing security among residents of Cully's MHPs. When the former owners of the Oak Leaf MHP put the park up for sale in 2016, residents of the park's thirty-four occupied homes feared that a developer would buy the park, shut it down, and build high-end housing (see figure 7.6). Instead, working with the Portland Housing Bureau, Living Cully bought the park and made substantial investments, including upgrades to infrastructure systems (e.g., electric, sewer, and roads), and replacement of degraded homes with new energy-efficient units. Living Cully then sold the park to the Society of St. Vincent de Paul of Portland, a local affordable housing and social services provider.[38] Chastened by the near eradication of so many affordable homes at the Oak Leaf MHP, Living Cully staff began advocating for zoning protections to ensure that existing MHPs across the city would be preserved as affordable housing. According to Living Cully organizers, they worked with residents to overcome the stigma associated with MHPs by "inviting policymakers to come and tour the mobile home parks and see their homes and see their communities."[39]

GOVERNANCE: CONFRONTATION, COALITION, AND INNOVATION FOR SUSTAINED SUPPORT

Living Cully's coalition structure has both strengths and weaknesses. The coalition members have strong relationships within the neighborhood that have enabled them to advocate effectively for marginalized groups. In the words of one Living Cully staff member, their effectiveness is rooted in "the level of trust that the community has in Living Cully staff," especially those members who are "living in the community."[40] Community embeddedness and trust enabled the participatory design of the Cully Park. Verde's Director of Strategic Partnerships, Vivian Satterfield, says, "Every single aspect of the park is infused with not just community-*informed* design, but actual *direct* community design." Designers of the natural play structures engaged neighborhood youth, and NAYA "brought together Indigenous people, enrolled tribal members, and urban Indians to really talk about how they wanted to design [the Inter-Tribal Gathering Garden]."[41]

The coalition structure of Living Cully has allowed members to act both individually and in coordination to advance their shared mission, taking advantage of shifting government priorities (e.g., ARRA green jobs funding), capitalizing on opportunities to purchase key properties, and advocating for policy changes. Cameron Harrington describes the coalition structure as "politically valuable and strategic," since it creates a "separate identity that can do things as Living Cully without implicating [coalition members, such as] Hacienda or NAYA or Verde." He goes on to describe how Living Cully can form "community organizing groups that have their own name and identity," such as CHAT, enabling those groups to use confrontational tactics, such as protesting at City Hall or at "a landlord's office downtown" or doing "street theater." The coalition structure creates space for more radical actions without jeopardizing coalition members' relationships with empowered institutions (figure 7.6).[42]

Living Cully adopts strategic confrontation to advance their advocacy goals, but they are also effective in building power through linkages with powerful institutions, including local government. Satterfield explains: "Our relationship with City Hall is one of power building and challenging power and saying 'This is what we need. This is what we want.' . . . People do take us seriously . . . and they take our phone calls, and they will assign staff to work on issues that we want to work on."[43] The coalition has also built successful partnerships with other local institutions, including neighborhood businesses in the Cully Boulevard Alliance and academic institutions such as Portland State University and Mt. Hood Community College, supporting useful research and educational opportunities for community members.

Living Cully has successfully adapted to changing government priorities, but reliance on shifting resources can present serious challenges. Reflecting on the purchase of the former Sugar Shack and the Oak Leaf MHP, Living Cully staff member Harrington notes, "The amount of work and luck and just confluence of circumstances that it took for either of those to happen point to the need for mechanisms that make it possible to replicate that kind of project, because we can't count on all of those circumstances and lucky partnerships coming together."[44]

Recognizing that inconsistent funding undermines community self-determination, Living Cully coalition members use their shared resources, networks, and political influence to create sustained sources of support for their work. The PCEF represents one such stable source for resources. Living Cully has also been an integral part of establishing an innovative community-led tax-increment finance (TIF) district in Cully. The nascent district could spur even more transformative change. Organizing to design and establish the TIF began with a city-wide effort called Anti-Displacement

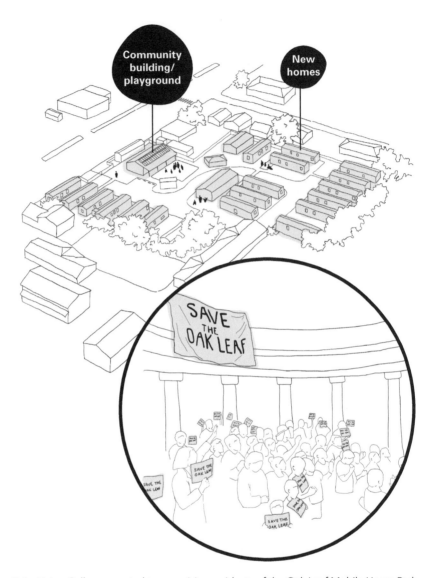

7.6 Living Cully was central to organizing residents of the Oak Leaf Mobile Home Park to preserve their affordable housing and avoid displacement. *Source*: Mora Orensanz.

PDX associated with the Portland Comprehensive Plan process in 2018. The TIF is intended to capture property-value increases to create dedicated funding to buy both residential and commercial property to preserve their affordability. The Cully TIF, unanimously approved by Portland City Council in November 2022, is projected to generate more than US$100 million over twenty years to "prevent displacement and benefit people of color and low-income people."[45] Proposal documents assert that the funds will "prioritize investments that remove properties from market-based ownership" and "permanently preserve" their affordability and access through ownership by "non-profits, public agencies, land trusts, resident-owned cooperatives, and other models of community ownership."[46]

Living Cully staff argue that they are "taking the tool of tax-increment financing and turning it on its head" by enabling "community ownership and community decision-making."[47] Program manager Harrington states that the "list of eligible investments will be focused on affordable housing and economic development investments that are community-led, and that explicitly create and preserve opportunities for small businesses, and for businesses owned by people of color." In designing the Cully TIF, Living Cully and the other members of the coalition aim to create "a new kind of governance structure where community members and organizations are making real substantive recommendations and decisions—over the entire life of the district, twenty years plus—about what projects are getting funded."[48] This governance and funding structure is meant to create a form of participatory budgeting that actively counters structural racism and historical and ongoing injustices against people of color.[49]

While the Cully community-controlled TIF district represents an exciting avenue to secure sustained support for Living Cully's anti-displacement activities, it is not yet clear how these efforts will relate to the coalition's environmentally focused work. TIF planning documents somewhat evasively states that the "Climate Action and Environmental Stewardship" guidance for TIF investments "will be developed in Phase 2 of the project."[50]

CONCLUSION: ENVIRONMENTAL WEALTH FOR EQUITABLE RESILIENCE

As a coalition of four NGOs rather than a single entity, Living Cully can face complicated organizational challenges. Adding further complexity, Living

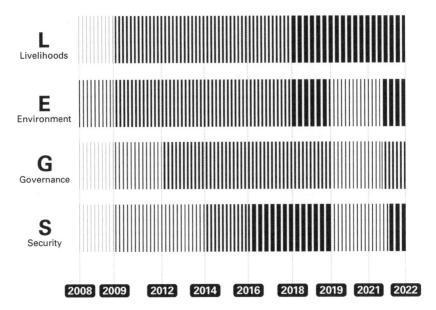

L
Livelihoods

E
Environment

G
Governance

S
Security

2008 2009   2012   2014   2016   2018   2019   2021   2022

7.7 Living Cully's work ties livelihood generation and security from displacement to neighborhood environmental improvement. *Source*: Smriti Bhaya. 2008: Great Recession, ARRA green jobs funding, Portland Sustainability Institute Ecodistrict Initiative; 2009: Living Cully, Cully Park coalition, and CEWP founded; 2014: City grants Verde development rights for Cully Park, Participatory design/planning begins; 2017: Former Sugar Shack bought and planning for Las Adelitas begins; 2016: Living Cully takes ownership of Oak Leaf MHP; 2018: Living Cully Energy Plan, PCEF founded; Cully Park opens; 2020: COVID-19 pandemic, emergency relief efforts overtake other priorities; 2021: Heat dome event; 2022: TIF proposal approved.

Cully often contributes to other coalitions for projects spanning affordable housing development, energy-efficient building retrofitting, green infrastructure building, and job training. Two elements tie these various organizations and activities together. First, all of the coalition's projects are rooted in the notion that "environmental wealth" can be an effective tool for fighting displacement and supporting the livelihoods of low-income people of color. Second, while their advocacy impacts territories beyond Cully, Living Cully is steadfastly focused on a single neighborhood. In Vivian Satterfield's words, the coalition is "impactful" because they have "all these partner organizations working together and really focused on a geographic area," allowing them to "go really deep in the neighborhood."[51]

Living Cully's work integrates all four elements of equitable resilience (figure 7.7). Their focus on building "environmental wealth" includes both pollution reduction, through efforts such as energy-efficiency retrofitting and brownfields redevelopment, and adaptation to climate hazards such as flooding and heat. The coalition pursues security against displacement by developing and preserving affordable housing and advocating for tenants' rights. They foster community power and self-governance through coalition building, linkages to powerful institutions, and sometimes-confrontational advocacy. Crucially, Living Cully has made supporting dignified livelihoods for low-income residents a central part of their work. Whether supporting residents in learning English and financial literacy, training and employing people in energy efficiency and green infrastructure construction, or mobilizing emergency relief during the COVID-19 pandemic, forging dignified livelihoods is at the core of Living Cully's concept of "environmental wealth" and thus at the core of their pursuit of equitable resilience.

Case 8

YERWADA IN SITU UPGRADING: PRESERVING AND GROWING PLACE-BASED LIVELIHOODS IN PUNE, INDIA

OVERVIEW

Informal settlements such as those in the Yerwada area of Pune, a city in western India, are critical sites for advancing equitable resilience. Beginning in the late 2000s, city leaders and civil society organizations undertook ambitious projects to upgrade thousands of self-built homes in Yerwada. The projects demonstrate an alternative to dominant models of "slum clearance" and relocation. Interventions led by two civil society teams, Maharashtra Social Housing and Action League (MASHAL) and Society for the Promotion of Area Resource Centers (SPARC)/Mahila Milan, used innovative governance and design strategies to ensure that upgrading reflected the needs of residents. While the projects supported residents' livelihoods through several mechanisms, the house-by-house in situ upgrading process limited progress on larger problems with public space and infrastructure. This case study draws on site visits, analysis of existing studies and project materials, and interviews with project participants, administrators, observers, and critics.

INTRODUCTION

In his bill of rights for housing, pioneering Indian architect Charles Correa lays out eight principles for housing in rapidly urbanizing settings,

8.1 Homes in Yashwant Nagar, a settlement in the Yerwada area of Pune that benefited from in situ upgrading, include a range of livelihood-supporting features, such as small storefronts and space for storage of materials and equipment. *Source*: Smriti Bhaya.

including (1) incrementality, (2) open to sky space, (3) equity, (4) disaggregation, (5) pluralism, (6) malleability, (7) participation, and (8) income generation.[1] While the in situ upgrading of homes in the Yerwada settlements address all of these principles to some degree, it is the last principle—creating housing to support dignified and sustained livelihoods—that is the focus of this chapter.

In cities across India, as in other rapidly urbanizing regions, self-built "informal" settlements constructed by poor people have long been treated as problems by authorities.[2] When they are not ignored and neglected,

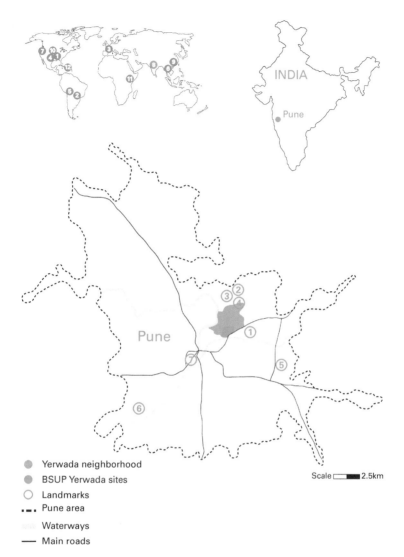

8.2 Location of the Yerwada In Situ Upgrading projects and other Pune landmarks. *Source*: Mora Orensanz. 1. Kalyani Nagar; 2. Airport; 3. Stone Quarry; 4. Viman Nagar; 5. Hadaspar; 6. Warje; 7. Pune Municipal Corporation.

these settlements, which are called by many names, including "basti" and "slum," are often bulldozed in the name of modernization and sanitation. Informal settlements are frequently located in areas that are hazard prone but well located for economic opportunity. Even when residents of cleared settlements are rehoused, new housing is often located on urban peripheries, far from residents' former homes, social networks, and livelihoods.

The Yerwada In Situ Upgrading projects in Pune, like the Paraisópolis Condomínios (see case 2) and the Baan Mankong upgrading projects in Bangkok (see case 6), represent an alternative approach, supporting residents' livelihoods by maintaining their locational advantages. The Yerwada upgrading projects emerged from a larger initiative supported by the Indian government's Basic Services for the Urban Poor (BSUP) program under the Jawaharlal Nehru National Urban Renewal Mission (JNNURM). This chapter focuses on in situ upgrading by two groups of prominent NGOs: first, a partnership between the Pune chapter of Mahila Milan and SPARC; and second, MASHAL.

The Yerwada neighborhood includes several self-built settlements in the northeastern part of Pune, a fast-growing city of about four million residents two hundred kilometers southeast of Mumbai (figure 8.2). In 2012, approximately one third of Pune's residents lived in the city's 564 informal settlements.[3] Yerwada represents one cluster of these settlements, which took shape in the 1960s when low-income people started settling the area to gain access to livelihood opportunities, including in nearby mines and quarries.[4] A census in 2008 estimated that the Yerwada settlements housed 87,400 residents.[5] Residents reported difficult conditions, with people "living in mud and patra [asbestos sheet] huts,"[6] with an average footprint of just thirteen to fifteen square meters (140–160 square feet). Few homes had toilets or piped water. The nearby Mula Mutha River periodically overflowed its banks, flooding low-lying areas.

The BSUP-sponsored projects in Pune set out to upgrade housing for nearly 205,000 households, but they started inauspiciously. Phase I included construction of some 1,300 units of housing in eight-story towers in the Warje and Hadaspar neighborhoods on Pune's outskirts. These first projects prioritized relocation of people from informal settlements "falling under riverbeds, green belts, road widening, hill top, hill slopes, [and] vitally required lands."[7] Shelter Associates, a housing advocacy group, reported that the BSUP relocation projects "failed to address the aspirations and nuances of the target community's culture" because "the communities for whom they were intended were not taken into confidence."[8] As a result, since families were reluctant to relocate away from their homes, communities, and livelihoods, the buildings sat empty for many months.

After these early failures, local leaders and advocacy groups devised an alternative in situ upgrading strategy for subsequent BSUP projects. In the end, roughly half of the city's BSUP budget went to upgrading homes and infrastructure in twenty-seven settlements rather than relocating residents.[9] SPARC and MASHAL led upgrading for a combined four thousand Yerwada households in seven and eleven sites, respectively (figure 8.3). Rather than relocating residents or clearing and rebuilding entire settlements to regularize their form, these NGO-led projects sought to rehouse people in upgraded *pucca* homes on the same parcels where their earlier *kuccha* or "temporary" homes had stood. The teams conducted extensive surveys in each settlement to assess household eligibility. Only households living in *kuccha* houses constructed of materials such as tin sheets and mud were eligible for upgrading under BSUP rules. Rigid program requirements that new *pucca* homes must be twenty-five square meters (270 square feet) complicated implementation and made the program less appealing to some eligible households. While many families living in houses smaller than twenty-five square meters were eager to upgrade to larger homes, accommodating these households on existing parcels presented challenges. Households with existing houses that were larger than twenty-five square meters frequently opted not to participate because they did not want to reduce the size of their homes. Participation varied from settlement to settlement, although one leader estimated that the project upgraded roughly 65 percent of dwellings in participating settlements.[10] Upgrades were funded through a combination of federal (50 percent), state (30 percent), local (10 percent), and beneficiary (10 percent) contributions. Each household paid roughly thirty thousand rupees (approximately US$400) for their upgraded home.

YERWADA AND EQUITABLE RESILIENCE

While the SPARC and MASHAL upgrading projects in Yerwada followed similar strategies, their implementation differed somewhat. SPARC partnered

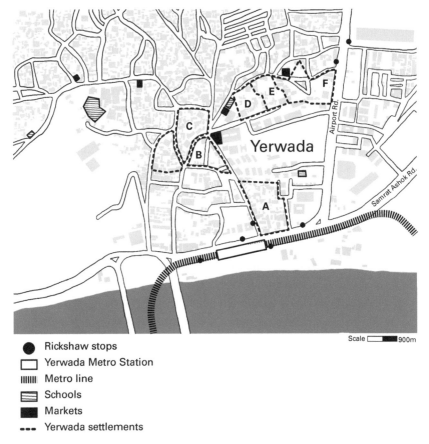

Yerwada

Scale ▭◼ 900m

● Rickshaw stops
▭ Yerwada Metro Station
▥ Metro line
▤ Schools
◼ Markets
▪▪▪ Yerwada settlements

8.3 Yerwada settlements. *Source*: Smriti Bhaya. A. Mother Teresa Nagar; B. Sheela Salve Nagar; C. Wadar Wasti; D. Bhatt Nagar; E. Netaji Nagar; F. Yashwant Nagar.

with the National Slum Dwellers Federation and Mahila Milan, a women's collective that has been active in Yerwada and across India for many years. The SPARC projects emphasized design innovation, developing models for incrementally upgradeable infill housing through extended participatory processes led by outside designers, first Filipe Balestra of Sweden-based Urban Nouveau and then Pune-based architect Prasanna Desai. MASHAL focused less on design innovation and more on creating a streamlined process for construction with small local contractors, maximizing flexibility and local economic benefits.

In both cases, project leaders sought to maximize livelihood benefits for Yerwada residents, including by employing local contractors and laborers, accommodating home-based livelihoods such as small storefronts, and avoiding relocation to preserve place-based livelihood networks. The Yerwada upgrading projects also reduced residents' vulnerability to environmental hazards through simple strategies such as building homes on small, raised plinths to avoid floodwaters. Some observers have critiqued the projects for focusing on individual housing upgrades rather than improving public space and infrastructure. Other critiques of the projects center on the failure of the city to grant residents official tenure rights. In both struggle and triumph, the Yerwada In Situ Upgrading projects offer insights on pursuing equitable resilience in informal settlements.

While our central focus is how the upgrading projects support residents' livelihoods, to understand these contributions first requires consideration of the projects' innovations in governance.

GOVERNANCE: CIVIL SOCIETY DEVELOPERS AND RESIDENT-DRIVEN DESIGN AND CONSTRUCTION

Prasanna Desai, a Pune architect who worked on the SPARC Yerwada upgrading projects, describes the project as "a completely different approach," departing from the top-down relocation and redevelopment models of "slum rehabilitation" by treating "the slum dwellers and their community" as "the clients" and "decision maker[s] of the entire process."[11] While their approaches differed somewhat, both SPARC and MASHAL sought to treat residents as clients whose needs should inform upgrading. In both cases, these NGOs acted as trusted intermediaries between state institutions and residents.

Through partnering with Pune Mahila Milan, SPARC built upon the collectives' reputation established over years of work in Yerwada, including constructing community toilets and initiating savings groups. Pune Mahila Milan's involvement built trust with residents and enabled ongoing advocacy and organizing, even after the end of the upgrading process.[12] The SPARC/Mahila Milan team started their efforts in seven Yerwada settlements in 2007, with several months of work surveying community needs, assessing household eligibility, and forming savings groups. In addition to conducing cadastral and socioeconomic surveys, the team used support from the Gates Foundation to hire Urban Nouveau to initiate participatory design activities early in the process.[13] Architect Balestra reported, "We used to arrive on site early morning, and map and document the various ways in which informal living happened in the settlement. We understood that each house, kuccha or pucca, was built by someone and their design input was crucial."[14]

Based on this initial research, Urban Nouveau developed three house concepts, each featuring a multistory reinforced concrete post and slab structure, with some areas open for future expansion (figure 8.4). They then refined these concepts through meetings with potential beneficiaries using both 1:50 scale models and a full-scale bamboo and cloth mock-up to allow residents to understand the proposed spaces better. Following these meetings, the team simplified the designs, moving forward with only one- and two-story models. Where existing houses had footprints that were too small to accommodate the programmatically required twenty-five square meters of space, the team grouped households together in taller multifamily clusters.[15] Narmada Vetaale of Mahila Milan reports, "The participatory design part of the process was an important success. We created house model exhibitions, went door to door."[16] Resident participation continued after the design phase, with residents often contributing to construction labor and suggesting changes to suit their needs.

The strict interpretation of "in situ" upgrading deployed in this project meant that residents were not only rehoused within the same community but also overwhelmingly rehoused on the same parcel of land where their former *kuccha* house stood. This approach had both advantages and disadvantages. Architect Desai, whose firm led the later design stages of the SPARC/Mahila Milan projects, says that parcel-by-parcel upgrading enabled the team to "retain the overall fabric of the slum in terms of existing street patterns and existing footprints," preserving important sites (e.g., shrines) and place-based networks. The team attempted targeted "re-blocking," urging owners of larger lots to "surrender a part of the plot in the larger interest of the community, making it possible to widen the existing streets for better accessibility and create community interaction spaces."[17] In practice, few such adjustments were realized, in part because the BSUP policy framework did not provide flexibility to facilitate re-blocking, for instance by incentivizing residents to give up some portion of their plot by allowing them to build taller structures.[18] Without substantial re-blocking, plot-by-plot upgrading maintained the existing settlements' form, for better and

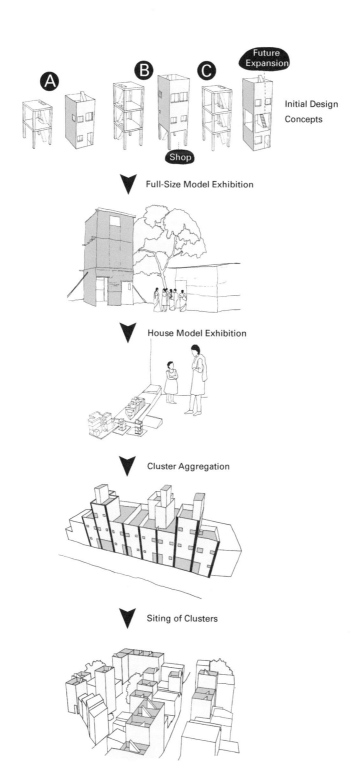

8.4 Participatory design process for SPARC/Mahila Milan upgrading. *Source*: Ipshita Karmakar.

for worse, preserving places and networks, but also retaining the settlements' extreme density, limited access to light and air, and constrained emergency services access.[19]

Keya Kunte, who worked on the SPARC upgrading, describes the tensions created by plot-by-plot upgrading, saying, "It's a very dense settlement so you're working with small space . . . the minute you go to upgrade . . . inevitably the roofs begin to touch each other." While some critics suggest that a master planning approach would have been preferable, starting with rationalized infrastructure and accommodating housing thereafter, Kunte counters that "people who needed housing wanted the housing to start first, because that was a bigger requirement."[20]

Project participants noted other challenges with SPARC's and MASHAL's upgrading, including organizational and financial difficulties encountered by NGOs unaccustomed to working as building contractors; quality control complaints, particularly in projects built by some subcontractors; and financial strain on the poorest residents to secure the 10 percent beneficiary contributions.[21] Project participants also lamented the slow and complex work of performing thousands of individual upgrading projects with individual clients on irregular and informal parcels. One observer noted that "the project had an enormous organizational task . . . Since no two units were the same because of varying plot sizes, each unit was treated as a separate project in itself."[22]

LIVELIHOODS: PRESERVING ACCESS, ACCOMMODATING HOME-BASED WORK, AND DIRECT EMPLOYMENT

The Yerwada settlements originally formed largely because they gave poor people access to economic opportunities in central Pune. Today, residents work in a range of fields, including construction, driving autorickshaws, domestic services, welding, vending, and stonework and mining in nearby quarries. The first round of BSUP projects, which aimed to relocate residents to distant suburbs ten to fifteen kilometers from Yerwada, struggled because, as SPARC reports, there was "no provision of capacity building or addressing livelihood issues."[23] The SPARC and MASHAL projects took a different approach, supporting residents' livelihoods not just by preserving locational advantages but also by accommodating home-based work and employing residents in the upgrading process.

IN SITU LIVELIHOODS: CAPACITY AND EMPLOYMENT

Both MASHAL and SPARC organized their upgrading projects to ensure that the residents would enjoy direct economic benefits. One observer noted that a "large part of the workforce of Yerwada works in the construction industry" and the project "managed to tap this workforce during the construction phase," providing "great economic benefits to the settlement."[24]

SPARC's partnership with Mahila Milan shaped their approach. Mahila Milan's own contracting business performed an estimated 40 percent of the work in the seven settlements in which they operated.[25] In some cases, they employed project beneficiaries in demolition and construction work, including for their own homes. While Mahila Milan hired local people, especially women, for upgrading work, some observers noted concerns about conflicts of interest, since residents could not choose their own contractors and Mahila Milan was not required to participate in competitive bidding to secure projects.[26] In other cases where SPARC employed outside contractors, different challenges arose, including contractors quitting because of slow payment, contractors struggling to complete work in the confined environment of informal settlements, and, occasionally, shoddy workmanship and materials.[27]

MASHAL adopted a somewhat different approach to contracting and hiring. While SPARC worked with Mahila Milan and a limited group of larger builders, some of whom were from other cities, MASHAL allowed beneficiaries to choose from a list of small local contractors, many of whom were based in Yerwada. In total, some 180 local contractors worked on MASHAL's upgrading projects, often in the neighborhoods in which builders themselves lived. This arrangement allowed beneficiaries to establish relationships with builders, creating a sense of mutual accountability and enabling residents to shape decisions throughout the building process.[28] MASHAL's more standardized housing design streamlined the building process. Whereas SPARC did many idiosyncratic cluster developments housing several households in larger buildings, MASHAL worked primarily with individual households. In 2015, administrators resolved that project beneficiaries themselves could act as contractors, further advancing local economic benefits.

INFORMAL ACCOMMODATION OF HOME-BASED LIVELIHOODS

Most SPARC and MASHAL Yerwada upgrading projects were not designed explicitly to accommodate home-based work. Even so, both groups' resident-engaged processes allowed for significant customization to support income generation. Mahila Milan's initial household surveys included questions about home-based work. According to one Mahila Milan official, prior to upgrading, few residents did home-based work such as sewing or basket making. Due to the poor condition of Yerwada homes, "manufacturers were reluctant to let the women in the settlement use equipment like sewing machines in the house." After upgrading, however, "more NGOs and manufacturers have started giving us home-based work." For example, "The women in Wadarwasti make baskets, do tailoring, run grocery or stationery shops and other such work from their homes."[29] A SPARC engineer reports that in approximately thirty houses, "which had previously accommodated shops in the ground floor, we modified the design to some extent so the upgraded house could also support a shop" (figure 8.5).[30] Even when the initial design was not specifically tailored to home-based work, flexible home designs have enabled many residents to make small changes to facilitate businesses, including installing roll-up doors for small storefronts.

The upgraded homes in Yerwada are small, but they accommodate residents' livelihoods through subtle features that allow for storage of materials and equipment in the immediate area around the homes. In many cases, new houses feature *otlas*, small verandas created by extended plinths and cantilevered second floor volumes. These spaces fill many livelihood-supporting functions, including facilitating storage of autorickshaws, vending carts, building materials, water vessels, cooking pots, and other equipment for household and commercial use.

ACCESSING THE CITY

Many Yerwada residents are employed in informal economies and formal jobs tied to nearby locations. Most residents work within just a few miles

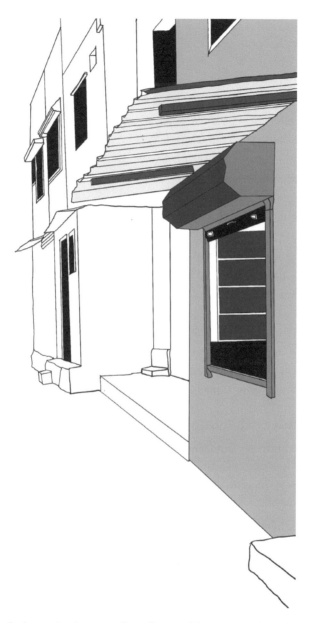

8.5 House-by-house in situ upgrading allows residents to continue improving their homes, including creating storefronts and other spaces for livelihood generation. *Source*: Ipshita Karmakar.

of their homes and rely on access to inexpensive transportation. The upgraded settlements have easy access to a major railway station, Pune's airport, the municipal corporation's offices, and nearby colleges and universities, all of which employ local residents. Many residents also work in the municipal produce market in Yerwada. After observing the failure of early BSUP relocation projects, SPARC leader Sheela Patel wrote, "Clearly, the preferred option was to upgrade habitat in the same location. For residents, this avoids the disruptions to livelihoods and social networks that relocation usually entails."[31] A 2020 study assessing perceptions of livability among residents in several Pune communities—including an informal settlement that was not upgraded, a relocation housing scheme, and one of the MASHAL upgraded Yerwada communities—found that "proximity to public transportation" and "proximity to employment opportunities" were among the most critical areas of satisfaction for Yerwada residents.[32]

The Yerwada upgrading projects also improved access to livelihoods for residents by removing the social stigma of living in *kuccha* houses. Project leaders and funders observed that, with this stigma removed, residents of upgraded houses enjoyed improved access to jobs and schools—benefits that especially helped young residents.[33]

Even as upgrading supported residents' livelihoods through direct employment, accommodating home-based work, and maintaining access to nearby jobs and amenities, project participants and observers also noted some shortcomings. The Yerwada projects did not include transitional housing or livelihood support for residents forced to relocate temporarily during upgrading. In some cases, residents had to pay rent or live with family members for multiple years as construction lagged. As in other upgrading schemes we profile in Thailand (case 6) and Brazil (case 2), residents of upgraded homes in Yerwada complained of increased living expenses associated with their new homes, including both the purchase cost and new expenses, such as electricity and water payments. Despite these challenges, the Yerwada upgrading projects demonstrate several ways that pluralistic, flexible, and participatory upgrading can enable residents to shape interventions to support their own livelihoods.

ENVIRONMENT: ADJUSTMENTS TO PLACE-BASED HAZARDS

According to project leaders, proximity to the Mula Mutha River floodplain helps explain why the BSUP funded upgrading in Yerwada. Some areas have suffered repeated river and rain inundation, including devastating floods in 1961 and 2019. In addition to flood risk, residents of Yerwada also face other hazards, including extreme heat, poor indoor air quality, public health threats from inadequate sanitation, and seismic risks, which are exacerbated in former mine and landfill sites. While the upgraded housing developed by SPARC and MASHAL does not include flashy green building or resilience features, the projects did incorporate strategies to reduce vulnerability for both individual households and the broader communities.

Upgraded houses feature elevated plinths lifting them above the street level, reducing flood vulnerability.[34] In Shanti Nagar, the city corporation built a new floodwall to keep floodwaters out of the settlement. Although drainage congestion caused waterlogging on the "dry" side of the new Shanti Nagar floodwall during heavy rains in 2019, simple flood-mitigation strategies deployed in upgraded houses largely spared them from damage (figure 8.6).

Other environmental resilience features include reinforced foundations engineered to resist damage from earthquakes.[35] Additionally, wherever possible, designers incorporated more and better windows in living spaces to improve indoor air quality and natural ventilation. The same overhangs that create livelihood-enhancing storage spaces between houses and streets also provide thermal and weather protection, shielding ground-floor spaces from sun and rain.

While SPARC's and MASHAL's upgrading overwhelmingly focused on individual households, they also improved community infrastructure in several cases, upgrading drainage, sanitary sewers, and roads. Each upgraded house included an indoor toilet. The program also provided fifteen thousand rupees (approximately US$200) for other households to install toilets, thereby dramatically improving sanitation conditions in the settlements.

Though the upgrades have reduced residents' vulnerability in several ways, critics point to lingering challenges. Observers have noted that the programs' focus on housing, rather than holistic urban upgrading, means

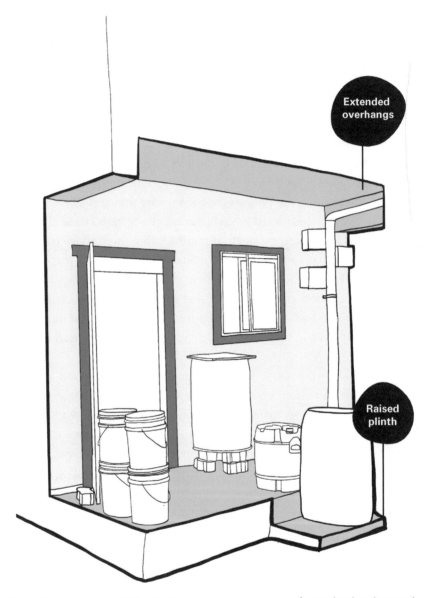

8.6 Simple environmentally adaptive measures improve comfort and reduce heat and flood vulnerability. *Source*: Ipshita Karmakar.

that collective services, including drinking water, are still inadequate.[36] Shelter Associates, a group that withdrew from Pune's BSUP projects because of conflicting visions, has leveled critiques against the house-by-house in situ strategy deployed by MASHAL and SPARC. They argue that the projects "actually worsen[ed] the situation [in] the slums from the point of view of safety and health" because "narrow streets remain and restrict access for the emergency services and restrict the evacuation of the residents."[37] They also fault the projects for rebuilding homes in configurations that limit both interior and exterior access to light and air due to extreme density. These critiques suggest that future in situ upgrading projects might prioritize targeted re-blocking to enable infrastructure upgrades while still rehousing residents near their original locations.

SECURITY: CONSOLIDATED COMMUNITIES WITH LINGERING INFORMALITY

The Yerwada upgrading projects improved security of tenure and safety for residents in several important ways, but they have still not achieved formal legal tenure security. Even so, residents often believe that state-sponsored upgrading improves de facto security against forced displacement (figure 8.7). When asked about security of tenure, project leaders cited the fact that there is no recent history of eviction in this area, stressing that each household received a certificate of occupancy, assigned to women, upon completion of the upgrading. These certificates allow residents to collateralize their homes to access loans. They also permit residents to sell their houses after five years of residence. According to NGO leaders, few residents—an estimated 5 percent—have exercised this right to sell upgraded homes.[38]

The de facto informal security of tenure enjoyed by residents of upgraded settlements is evident in the substantial investments that families have made to improve and expand their homes (figure 8.8). One study describes these resident-driven improvements, including "vertical expansion as well as cantilevering of the unit . . . improving and adding toilets, storage spaces and . . . improving the quality of life in the living and sleeping spaces." Upgrades go beyond such functional matters to include "beautification in the form of tiling, painting and furniture."[39]

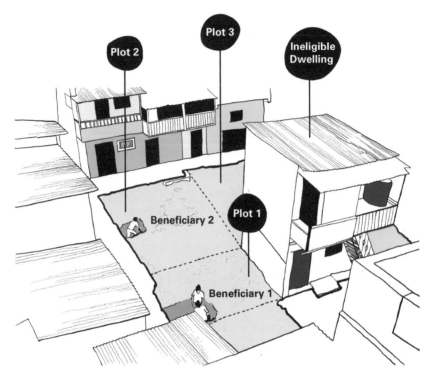

8.7 The upgrading process allowed eligible residents to upgrade their homes on the same site where they had previously lived. *Source:* Ipshita Karmakar.

Even with increased perceived security, residents and organizers had hoped to achieve formal tenure rights after upgrading, and that recognition has not been granted. Residents of the distant BSUP relocation sites at Warje and Hadaspar received ninety-nine-year leases on their new homes, but residents in in situ upgraded homes have no such security. Both MASHAL and SPARC called for the formation of cooperative housing societies to help secure residents' collective tenure rights,[40] but these institutions have yet to be formed, and residents continue to rely on informal and partial security of tenure.[41] Further, the Yerwada settlements have yet to be "de-notified," meaning that they are still considered by the government to be temporary. This status makes the settlements eligible for further upgrading assistance, but it also leaves residents in a precarious position with respect to their tenure rights.

8.8 Encouraged by increased perceived security, residents have continued upgrading their homes. *Source*: Ipshita Karmakar.

CONCLUSION

The in situ upgrading projects led by the MASHAL and SPARC/Mahila Milan teams in Yerwada represent a substantial shift from the dominant model of "slum clearance" and relocation. The projects provide crucial insights for pursuing equitable resilience, particularly in their governance and participatory design and in their layered approaches to supporting residents' livelihoods (figure 8.9).

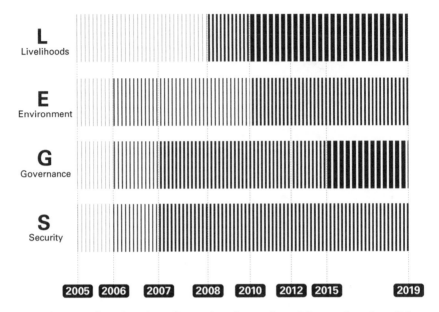

8.9 The Yerwada In Situ Upgrading projects have substantially contributed to all four dimensions of equitable resilience, although residents lack full recognition of their tenure security. *Source:* Smriti Bhaya. 2005: JNNURM and BSUP initiated; 2006: Procurement begins; 2007: Community engagement started; 2008: Initial design process initiated; 2010: Construction commences; 2012: Floodwall constructed; 2015: Resolution 420 initiated, allowing residents to act as their own contractors; 2019: Major flood without major damage.

The program's commitment to house-by-house in situ upgrading preserved place-based social and livelihood networks. SPARC, MASHAL, and Mahila Milan acted as intermediaries between inflexible state bureaucracies and local people, enabling residents to shape the projects and customize their homes to meet their needs. With this pluralistic and resident-driven model, the projects supported residents' home-based livelihoods and secured direct economic benefits through employment in the upgrading process.

These projects made substantial contributions in each of the four dimensions of equitable resilience. Yet, they also had several shortcomings. In the realm of governance, both SPARC and MASHAL struggled to manage the complexities of contracting, payments, and quality control on thousands of individual upgrading projects. The lack of temporary housing and livelihood support during construction, the required financial contributions from beneficiaries, and the increased costs of living in upgraded homes have presented significant hardships for poor residents. Finally, and crucially, while upgrading improved residents' de facto security, they still do not have formally recognized tenure rights.

Because the SPARC- and MASHAL-led Yerwada projects emphasized house-by-house in situ upgrading, they missed some opportunities to improve conditions through investments in public space and infrastructure. Yet, realizing such public improvements would have required more than just further negotiation among households and design innovation by project leaders. It would also have required government administrators to provide greater flexibility and incentives to facilitate re-blocking. Reporting on the trade-offs between household-level upgrading and public realm benefits, one observer notes, "The one-on-one meetings proved to be very successful in bringing about a consensus on design of the individual dwelling units but did not go further to address larger issues." As such, "the programme failed in providing adequate infrastructure, street and public-space development, waste disposal systems and health and educational facilities."[42]

In both its victories and shortcomings, the Yerwada upgrading projects demonstrate valuable insights for equitable resilience. Focusing on household-level improvements and preservation of place-based assets is essential to equitable resilience, although such an approach may fail to deliver essential improvements to public spaces and infrastructures.

Case 9

FROM DAFEN VILLAGE TO DAFEN OIL PAINTING VILLAGE: SELECTIVE ADVANCEMENT OF LIVELIHOODS IN SHENZHEN

OVERVIEW

Dafen is arguably the most celebrated of the hundreds of villages engulfed by urbanization in Shenzhen in China's Pearl River Delta. The once-impoverished settlement has twice been reinvented: first, in the late 1990s and early 2000s as the world's leading supplier of reproduction oil paintings, and then, more recently, as an exemplar of state-led culture-based economic development. In contrast to most other case studies in this book, the transformation of Dafen village into the economic powerhouse of Dafen Oil Painting Village is very much a top-down vision. It has been substantially led by outsiders, even as it has brought considerable wealth to the original villagers. The initiative, which includes efforts to revitalize livelihoods, investments in environmental infrastructures, and affordable housing for migrant painters, is a rare Chinese example that embodies some of the principles of equitable resilience, made possible in part because urban villagers retain unusual forms of collective land tenure. However, interviews with two dozen residents and art workers during visits in 2016 and 2022, coupled with review of village documents and assessments by scholars and journalists, suggest that early hopes for equitably resilient development have remained largely unfulfilled.

9.1 Dafen Village surrounded by recent urbanization. *Source*: Urbanus Architecture and Design.

INTRODUCTION

CHINA'S URBAN VILLAGES

Building off of the rural impetus of the Chinese Revolution and Mao Zedong's celebration of village life, China developed separate land governance classifications wherein the state owned urban land while village collectives owned rural lands. This distinction meant that, as cities expanded into their hinterlands, villages became islands of collectively owned land in the sea of state-owned urban land. Although these settlements are known as "villages in the city" (*chengzhongcun*), there is little that remains village-like in their appearance or economy.[1]

Shenzhen stands as an extreme example of village urbanization. As a centerpiece of post-Mao economic liberalization, in 1979, Guangdong's provincial government transformed the predominantly rural Bao'an County, just north of Hong Kong, into Shenzhen municipality. A year later, Premier Deng Xiaoping's government designated the area closest to the border as China's first special economic zone (SEZ), opening the area to industrial manufacturing and foreign investment. In the decades that

9.2 Dafen Village location in Shenzhen, China. *Source*: Mora Orensanz.

followed, scattered villages that once housed about three hundred thousand people exploded into a sprawling metropolis of twenty million (figure 9.2).[2] Although Dafen initially stood outside the SEZ, that zone was officially extended in 2010 to encompass all of Shenzhen's outer precincts, including Dafen.

In many villages, including Dafen, traces of older built environments, including a traditional ancestor hall, remain. However, most villages, beginning with the most centrally located, have followed a path of urbanization that can be roughly broken into four stages. The rural buildings that might be termed "stage 1" were initially replaced by dense, low-rise settlements ("stage 2"). In the third stage, village lands in many settlements were divided into ten-by-ten-meter parcels out of which sprung dense grids of seven- to ten-story buildings, often placed so close together that balconies almost touched, spawning the term "handshake buildings." These buildings often housed migrant workers who lacked the *hukou* (household registration) necessary to obtain housing on the official non-village market. While village collectives typically led these early redevelopment efforts, subsequent redevelopment has often led villagers to sell their land to private developers, eager to build high-rise housing and upscale malls (Stage 4).

The transition from stage 3's dense grids of migrant worker housing to stage 4's high-end superblocks often makes villagers quite wealthy but reduces the amount of centrally located affordable housing. Villagers enjoy financial windfalls, but migrant renters—and the urban economies that depend on their labor—lose out. Urban villages exist in many parts of China, but these communities and their tumultuous redevelopment have been especially consequential in Shenzhen, where rural migrants without urban *hukou* are estimated to comprise the majority of residents.[3] One housing survey in 2013 found that self-constructed and co-constructed housing in Shenzhen's urban villages totaled more than 260 million square meters—fully half of the area devoted to housing in the entire city.[4]

As the Shenzhen-based urban ethnographer Mary Ann O'Donnell has written, Shenzhen's urban villages "have been the architectural form through which migrants and low-status citizens have claimed rights to the city." These dense settlements "sustained the city's extensive grey economy, including piecework manufacturing, spas and massage parlors, and cheap consumer goods." Moreover, with some easing of *hukou* restrictions, "urban villages have been the form through which migrants have had access to social services, including schools and medical clinics." In short, she argues, these places have "provided informal solutions to boomtown conditions." Despite their importance to the city's economy and development, the Shenzhen municipality legally abolished villages in 2004 and began implementing plans for wholesale redevelopment in 2007.[5]

FROM DAFEN VILLAGE TO DAFEN OIL PAINTING VILLAGE

Dafen's urbanization broadly follows the sequence of other urban villages in Shenzhen, although the story of Dafen is also marked by distinctions with implications for improving livelihood prospects for both migrant workers and original villagers.

In its pre-urban state, Dafen village housed about three hundred people of Hakka descent, surrounded by agricultural fields and waterways. After 1979, things changed dramatically. In the 1980s, Dafen village attracted migrant workers, "who came to the fringes of a booming city seeking to work as painters and to secure a modest livelihood as independent painters." The village nurtured opportunities for "social mobility through self-employment and entrepreneurialism," especially for those who opened their own galleries, frame shops, or art supply stores.[6] By the end of the 1980s, migrants from elsewhere in Guangdong Province, often with little education or training in art, began to cluster in Buji, a small settlement one and a half miles north of the original SEZ. There, they found low rents and few regulations—factors that also attracted entrepreneurs seeking to establish art businesses close to Hong Kong.[7]

Dafen village lay just beyond Buji. Its growth as a mecca for art production commenced in 1989 when Hong Kong oil painting businessman Huang Jiang rented village space to establish his first factory. According to Dafen artist Zhou Yongjiu who himself arrived in 1991, Huang wanted "a remote place, cheap and large" and was charmed by Dafen's tile-roof houses. Huang and Zhou helped launch a system of apprenticeship, enabling generations of new migrants to learn oil painting production. Thirteen Zhou-trained apprentices still work in the radically transformed village. Three decades earlier, he remembers, Dafen had only a single paved road, lined solely with "grocery stores."[8] Gradually, the local government eliminated nearly all shops that were not engaged in painting-related industries. Dafen grew exponentially, with an ever-larger number of painting production facilities, some employing hundreds of painters, each completing as many as twenty canvases daily. Huang's original factory and others that joined it in Dafen became an international phenomenon. Aided by middlemen in Hong Kong and elsewhere, Dafen painters soon generated huge numbers of oil paintings sold by retailers in the US

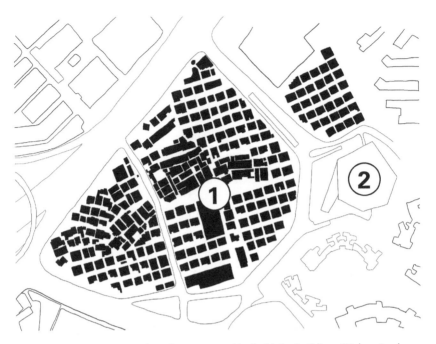

9.3 By the early 2000s, Dafen village was a grid of mid-rise buildings (1), housing large numbers of migrant artists and galleries. The Dafen Art Gallery (2) opened in 2007. *Source*: Mora Orensanz, based on an image from Urbanus.

such as K-Mart, Walmart, JCPenney, and Target.[9] By the mid-1990s, savvy villagers recognized that growing demand for their space offered significant opportunities for profits.

Urbanization did not take place smoothly in Dafen. In the early 2000s, government authorities began developing master plans for the village, which included replacing the irregular layout with standardized ten-by-ten-meter plots (figure 9.3). Many villagers resented the plans once they realized that they included substantial forced demolition. Having already lost access to nearby fishing grounds, villagers worried that they lacked the skills to thrive in the context of rapid urbanization. As first detailed in a Chinese-language account of *The Rise of Dafen*, they "blocked the bulldozers and surrounded the demolition team." After this tumultuous start, the local government established a "Leaders' Team on the Environmental Transformation of Dafen," bringing together painters and village committee members for a series of meetings. The Shenzhen municipal government, eager to redevelop the village, cast one set of community members against others in a "confrontation between the 'heroic role model villagers'" who went along with redevelopment and the "stubborn villagers" who resisted.[10] In the end, those judged heroic by the government won out, and the demolition proceeded.

With its rebranding as Dafen Oil Painting Village in 2004,[11] development accelerated, and art galleries proliferated. Villagers developed multistory rental housing to accommodate ever-larger numbers of migrants. Former peasant agriculturalists and fisherfolk became wealthy urban landlords. In 2004, when Dafen was formally integrated into the municipality of Shenzhen, the Dafen Village Corporation and its management office superseded the village committee as the village's primary governing authority.

With new rental apartments and commercial spaces, Dafen attracted artists from distant provinces, who increasingly arrived with more training in oil painting. Dafen's development caught the attention of Shenzhen officials who sought to make the village part of China's push for "cultural industries," beginning with the 10th Five-Year Plan of 2001–2005. Dafen's emergence as a painting hub also resonated with global trends emphasizing cultural development, framed in Europe as Cities of Culture and by UNESCO's Creative City Network, and by scholars touting the rise of the "creative economy" and the "creative class."[12]

Beginning in 2004, the year that the Chinese Ministry of Culture designated Dafen village as the country's first model cultural industry, Shenzhen officials started marketing Dafen nationally and internationally. By 2006, Buji subdistrict propaganda director Ren Xiaofeng praised Dafen as a "miracle in the development history of the world culture industry." The Oil Painting Village showed how "the new socialist market economy system" enabled China to compete with the US, Japan, and Korea in the export of cultural products—a "brilliant" counterpoint to "so many 'invasions' of foreign goods." Ren co-edited a book—bilingual in English and Mandarin— extolling the achievements of dozens of talented migrant artists whom local officials bestowed with the honorific "special villagers."[13] Government officials relaxed *hukou* restrictions for such preferred denizens, giving them access to local services, including schools. More ideologically, they extolled the concept of the artist-worker as a demonstration of the capacity of "disadvantaged, rural-born individuals" to build productive urban lives

9.4 Dafen emerged in the 1990s as a mecca for migrant workers copying famous oil paintings. *Source*: Mora Orensanz, based on a photograph from Urbanus.

through "self-perseverance, hard work, and self-skilling."[14] Even as some Western critics dismissed Dafen's paintings as derivative copies made by sweatshop workers, Chinese journalists and government propaganda celebrated creative peasants whose hard work enabled them to participate in a global economy (figure 9.4).[15]

In support of these migrant livelihoods, and to support tourism-based economic development, the local government upgraded sewage and water services starting in 2001, repaved streets and alleys, invested in larger commercial buildings, and, in 2011, added a Dafen stop on the Shenzhen Metro.[16] Other key projects included the Oil Painting Trading Square, a building with gallery space, stores, and rental apartments sponsored by the district government, an auction house, and extensive streetscape improvements, showcasing a variety of sculptures, fountains, and cafés. The Dafen Art Museum and its six-thousand-square-foot art plaza, both largely dedicated to nonlocal artists, replaced four older village buildings in 2007.[17]

In 2010, Shenzhen's pavilion in the "Best Urban Practices Zone" of the Shanghai World Expo highlighted Dafen's achievements, casting the village as a microcosm of the city's creative growth. State-controlled media praised it as "the most beautiful village in Shenzhen" and extolled it as showcasing "how cultural development can flourish in an outlying area."[18] In preparation for the Expo, the Dafen Museum of Art assembled five hundred artists with easels on the museum's plaza. Curators provided each painter with thirty-by-twenty-centimeter rectangular digital images, largely monochrome, to be rendered in oil on canvas. The resulting 999 panels were mounted on an exterior wall of the Expo pavilion—seven meters high and forty-three meters wide. The consequent collaborative work depicted the central visage of the *Mona Lisa* and was nicknamed the *Dafen Lisa*.[19]

Since the Expo, development surrounding Dafen Oil Painting Village has accelerated. In 2018, the district government commenced construction of the Shenzhen Dafen Oil Painting Industrial Park adjacent to the old village. The facility replaced an industrial area with luxury housing, commercial development, offices, hotels, and schools. This development, known as Mumianwan (recalling the name of a now-demolished urban village), also displaced 3,300 households.[20] Just beyond the boundaries of Dafen Oil Painting Village, the Village Corporation launched new commercial development and the skyscraper of the Sky Point Hotel, as well as additional development with shops and eateries, including McDonald's, KFC, and Starbucks, anchored by Walmart. In the place where a village once produced paintings to be sold in Walmart stores across the globe, Dafen itself now includes a Walmart (figure 9.5).

DAFEN AND EQUITABLE RESILIENCE

The saga of Dafen, with its lofty aims, tumultuous urbanization processes, and entrenched economic inequalities, offers insights into the pursuit of equitable resilience. In contrast to other forms of urbanization in China, the combination of location and collective property rights enabled urban villages to offer potential respite from brutal state-led gentrification and displacement—at least temporarily.

In addition to providing affordable housing essential to the economic health of a growing region, China's urban villages may also hold an unusual capacity for supplying more environmentally sound accommodations to low-income households. This is especially important, since

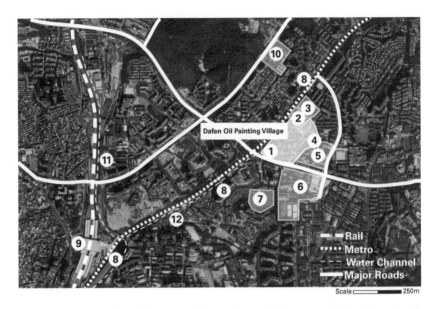

9.5 Since 2004, Dafen Oil Painting Village is increasingly surrounded by dense development sponsored by the village corporation and the municipality. *Source*: Smriti Bhaya. 1. Dafen Oil Painting Village; 2. Commercial buildings built by the village corporation, anchored by Walmart; 3. High-rise commercial buildings built by the village corporation; 4. Art museum; 5. Residential and commercial buildings built by the village corporation; 6. Commercial and residential development built by the Longgang District government; 7. Neighboring urban village; 8. Metro station; 9. Train station; 10. Dafen New Village; 11. Buji River; 12. New drainage channel.

Shenzhen is considered one of the most vulnerable coastal cities in the world to flooding and climate change impacts.[21] Nonetheless, at least among Dafen residents, rising seas and falling rain are often less-pressing concerns than rising and falling livelihoods.

LIVELIHOODS: PAINTING THE WAY TO ADVANCEMENT, POINTING THE WAY TO INEQUITY

SHENZHEN'S EMPLOYMENT DRAW

Dafen Oil Painting Village is one small enclave whose story is inseparable from larger regional urbanization. Shenzhen's economy has long relied on low-cost labor, often enabled by housing and production in urban villages. Today, as the economy shifts toward technology and innovation, urban villages provide rare affordable commercial, startup, and industrial space. Dafen, although specialized to serve the "art industry," has been one prominent variant in efforts to deliver both affordability and economic development.

ACCOMMODATING EMPLOYMENT IN DAFEN

Dafen, like other urban villages, has been explicitly built as an engine of job creation. Yet, accommodating jobs has also entailed building subsidized housing, enhancing infrastructure, and providing public space to attract visitors and market art. One estimate claims that, at its peak, Dafen was responsible for 60 percent of global production of oil paintings.[22] Some of the larger painting "factories" utilized assembly-line practices where, in order to fulfill large orders, each painter completed only small parts of each painting in a sequence.[23] Even so, studies suggest that work-for-hire painters often regarded such art production as "far more independent and appealing than (actual) factory work in South China."[24] Compared to the alternatives, life as an artist worker in Dafen offered substantial benefits to migrants.

The state-run *China Daily* has repeatedly celebrated Dafen's artists for sustaining their livelihoods, despite global economic challenges. When "the 2008 financial crisis quashed foreign demand for reproductions," painters responded by shifting from copy work (*diduan henghua*) to "original artwork" (*gaoduan yuanchuang*) and took advantage of the burgeoning domestic art market. Dafen-based merchant Fen Jianmei estimates that Dafen provided 70 percent of the interior décor for China's new five-star hotels. In this rosy view, even COVID-19 did not defeat the resilience of Dafen artists, since many had been shifting to online sales for many years and some experienced a surge of orders during the pandemic. For greater engagement with the buying public, some artists began documenting and live streaming their painting through TikTok and other platforms. As of 2021, *China Daily*'s correspondent reported "2,500 digital shops for Dafen businesses," many with annual sales between three and thirty million yuan (US$450,000–$4.5 million).[25] As of 2022, Dafen's main planning agency estimates that eight thousand people are either employed as painters or undertake other art-related work, including in 1,200 galleries and studios.[26]

Despite such robust numbers, many of Dafen's migrant workers struggle financially. Rents increased rapidly after 2004, and Dafen's growing fame encouraged development of high-end housing nearby, putting pressure on prices.

While local "households-cum-landlords and the village committee" welcomed increasing rents and profits,[27] migrant renters often suffered. With studio rents out of reach, two hundred painter-workers resorted to renting outdoor "wall galleries." Despite their popularity with visitors and some artists, Dafen officials eliminated these galleries in 2019. Officially, the new ban aimed to improve firefighting access, but many artists suspected that the wall galleries had come to represent unwanted competition for landlords.[28] Geographers June Wang and Si-ming Li describe contemporary Dafen as a competitive environment of eroded trust and consequent "individual self-entrepreneurialism." After the villagers largely relinquished their authority over their former territory after 2004, painters regarded them as "useless." The Dafen Management Office strains to manage the conflict.[29]

Interviews with twenty-one artists and business owners conducted in 2022 confirm these tensions. One artist, working in Dafen since 1999, blames the village landlords for excessive rent increases, noting, "If he increases the rent and you don't agree, he will kick you out." Rent for his gallery space has risen dramatically, especially after completion of the Metro stop: from 1,200 yuan (US$180) in 1999 to almost 9,000 yuan (US$1,340) in 2022. Another longtime artist/resident comments, "The landlord, of course, doesn't live here, so they don't care if we have some trouble." Others agreed. During the pandemic "some landlords haven't lowered the rent, even a penny"; "We are all struggling and waiting [since] people like us are unable to do other things"; "Landlords in the village are doing nothing" and are "united" in setting rents.[30]

Some artists began offering painting lessons and other activities to children and tourists to replace lost sales revenue (figure 9.6). Pioneering artist-turned-dealer Zhou Yongjiu says he initiated this practice in 2014, but largely abandoned it in 2017 after others joined in and drove down prices. Visiting participants pay little and often depart Dafen carrying only their own creations. Some professional painters see this "experience" economy as replacing more lucrative sales. One old-timer who reluctantly joined this practice is ashamed that it has turned him into a shouting hawker, akin

9.6 To seek additional income, some Dafen artists turned to hawking outdoor "painting experiences." *Source*: Mora Orensanz, based on a photograph by Gan Xinyue.

to "a vendor selling sweet potatoes." "Reality," he laments, "forces us to be like this now . . . In these last two years, even masters cannot make a living."[31]

UNEVEN CAPACITIES: DIFFERENTIATED WORK, DIFFERENTIAL WELCOME

Within Dafen village and its immediate environments, some artist-led households have benefited much more than others from the painting economy. Sensitive to critics who decried Dafen's artists as mere copyists, after 2004, the local party-state introduced policies and subsidies to selectively attract "original artists." Many of these new arrivals graduated from art academies or the art departments of teacher's colleges or came as retirees from other arts-related professions.[32] Local officials granted Shenzhen *hukou* rights to artists who could demonstrate special talent by passing an exam (including art history knowledge, sketching, and oil painting prowess) offered annually from 2005–2007. Between 2008 and 2010, they selectively offered two-year rent waivers for gallery space to those deemed "decent art dealers." Starting in 2008, authorities allocated most places in a 268-unit public rental housing project to "quality" artists.[33]

In 2012, the Longgang District Housing and Construction Bureau promulgated specific guidelines for artists seeking public rental housing. They sought to provide a "good living and working environment" for three distinct groups: great painters (an honorific term), painters with good skills, and less-skilled artisans. Preferences for housing, in turn, depended on official classifications,[34] including membership in the highly selective Dafen Artists' Association, established in 2009. This organization supplemented the existing Dafen Art Industry Association, which focused more on the business aspects of art.[35] Being a member of these associations also enabled artists to access other government benefits.

To qualify for Dafen's subsidized public rental housing, applicants had to demonstrate that they had some degree of arts training, lacked access to other affordable housing, complied with the (then-existing) one-child policy, did not have a criminal record, would "obey the administrative management," and would contribute "work that is beneficial to the development of Dafen's cultural industry." The government listed residency requirements for different sizes of apartment, explicitly prioritizing those who had received "preferential *hukou* registration"; won art prizes at the national, provincial, or city level; and/or "made outstanding contributions" to Dafen's development. Rents initially ranged from about 400 to 1,100 yuan/month (US$60–US$165) but had doubled as of 2022. The official status hierarchies imposed upon artists seeking state subsidy led one artist to state that access to one-, two-, or three-bedroom apartments was directly proportional to having city, provincial, or national status as an artist.[36] By some accounts, such state-led programs to reward subsets of migrant artists further entrenched a neoliberal logic that "welfare is for the worthy."[37]

The status hierarchy from artisan/copyist to painter to famous artist also manifests itself in the placement and visibility of work and gallery space in the village.[38] One planning official describes this as a "chain of contempt" in which those artists deemed less "original" are disdained and high-end galleries along the main roads have "a sense of superiority over those who hide in alleys in Dafen." Dafen officials now ban factories completely, and even frame makers have largely moved outside the village. As Dafen itself shifts from a site of production to one of consumption, many artisans moved away from the village to less expensive nearby areas.[39]

For decades, overcoming its early reputation for cramped conditions, Dafen has supported the upward mobility of rural-to-urban migrants. From a distributional perspective, however, it seems clear that the largest livelihood benefits have gone to two other constituencies: the original villagers, who profit from their newly valuable property, and the bosses and entrepreneurs, who parlayed village-generated products into a global network of wealth generation.

These patterns raise questions about whether Dafen's innovations in livelihoods can be considered equitably resilient. On the one hand, formerly poor villagers gained once-unimaginable wealth in the course of a single generation, and Dafen was a gateway to greater prosperity and better conditions for many migrant artist workers. On the other hand, as in so many other places, Dafen is clearly a site of massive wealth extraction that has magnified economic inequities.

SECURITY: SEEKING A STABLE PLACE TO PROSPER

The tale of Dafen, although often framed to emphasize its contributions to livelihoods, has also been a quest for security. For original villagers, security entails parlaying their collective tenure rights into economic gain. For poor migrants, security is tied to *hukou* privileges to remain in Shenzhen.

As urbanization around Dafen Oil Painting Village has intensified and shifted toward higher-end constituencies, the village's affordability has been threatened. The tensions between migrant renters and their villager landlords, rooted in differential tenure status, have only increased. With their incomes now assured, most of the original villagers who gained so much from their collective property rights have moved elsewhere. By 2011, approximately two hundred of the remaining villagers had decamped to Dafen New Village (figure 9.7), a gated community of villas—bounded by a barbed wire–topped wall—a short distance from the gridded warren of artist spaces in the old village.[40] Others moved further afield, still drawing

9.7 The original villagers have largely decamped to a luxury gated community, located a few minutes' walk away from their old homes. *Source*: Mora Orensanz, based on a photograph by Gan Xinyue.

significant income from rental property in Dafen. Village owners also typically exacted a share of the hefty transfer fees that would-be renters of gallery space often paid to former tenants and "tea money" (*hecha fei*)—an "arbitrary charge for giving the lease to specific tenants."[41] The new wealth of the villagers enabled them to restore their village ancestral hall. Yet, it now stands anomalously in the core of a settlement that is no longer integrated with their daily lives. Villager security is financially guaranteed by the returns from their ownership rights and architecturally signaled by the elaborate walls and gated entrance to their new enclave.

GOVERNANCE PARTNERSHIPS: STRUGGLES TO LINK INDIGENOUS VILLAGERS AND LOCAL GOVERNMENT

Although Dafen's Village Management Office, established after 2004, is not a formal level of government like the Buji subdistrict or the Longgang District, it serves as a key unit of governance. The once-rural village collective now operates as "a holding company, with the shareholders being the individual households of the village." The board manages properties within Dafen while also providing social services, including oversight of subsidized rental housing.[42]

As the original villagers have moved out, the Management Office has focused increasingly on economic development and marketing. Responsibility for planning in much of Dafen has been "handed over" to China Resources (CR) Land. CR Land focuses on placemaking, seeking to invest in "the feeling" of Dafen, including enhancements to the tourist experience. CR Land announced plans to renovate and expand the Dafen Art Museum to include rooftop galleries. Elsewhere in Dafen, they seek to add more services and cafés, improved streetscapes and wayfinding, water plazas, tree plantings, and a multistory parking structure to facilitate access for visitors arriving by car. One explicit goal is to create new places that will encourage younger visitors to "check in" by posting on social media. Akin to the government-established hierarchies used to allocate subsidized housing, CR Land employs a multilevel framework to categorize artists based on "rules of the contemporary art market."[43]

Recent assessments cast the story of Dafen as an instance of state-led gentrification that has harmed the village's painter workers. Social scientist Karita Kan sees Dafen's trajectory as akin to "building SoHo in Shenzhen" but with a greater role for the state. The shift from mass-produced art in the 1990s to a twenty-first-century focus on tourism and consumption has been prompted by the particular characteristics of the Chinese land tenure system. While the village collective tenure regime initially nurtured informal practices and protected original villagers from displacement via speculation, subsequent state-led gentrification has displaced many of the migrants responsible for Dafen's prosperity.[44] Hence, tenure security and governance structure became intertwined.

Geographers Wang and Li emphasize the "market mind-set" that "re-territorialized" the village into a state-controlled "cultural cluster."[45] Art historian Winnie Wong, whose book, *Van Gogh on Demand*, situates Dafen within larger global art practices and markets, sees this reassertion of state control in a different realm. She describes a government-led campaign to regulate Dafen as a "model bohemia"—"a place where the political ideals of artistic freedom and rebelliousness from the market is remade into a state-sanctioned creative industry." In short, the goal has been to create a productive zone stripped of artistic criticality, a bohemian vibe that is devoid of "any avant-garde challenge to social and political norms."[46] Shenzhen-based planning professor Gan Xinyue, however, resists the

9.8 Dafen's management implemented major upgrades to drainage, hoping to sustain tourism. *Source*: Mora Orensanz.

state-led gentrification charge, observing that, in Dafen, "all kinds of artists can survive," and that such artists have pragmatically adapted to the market in a more self-empowered "bottom-up" manner.[47]

ENVIRONMENTAL RESILIENCE: BELATED PROTECTIONS FOR NEW CONSTITUENCIES

Shenzhen, including its former villages, faces a range of climate impacts. Residents blame rampant development surrounding the village for exacerbating flood risk by increasing impermeable surfaces, obstructing drainage, and elevating surrounding land, rendering the original village a local low point. Recent years have brought both significant warming and increasingly common devastating rainfall and floods in the Pearl River Delta.

Most investment in Dafen's environmental infrastructure occurred only with increasing government investments in tourism and economic development (figure 9.8). Recent improvements have included the Dafen Shui

Flood Diversion Project, which commenced in 2017 and includes more than a kilometer of channels and tunnels to divert flood water from the villages of Dafen and Mumianwan to the Buji River (see #12 on figure 9.5).[48] While some regard these projects as bringing relief to the riskiest areas, many local merchants and painters blame construction of drainage and road improvements for disrupting access to their businesses. As one Dafen artist put it, "the government has allocated a lot of money [to infrastructure] but we painters have lost our incomes."[49]

Beginning with a 2014 directive from the central government, city officials in Shenzhen and other Chinese cities have sought to implement "Sponge City" strategies to relieve pressure on "grey infrastructure" by building new "blue-green infrastructures" to manage stormwater through landscape processes.[50] In Dafen, however, traditional engineering approaches continue to dominate, even as new high-density development in former agricultural areas exacerbates flooding. Rather than addressing broader changes in land use and development, government entities often blame inadequate village infrastructure for flooding, providing a rationale for ever-more-intensive megaprojects. Meanwhile, the ahistorical and narrow focus on managing flood hazards as technical infrastructure challenges blinds the city to the uneven vulnerability of low-income groups.

CONCLUSION: THE LIMITS OF EQUITABLY RESILIENT LIVELIHOODS

The saga of Dafen village reveals multiple challenges to pursuing equitable resilience. The villagers clearly facilitated improved livelihoods for themselves, and also arguably improved prospects for thousands of migrant art workers. Still, the various aspects of equitable resilience did not proceed smoothly (figure 9.9). Most prominently, the environmental gains and protections against flooding have come only after gains to livelihoods. Enhanced environmental safety and streetscape amenities seem motivated more by the pursuit of tourism and high-end housing than by a desire to improve conditions for poor migrants. By pricing out many of the early beneficiaries of the village's economic rise, recent development projects have made Dafen less equitable. Dafen's original villagers parlayed their collective tenure rights into great community and individual

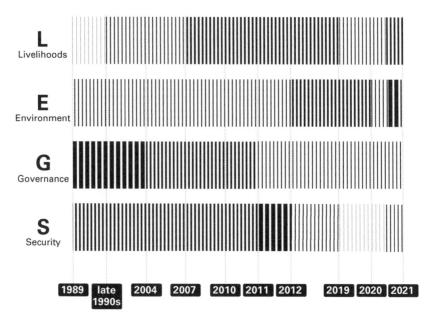

9.9 Dafen's shift from an urban village to a more corporate oil painting village has sustained livelihoods but undermined collective governance. *Source*: Smriti Bhaya. 1989: First Dafen art factory established; Late 1990s: Village densifies into mid-rise grid of apartments, galleries, and shops; 2004: Dafen village committee becomes Dafen corporation; 2007: Dafen Museum opens; 2010: Shanghai World Expo features Dafen; 2011: Last original villagers move to Dafen New Village; 2012: Public rental housing opens for artists; 2019: Wall galleries outlawed; 2020: COVID-19 pandemic begins; 2021: Dafen Shui Flood Diversion completed.

wealth, only to be co-opted by the developmentalism of local and national governments.

Although the villagers' collective tenure arrangements proved to be a profitable asset, they have seen their self-governance authority eroded. While Dafen's villager landlords enjoyed enough secure tenure to extract significant wealth from their urbanizing village land, they were displaced to the nearby New Village. Further, the villagers lack control over their environment in significant domains. They are landlords and stakeholders of the Village Corporation but have been relegated to ever-smaller roles in directing redevelopment. Moreover, tensions between villager landlords and their artist tenants have often undermined community solidarity and equity.[51]

Even so, Shenzhen's remaining urban villages play a vital role in the city. If villages are eradicated, the physical, social, and administrative anomaly that has allowed them to contribute to economic growth and urbanization will be lost. Low-income workers will lose access to large parts of Shenzhen, harming the city's overall resilience. Balancing Shenzhen's need for low-cost housing against the economic windfalls of wholesale redevelopment, more recent policy shifts promote in situ models of upgrading.[52]

With state-led uneven allocation of benefits, Dafen's resilience has become increasingly unequal; the village has become an environmentally safer place, but it is home to a more affluent population. Collective tenure rights catalyzed great wealth for three hundred villagers and enhanced livelihoods for many migrants, but neither villagers nor migrant tenants have much control over how their community is managed. In the end, it seems, equitable resilience does not go very far without substantial self-governance.

LEARNING FROM THREE STRUGGLES FOR EQUITABLY RESILIENT LIVELIHOODS

An urban village of entrepreneurial Chinese artists, a diverse Portland neighborhood with robust community organizations, and an upgraded Indian informal settlement may seem to hold little in common. Still, taken together, the stories that unfold in these three places as outlined in this section reveal key aspects of livelihood generation in the face of a changing climate, surging inequity, and demographic change. The residents of Dafen, Cully, and Yerwada value their homeplaces in part because these neighborhoods provide affordable bases from which to access the networks, services, and employment necessary to survive and thrive.

These cases collectively highlight several challenges. First, the recent trajectories of these places show how hard it is to invest in meaningful neighborhood improvement without courting gentrification—green or otherwise. In Dafen, for instance, enhanced flood-prevention spending makes the urban village environmentally safer but is also likely to trigger further erosion of affordable housing. Many of Living Cully's projects, by contrast, aim to build low-income residents' "environmental wealth" while explicitly fighting displacement, including through affordable housing provision and preservation, tenant organization, and emergency COVID-19 relief. The Yerwada settlement reduced hazard risks while recommitting to serving existing residents through in situ upgrading.

Second, all three cases include efforts to help residents build market-able skills and direct employment opportunities. Dafen's painting industry provided an economically viable path upward for thousands of migrants from rural areas. Similarly, the Yerwada upgrading projects included capacity building through Mahila Milan's training and contracting programs and MASHAL's preferential contracting with local builders. Building livelihood capacity is at the heart of Living Cully's work combating poverty and displacement through training and employment for well-paying "green" construction and landscaping jobs.

Third, these cases problematize the uneven distribution of benefits from neighborhood investment. Dafen is the most extreme case. While the village painting industry has provided livelihoods for many thousands of migrant art workers, the original villagers profited extravagantly as their land became coveted for urbanization and entrepreneurs from the arts and tourism industries have effectively extracted enormous value from the village's artists. Conversely, the Cully TIF is explicitly intended to create a mechanism by which low-income residents and people of color can enjoy benefits—rather than threats—from increasing neighborhood property values.

Fourth, these pursuits of equitable resilience demonstrate a persistent tension between local initiatives and powerful forces located beyond the bounds of any single neighborhood or district. Living Cully emerged as a coalition of neighborhood organizations seeking support and investment from higher levels of government and other organizations, whereas Yerwada's transformation depended on government funding channeled through NGOs committed to sustained partnership and community engagement. Dafen exhibits yet another sort of governance structure: one that commenced with substantial local village control but evolved into a more corporate form of management, with increasing municipal oversight.

These three examples make clear that building equitable resilience through livelihoods does not proceed in a linear manner. Livelihood gains—just like environmental protection and security of tenure—may wax and wane, depending not only on larger economic policies but also on government priorities and the fortunes of local organizations that advocate for the least advantaged (table III.2).

Table III.2

Multiscalar analysis of livelihood cases

	Living Cully	Yerwada Upgrading	Dafen Urban Village
Individual/ Household Capacity	Skill development and employment programs	Reduced stigma; local contracting; construction skills training	Skill building through apprenticeship, but hyper-competitive and hierarchical market
Project Accommodation	COVID-19-motivated technology program for remote school and work	Customization to accommodate storefronts and other commerce; outdoor storage	Arts and arts-related work in home-based production, wall galleries, and studios
City/Region Access	Affordable housing preservation near job centers	Maintains access to nearby jobs and transportation	Supports affordable housing needs of region; linked to Metro

Ultimately, these cases reveal that pursuit of equitable resilience entails more than provision of secure housing and livelihoods in environmentally safe places. It also matters who exercises control over development processes and who has a voice in how communities are managed. These matters of legitimacy and equity are centrally questions of governance—the focus of this book's final section.

IV.
GOVERNANCE

EQUITABLY RESILIENT GOVERNANCE: SELF-EFFICACY, CONTROL, AND INFLUENCE

Even if people live in settlements with reduced hazard vulnerability, improved security against displacement, and stable and dignified livelihoods, planning and design interventions are unlikely to reflect community priorities or be sustained in the long term unless residents can shape design and operations decisions. This part focuses on empowered self-governance, the last of the four LEGS dimensions of equitable resilience. However, the position of governance as the final dimension to be discussed here does not suggest that it is less important or should come sequentially after the other three dimensions in implementation. Rather, as our case studies make clear, enabling people to shape the decisions that impact their lives is essential to ensuring that environmental, livelihood, and security gains are sustained and that they reflect the values and priorities of impacted people.

Inclusive governance can ensure that environmental hazard mitigation is informed by diverse forms of experiential and technical knowledge. Governance is inextricably intertwined with questions of security from displacement. Communities around the world, including several featured here, are improving their security through empowered self-governance, often linked to shared ownership of land and housing. Finally, movements to build democratic and solidarity-based economic structures show that improved self-governance can advance stable and dignified livelihoods.

In highlighting self-governance as a central pillar of equitable resilience, we embrace a capacious definition of governance that goes beyond the actions of governments to include a wide array of decision-making institutions and practices across scales.[1] Climate change presents difficult urban governance challenges. The scalar complexities, uncertainties, and uneven nature of climate impacts require governance strategies that are multiscalar, flexible, and inclusive, enabling often-marginalized groups and forms of knowledge to shape decision making.[2]

The case studies in this section and throughout the book demonstrate the range of actors that can advance equitable resilience in the face of climate change. Essential contributions can come from government entities, ranging from national agencies such as the US Army Corps of Engineers (Caño Martín Peña, case 12) and CODI (Baan Mankong, case 6) to local public-school officials (Paris OASIS, case 3). Experts from private design, planning, and engineering firms play an important role in shaping many projects, including the Gentilly Resilience District (case 1), Yerwada upgrading (case 8), and the Kibera Public Space Project (case 11). In many cases, NGOs can serve as essential bridging institutions, including locally oriented groups such as Living Cully (case 7) and groups operating at the national or international scale such as Habitat International, which played a role in several of the featured interventions (cases 5, 7, and 8). Most essential to ensuring that interventions benefit disadvantaged residents, several of the cases that we highlight were substantially driven by residents themselves, both through formal ownership and governing structures such as resident cooperatives (cases 4 and 6) and community land trusts (case 12) and through more or less formal advocacy efforts that can effectively pressure governments and other empowered institutions to attend to critical needs (e.g., cases 2 and 7).

This chapter discusses how governance relates to urban design and planning more generally, including both consideration of who is empowered to make decisions about the form and operation of built environments and how design methods and strategies operate as mechanisms for governance. We then discuss governance in the context of extreme events and climate hazards, including how empowered groups use disruptions to undermine democratic structures. We also note strategies for advancing self-governance amid hazard-related upheavals. Finally, we present a framework for empowered self-governance for equitable resilience across scales before introducing three case studies.

GOVERNANCE, PLANNING, AND DESIGN

Governance and decision making are integral to urban planning and design. Complex city-making processes link designers and planners to a variety of other participants. A "just" approach to urban design requires inclusive processes that enable shared agency and management of places.[3] In this section, we briefly discuss the evolution of governance within the field of planning, from top-down rationalism and high modernism to participatory practices that aspire to redistribute power.

Since at least the mid-twentieth century, critics have raised alarms about top-down planning that does not reflect the values of residents, from "rational" planning in US cities to grand nationalist visions for planned capital cities. Such heavy-handed interventions inspired calls for greater consideration for the experience of ordinary residents.[4] Jane Jacobs urged preservation of existing socio-spatial relations in mixed-use, mixed-income neighborhoods.[5] Paul Davidoff argued that planners should explicitly work on behalf of marginalized residents.[6] Horst Rittel and Mel Webber challenged the assumption that expert-driven planning was appropriate in confronting the "wicked problems" of urban settlements.[7] With the "communicative turn" in planning theory and the embrace of "deliberative democracy," many scholars and practitioners regarded the principal task of planning not as devising solutions to concrete problems but rather as enabling diverse groups to deliberate in search of shared ambitions.[8]

Out of these critiques of top-down planning, many models for small-scale governance have emerged. Political theorists have examined case studies of "empowered participatory governance" in diverse settings—from neighborhood councils in Chicago to participatory budgeting in Porto Alegre, Brazil, to local government reforms in India—identifying key principles for success, including a focus on specific tangible problems, involvement of ordinary people, and deliberative development of solutions. Institutionally, these strategies require devolution of decision making to local scales, linkages between different local units and between local units and higher levels of government, and a desire to infiltrate and change existing state institutions through the insights produced by participatory processes.[9]

Design practitioners have also sought to democratize the formation of built environments through participation and user control. Advocates have refined strategies for participatory design, including "charrettes" in which designers solicit ideas from interested publics.[10] While some forms of participatory design naively ignore power imbalances within communities, others explicitly acknowledge and counteract inherited inequities, including Randy Hester's model of power mapping and "design for ecological democracy."[11] Some advocate going beyond charrettes to advance "community-driven design," in which the will of designers becomes secondary to that of people impacted by interventions.[12]

Some projects, especially in rapidly urbanizing settings, invite residents to shape places directly. In the 1960s and 1970s, John Turner advocated for "dweller control of the housing process," arguing that governments should support rather than oppose efforts by poor people to build their own informal communities.[13] Work by Turner and others led international development institutions to embrace the "site and services" model for low-cost neighborhoods. High-profile examples of this approach, wherein governments or other actors lay out parcels (sites) and provide basic infrastructure (services) leaving people to build housing as and when they can, include Balakrishnan Doshi's Aranya project in Indore, India.[14] In the 1980s, Dr. Akhtar Hameed Khan initiated the Orangi Pilot Project, through which residents of an underserviced area of Karachi, Pakistan, built their own water and sewer infrastructure.[15] More recently, Alejandro Aravena's firm, Elemental, designed "half houses," enabling people to control the design and construction of portions of their own homes.[16] "Self-build" or "auto-construction" models are most associated with rapidly urbanizing regions of the Global South,[17] but designers in the Global North have also experimented with these methods, from John Habraken's "supports"[18] to self-help housing in the US[19] and a self-built community in Trondheim, Norway.[20]

Participatory governance, whether focused on initial designs or ongoing decision making, has long been a topic in city planning. Soon after community engagement became mandatory in many US government-funded urban development projects, Sherry Arnstein's classic paper on the "ladder of citizen participation" argued that superficial participation often granted participants little actual power, amounting to little more than manipulation and co-optation.[21] Commentaries on recent participatory design efforts describe how Post-it Notes and colored dots used to solicit user input in charrettes frequently become an aestheticized performance divorced from decision-making power.[22] Research on participatory international development levels similar critiques, pointing out that planners frequently use the rhetoric of participation while remaining firmly guided by outside technocratic expertise.[23] In such circumstances, participation becomes a symbolic process, shoring up neoliberal hegemony rather than enabling genuine community control.[24]

Such critiques of superficial participation have led to calls for more radical interventions. Critical scholars argue that the definition of "formal" and "informal" urbanization represents an ideological battleground that excludes many people from shaping urban life.[25] Some call for "insurgent planning"[26] and recognition of "insurgent citizenship"[27] through which marginalized people can exercise their "right to the city." Exercising such rights can take many forms, from tenant organizing and the creation of "counter plans"[28] to occupation of public space and unauthorized squatting and settlement.

Community control of land and property through shared ownership is one means of extending participation into genuine power redistribution. Movements for community control have surged in cities around the world in response to mounting housing unaffordability, gentrification, and displacement. Strategies include cooperatives, community land trusts, and other structures for operationalizing "solidarity economy" principles in urban land and property.[29] We discuss shared ownership under the headings of both "security" (part II) and "governance" (part IV) because these models link tenure security with participatory decision making. Shared land ownership and governance has been the norm for much of human history, and as previously noted, this has continued into the present day.[30] While colonization and globalization have imposed privatized land tenure around the world, advocates are increasingly looking to models that link community ownership to democratic self-governance.

GOVERNANCE, DISASTER, AND RECOVERY

Destructive events, from armed conflict to hurricanes, bring particular governance challenges. Opportunistic actors can take advantage of disruptions, creating a "state of exception"[31] to justify changes to preexisting rights regimes and democratic governance norms. One well-documented instance of this phenomena is the surge of "disaster capitalism" that followed Hurricane Katrina[32] when government and private-sector actors systematically undermined many aspects of New Orleans's welfare state by destroying public housing, restructuring public education, and suspending worker protections.[33]

Both neoliberal privatization and authoritarian control can usurp the popular will of people impacted by disasters. Researchers have suggested many models for integrating self-governance into pre- and post-disaster planning. Reviews of participatory recovery from disasters, including tsunamis in India, earthquakes in China, and hurricanes in the US, suggest that facilitating participation by disaster-impacted communities is slow, politically complex, and prone to abuse, but that it can deliver long-term benefits, informing sound recovery decisions and building trust between community members and with outside actors.[34]

CLIMATE CRISES AND GOVERNANCE

The mounting impacts of climate change challenge the assumptions underlying urban governance and the governance of hazard risk, preparation, and recovery. As national and international climate governance fails to reduce greenhouse gas emissions significantly or coordinate effective adaptation, there is increased attention on subnational scales, including cities and regions, as sites of climate action.[35]

Mirroring research in critical disaster studies, research on urban climate change impacts also reveals that opportunistic actors can use real or anticipated disruptions to undermine democratic governance. The financialization of urban climate risk through instruments such as flood insurance and municipal bond rating can sharply limit the agency of democratic institutions and shift the burdens of adaptation, harming already disadvantaged households and municipalities.[36] Apocalyptic climate projections can quash debate and depoliticize decisions about adaptation.[37] Threats from climate change can be used to justify restructuring settlement patterns without democratic deliberation, reconfiguring the economic base of rural areas or razing low-income settlements to make way for green infrastructure.

Concerns over how city leaders make climate adaptation and mitigation decisions have led to increased scrutiny of "urban climate governance," which Isabelle Anguelovski and JoAnn Carmin define as "the ways in which public, private, and civil society actors and institutions articulate climate goals, exercise influence and authority, and manage urban climate planning and implementation processes."[38] IPCC reports state that current governance practices are "ineffective to reduce risks, reverse path-dependencies and maladaptation, and facilitate climate resilient development."[39] Efforts to reform urban climate governance often focus on three primary areas: (1) multiscalar governance, (2) adaptive governance, and (3) inclusive practices to insure that perspectives from often-marginalized groups are incorporated.

MULTISCALAR AND MULTILEVEL GOVERNANCE

City governments are becoming important players in climate action, but the nature of climate change requires actions on both larger and smaller scales. Greenhouse gas emissions do not honor municipal boundaries. Carbon dioxide emitted in Dallas can contribute to cyclones and rising sea levels in Chennai. Similarly, floods, heat waves, droughts, fires, and smoke plumes do not respect the administrative boundaries of cities, states, or nations. The mismatch between administrative boundaries and landscape and climate processes has long challenged effective and equitable environmental planning. Much as watershed-based planning and other ecologically informed strategies create structures to manage the impacts of development and pollution across municipal and state boundaries, climate change is spurring new regional planning efforts to coordinate adaptation and mitigation.[40] Climate researchers advocate for "multilevel governance" to overcome the scalar challenges of "traditional distinctions between local, national and global environmental politics."[41] The IPCC's Sixth Assessment Report reinforces this, arguing that "lack of coordination between governance levels and disagreement about financial responsibility" are key barriers to adaptation.[42]

ADAPTIVE GOVERNANCE

In addition to challenging the scalar conventions of planning, climate change forces planners to consider how governance regimes can adapt to the compound uncertainties of climate change and urbanization. Since the 1970s and 1980s, many ecologists and environmental policy researchers have advocated for "adaptive management" of environmental systems to facilitate "social learning" and flexibility in addressing socio-ecological challenges.[43] While some critics regard adaptive management as underestimating the importance of uneven power and subject to capture by empowered interests,[44] the scale and complexity of climate change impacts have motivated calls for new forms of adaptive urban climate governance.[45] Emily Boyd and Sirkku Juhola succinctly define adaptive governance as "decision making systems comprising formal and informal institutions and social networks that are able to adapt in the face of uncertainty."[46] A recent IPCC Assessment Report calls for "build[ing] adaptive governance systems that are equipped to take long-term decisions," driven by "processes of social learning."[47]

INCLUSIVE GOVERNANCE

Research on urban climate governance increasingly emphasizes the need for inclusion and pluralism, engaging perspectives that have often been marginalized from planning processes. The Sixth Assessment Report from the IPCC calls for "inclusive and accountable adaptation"[48] driven by "increasing community participation and consultation."[49]

This augmented focus on inclusive governance aligns with the ascendence of "transformative adaptation" and "climate justice" goals. Linda Shi and Susanne Moser argue that transformational adaptation requires "deliberately and fundamentally changing systems to achieve more just and equitable outcomes."[50] Such calls for equity-centered adaptation are explicitly linked to reforms in unjust structural conditions. Scholars describe three dimensions of climate justice: distributive justice, which considers the uneven allocation of costs and benefits of climate action; procedural justice, which demands that decision-making processes incorporate perspectives from those most likely to be impacted; and recognitional justice, which requires respect for the perspectives and historical experiences of impacted communities.[51] Acknowledging the importance of spatial settings for the enactment of justice, others advocate for interactional justice as a fourth dimension, emphasizing the need for multiple publics to shape the production and use of the public realm.[52]

Inclusive governance for justice-centered adaptation requires more than superficial participation. The IPCC calls for "transformation of ways of knowing, acting and lesson-drawing to rebalance relation between human and nature."[53] In widening the epistemologies deemed relevant to adaptation, the IPCC explicitly calls for "integration of Indigenous Knowledge and Local Knowledge, focusing on the most vulnerable."[54] Much as Susan Fainstein observes that "participation [in resilience planning] without financial resources is an empty promise,"[55] the IPCC recognizes that participation alone is not enough and that marginalized communities also need "the resources" and "legal basis" to act.[56]

Development organizations and governments are increasingly attempting to make climate governance more inclusive. Such efforts are often labeled "community-based adaptation" (CBA)[57] or "locally led adaptation," in which local people have greater agency in assessing needs, designing adaptation, and determining success.[58]

DESIGN AND CLIMATE GOVERNANCE

Calls for multiscalar, adaptive, and inclusive urban climate governance highlight the degree to which conventional infrastructure planning and spatial policy have been just the opposite: uniscalar, rigid, and narrowly technocratic. Geographer Kevin Grove describes the transition from previous generations of environmental infrastructures guided by a modernist "will to truth" to a newer generation of interventions emerging from a "will to design," viewed as "a desire to pragmatically and collaboratively engage with complex phenomena from a position of necessarily partial and limited knowledge."[59] Grove and others highlight how the rise of "resilience" as a framing concept and normative aim in environmental planning and governance has increased the centrality of design methods. Instead of infrastructure projects planned on a narrow project-by-project basis, this new mode encourages analysis and planning across scales. Where earlier projects used historical data to craft solutions that were presumed

to be stable, if not permanent, the will to design calls for flexibility, itera-
tion, and learning over time. Where modernist planning and engineering
valued technical knowledge to the exclusion of other ways of knowing,
planning for urban climate resilience must be open to a broad array of ways
of understanding and valuing places.

RESILIENT GOVERNMENTS WITHOUT
EQUITABLY RESILIENT GOVERNANCE

Urban design and development interventions can also serve as signals
of progress, providing spatial and material evidence of leaders' priorities,
especially after major disasters. Designed symbolic moments can be indicators of resilient governments but may fall short of equity-enhancing self-
governance. Government officials often seek to exert control over recovery
narratives, eager to project a bright future. Following the devastation of
Katrina, city leaders eagerly renovated the Louisiana Superdome—whose
damage had been a highly visible symbol of the disastrous response to the
hurricane and flooding—in time for the 2006 New Orleans Saints football
season. By contrast, other critical facilities, including public housing and
schools, lagged many years behind.

As we have noted throughout this book, resilience interventions often
fail when they run counter to the interests, desires, and priorities of
impacted people. For instance, when Indigenous Paiwan people in Taiwan
were relocated from their mountain village of Makazayazaya to a new site
following Typhoon Morakot in 2009, residents say that the government
"just built it how they wanted it" without regard to the needs of impacted
people. Further, people relocated to the new village were not allowed to
customize their homes to suit their needs.[60] However, following this same
typhoon in the same region, the story of Ulaljuc Village in Taiwu Township
outlined in part III demonstrates that it is possible to attend to the desires
and needs of impacted people while still mitigating environmental risk.
This is the difference between governance that aims to deliver resilience
and pursuit of equitable resilience that accords impacted people the power
to act.

EQUITABLE RESILIENCE THROUGH SELF-GOVERNANCE ACROSS SCALES

Enhancing self-governance among disadvantaged people is central to equitable resilience. As with the environment, security, and livelihood dimensions, governance for equitable resilience requires attention and action across scales. The IPCC's Sixth Assessment Report argues that "climate governance will be most effective when it has meaningful and ongoing involvement of all societal actors from the local to the global levels."[61] Given our focus on local planning and design interventions, we focus less on national and global scales, defining three primary scales of action: the individual or household, the community or project, and urban or regional scales. Planning and design interventions can advance self-governance through action across these scales by enhancing (1) *self-efficacy* (individual/household scale), supporting residents in developing their individual and collective efficacy in addressing community priorities; (2) *control* (project/community scale), providing structures for residents to shape interventions and ongoing community decisions; and (3) *influence* (city/region scale), creating structures for residents to advocate for themselves with respect to larger institutions, including governments.

CONTROL: SELF-GOVERNANCE FOR EQUITABLE RESILIENCE AT THE PROJECT OR COMMUNITY SCALE

Several recent projects have experimented with participatory design methods that incorporate diverse perspectives in adapting to climate threats. Following Hurricane Isaac's devastation in South Louisiana in 2012, the state government, in partnership with several NGOs and private design and planning firms, received a long-term disaster recovery grant from the US Department of Housing and Urban Development for the Louisiana's Strategic Adaptation for Future Environments (LA SAFE) program. Unlike high-profile resilience-focused design competitions, LA SAFE embedded design professionals in extended community planning processes in six impacted parishes. Over several months in 2017, architects, urban designers, landscape architects, and planners participated in dozens of gatherings with more than three thousand residents. In early meetings, designers used their visualization and analytical skills to facilitate discussions about climate threats and adaptation. Only after several rounds of meeting and discussion did designers present adaptation strategies, including both small "catalytic" projects to be carried out with dedicated funding and larger proposals for future intervention.[62]

Over the course of six months in 2015, WE ACT, a grassroots environmental justice organization, led a community-based planning process to develop the Northern Manhattan Climate Action Plan for an area of New York City that is home to large Black and immigrant communities. In addition to WE ACT and other grassroots organizations, the process included several city agencies, environmental and health-focused NGOs, political leaders, and faculty and students from New York City design schools. The group generated recommendations across four areas: energy democracy, emergency preparedness, social hubs, and public participation. The plan's approach to climate resilience incorporates not just strategies such as green infrastructure and flood barriers but also community control of land, food, banking, housing, energy, media, and industry.[63]

Barbara Brown Wilson's 2018 book *Resilience for All* includes case studies of several community-driven design initiatives that used design to advance community resilience among vulnerable groups. The cases that Wilson profiles are primarily small in scale, including a bayou restoration plan by the Gulf Coast Community Design Center and bicycle wayfinding by Living Cully (see case 7). The community-driven design approach that Wilson advocates is less centered on major planning and infrastructure projects and more focused on how small community-driven interventions can plant seeds of deeper change, shifting who has the power to act in the face of climate change and other threats.[64]

Participation in design and planning is a necessary first step to ensuring that resilience interventions respect the views of impacted communities, but it is equally important to establish mechanisms for shaping ongoing decision making after interventions are done. The early phases of the Baan Mankong program in Thailand (see case 6) successfully linked participatory design to ongoing decision making, using collective control of land as the basis for community governance.

In other instances, participatory governance need not be linked to inclusive design. Cooperatively owned enterprises in a range of sectors, from energy and agriculture to housing and transportation, are advancing climate action.[65] In the ROC USA model (see case 4), residents of existing manufactured home parks take on cooperative ownership and, through democratically elected boards, make decisions about infrastructure, common spaces, and other matters that shape vulnerability to climate change and other threats.[66] In addition to the limited equity housing cooperative model deployed by ROC USA, there are many other models for shared ownership and community governance in housing. The LILAC (Low Impact Living Affordable Community) project in Leeds in the UK uses a mutual home ownership society model to create "missing middle" housing designed to balance affordability with environmentally progressive features, including energy efficiency and on-site stormwater management.[67]

Community land trusts, such as the Fideicomiso Caño Martín Peña profiled in this section (see case 12), also join shared ownership of land to community governance. CLTs are typically governed by a tripartite board of directors, including equal representation from residents housed on CLT land, local leaders and experts, and members of the wider community. This three-part governance structure is intended to encourage decision making that benefits not only residents but also the broader community.[68] As the CLT model spreads globally, it reaches groups seeking to reduce threats from climate change in many urban areas.[69]

Numerous forms of empowered self-governance are linked to shared ownership. Yet, it is also possible for disempowered communities to organize, struggle, and claim power without formal ownership of land or housing. In the US and elsewhere, residents of public housing have organized to influence decisions that shape their lives.[70] Residents of informal settlements in urbanizing regions have also organized successfully to protect their right to remain and to fight for upgrades. Ciliwung Merdeka is a grassroots organization that advocates on behalf of residents of settlements along the Ciliwung River in Jakarta, Indonesia. The organization, whose name means "sovereign Ciliwung," has used a range of strategies, including participatory design to advocate for improvements for settlements imperiled by both rising floodwaters and eviction.[71]

Participation in design and decision making can advance community self-determination, especially when linked to community ownership. However, even the most elegant and well-intentioned project-based participatory interventions deliver few benefits if not coupled with influential linkages beyond the community and improved self-efficacy at the individual and household level.

INFLUENTIAL LINKAGES

Access to outside resources, power, and knowledge is often essential to improving local conditions. Urban sociologist Robert Sampson emphasizes the importance of "vertical linkages" for supporting the "collective efficacy" of communities in addressing shared problems.[72] Relatedly, researchers define "bridging" social capital as linking people, groups, and institutions that support community well-being, including resilience in the face of climate change and other hazards.[73]

In confronting complex environmental challenges, including those associated with climate change, multiscalar or multilevel governance is essential in linking local actors to external information and resources.[74] The IPCC's Sixth Assessment Report states, "Governance practices for climate resilient development will be most effective when supported by formal (e.g., the law) and informal (e.g., local customs and rituals) institutional arrangements providing for ongoing coordination between and alignment of local to international arrangements across sectors and policy domains."[75]

Several of the cases that we highlight illustrate the importance of strategic linkages to outside power and resources. Many resilience benefits of the ROC USA model (case 4) are associated with increased linkages that enable residents to access government and philanthropic resources. For instance, a ROC in Minnesota used state grants to partially fund a new tornado shelter and community center, and a ROC in Montana connected to city water and sewer infrastructure as their own systems failed under stress from deferred maintenance and climate change.[76]

In a radically different setting, the Baan Mankong program in Thailand (case 6) also improves linkages to outside resources. Baan Mankong projects in every city and region are driven by networked governance linking target communities with other institutions, including local and national government agencies and academic institutions. One interviewee described the Baan Mankong model as analogous to a power outlet, creating a point of connection through which marginalized people can access support, from low-cost loans and job training to infrastructure upgrading.

SELF-EFFICACY

Even when interventions include participation in design and ongoing decision making and strengthen linkages to influential outside knowledge, power, and resources, realizing the benefits of empowered self-governance requires that people have the skills and confidence to shape their own settlements. The same structures that create uneven burdens of climate vulnerability for urban poor communities can also strip people of agency and confidence. Therefore, it is especially important that any intervention to improve climate resilience among disadvantaged residents supports latent self-governance capacity and reinforces the idea that excluded people can and should make claims on the state and other institutions to improve conditions in their settlements.

Tactical urbanism, a recent movement within urban planning and design, explicitly seeks to broaden agency for changing urban environments. Proponents suggest that low-cost, often temporary, interventions—from street-side parklets to DIY traffic-calming planters and guerilla crosswalk painting—represent a new mode of democratized city making. Critics argue that these strategies remain exclusionary, as they tend to empower already-privileged groups to reshape the urban realm while leaving minoritized communities stripped of agency.[77] Countering this, Wilson's *Resilience for All* advocates for community-driven tactical interventions as a means to build coalitions and expand agency in communities harmed by exclusion, poverty, and violence.[78] In some cases, community-driven interventions can create or reinforce collective narratives and shared histories, which can support community self-governance, especially in the wake of major disruptions.[79] The Inter-Tribal Gathering Garden at Cully Park in Portland, Oregon, is an example of such an intervention, creating a space for celebrating and reinforcing community bonds among Native people and between Native residents and their non-Native neighbors (see case 7). The public art component of the Gentilly Resilience District in New Orleans is similarly intended to promote shared understanding among residents of flood-vulnerable neighborhoods (see case 1).

In other cases, interventions to support self-efficacy can more directly build skills for leadership and community organizing. The ROC USA model of community land ownership and self-governance includes training and peer-to-peer networking for cooperative board members (see case 5). Housing and infrastructure improvements supported by Baan Mankong in

Thailand often start with establishing community savings groups. These groups provide structures for budgeting and savings to enable community contributions to land and housing costs. Savings groups are also intended to create solidarity and build skills among potential resident leaders (see case 6).

Strengthening self-efficacy is essential to self-governance. However, as in the other dimensions of equitable resilience, overreliance at the household and individual scale can be problematic, reinforcing the neoliberal privatization of risk. For example, property buyout programs underlying managed retreat frequently focus on the household scale, disregarding the essential function that community ties play in supporting resilience.[80] People in marginalized communities do indeed have enormous capacity to solve problems and cope with climate change threats. However, focusing on these smaller scales of analysis and action can obscure the impact of vulnerabilization—the structural drivers of unequal vulnerability that place the heaviest burdens of climate change on already disadvantaged people (table IV.1).[81]

THE CASES: PARTIAL SUCCESS IN EQUITABLE RESILIENCE THROUGH SELF-GOVERNANCE

The case studies that follow feature three distinctly different efforts to support empowered self-governance among disadvantaged communities facing threats from climate change. As with the cases in the other sections, the accomplishments in these cases go beyond governance, also making strides in the environment, livelihoods, and security dimensions of equitable resilience. Also like the other cases throughout the book, these cases are not perfect. No actually existing planning and design intervention is perfect, especially in the context of the complexities, dynamism, and uncertainty of climate change. In each case, the projects demonstrate both the transformative potential of community self-governance and the fragility and limitations that beset efforts to struggle against unjust status quo conditions.

First, in case 10, we discuss the Thunder Valley Community Development Corporation, which is undertaking an ambitious effort at community building and economic empowerment on Lakota tribal territory at

Table IV.1

Three scales of equitable resilience through self-governance

Scale	Equitable Resilience Principle	Question	Examples
Individual/ Household	Self-Efficacy	Does the project support residents in developing individual and collective efficacy in addressing problems?	ROC USA leadership training; Baan Mankong savings groups; Native Gathering Garden in Cully Park; Gentilly Resilience District public art
Project/ Community	Control	Does the project provide structures for residents to shape design interventions or ongoing management?	LA SAFE; WE ACT Northern Manhattan Climate Action Plan; LILAC housing cooperative; Baan Mankong community design process; Ciliwung Merdeka in Jakarta
City/Region	Influence	Does the project enable residents to advocate for themselves and others with respect to larger institutions, including governments?	ROC USA linkages to government, philanthropy, commercial lending, and peer-to-peer networks; Baan Mankong as a "power outlet" for continued support

the edge of the American Great Plains. Their work explicitly addresses the generational trauma of land theft and genocide perpetrated against Native people by settlers. Operating within the complex institutional landscape of overlapping tribal and outside governments, the Thunder Valley CDC is building both a literal community of homes and a customized approach to "self-sovereignty." Next, in case 11, we profile the Kibera Public Space Project led by the Kounkuey Design Initiative, a planning and design firm partly headquartered in Nairobi, Kenya. In eleven projects over more than fifteen years, KDI demonstrates powerful methods for community-driven design along with long-term project ownership and stewardship.

Interventions have created new public amenities, including public plazas, market spaces, and toilets in flood-vulnerable sections of Kibera, one of Africa's largest informal settlements. Finally, the last case study of this section and the book (case 12) focuses on the ongoing transformative work in the communities surrounding Caño Martín Peña in San Juan, Puerto Rico. The effort, led by a coalition of community organizations and ENLACE, a government-sponsored planning entity, uses the structure of a community land trust to guide restructuring and upgrading of several flood-prone channel-side neighborhoods. The CLT, Fideicomiso de la Tierra del Caño Martín Peña, which was established and granted ownership of significant land holdings through an act of the territorial legislature, has enabled ongoing managed retreat, shifting people away from vulnerable areas and enabling improvements, including channel dredging, new sewer infrastructure, and additional green spaces.

Following the three case studies, this section closes with a brief conclusion drawing out common themes from the three cases and pointing toward areas of future research and transformative practice.

Case 10

THUNDER VALLEY: REGENERATING SELF-SOVEREIGNTY ON THE PINE RIDGE RESERVATION

OVERVIEW

The Pine Ridge Reservation, one of the most impoverished areas in the US, faces new challenges from an increasingly harsh and unpredictable climate. In response, members of the Oglala Lakota Tribe established the Thunder Valley Community Development Corporation (TVCDC) in 2007 to develop a community-led model for housing, ecological and cultural regeneration, and infrastructure. TVCDC is building a thirty-four-acre development with housing and facilities for training, education, and agriculture. Drawing upon Indigenous governance traditions, TVCDC seeks to enhance environmental well-being, generate livelihoods, and secure tenure through a Lakota-inflected approach to homeownership. Based on interviews, reviews of planning documents and a site visit, it seems that TVCDC's achievements are well underway, although challenges remain.

INTRODUCTION

The Pine Ridge Reservation, spread across three counties in South Dakota and a tiny extension in Nebraska, is the imposed homeland of the Oglala Lakota Nation (figure 10.2). The Reservation's residents have some of the lowest incomes and highest unemployment rates in the US. People who

10.1 New homes and community facilities under construction by the Thunder Valley Community Development Corporation. *Source*: Thunder Valley Community Development Corporation.

live on the Reservation disproportionately die young, and many struggle with addiction. Pine Ridge is an unforgiving landscape. Temperatures regularly fluctuate by nearly 140°F. Recent years have brought drought and wildfires in 2011, severe flooding in 2012, tornadoes in 2016, and hailstorms in 2018. In March 2019, a "bomb cyclone" wreaked havoc on the Reservation, followed by the protracted lockdowns and health threats of the COVID-19 pandemic.

These recent crises extend a much longer history marked by marginalization and displacement. Driven by conflicts with other Indigenous groups and the quest for abundant buffalo herds, the ancestors of today's Oglala Lakota migrated westward from the woodlands of present-day Minnesota to the Black Hills, arriving by 1775. White settlers soon followed, bringing disease and violent resource conflicts. In 1868, the Fort Laramie Treaty allocated to the Lakota a vast Reservation encompassing parts of the Dakotas and four other states. In 1889, the federal government reneged, confiscating 7.7 million acres (including the sacred Black Hills) and confining the Oglala to 2.7 million acres on the Pine Ridge Reservation.[1] A year later, government troops massacred more than three hundred Lakota men, women, and children near Wounded Knee Creek.

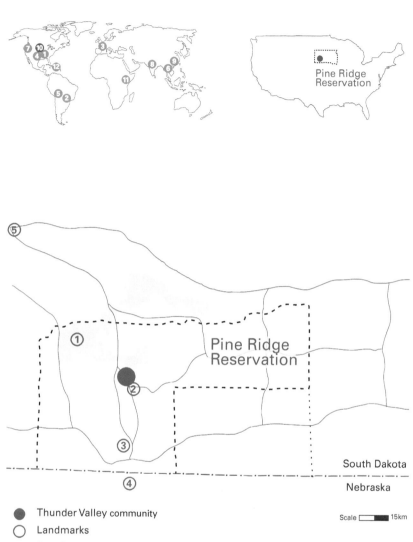

10.2 Site of Thunder Valley CDC on the Pine Ridge Reservation, South Dakota. The Reservation also includes pockets of trust and tribal lands located in the quadrant of territory southeast of the main portion. 1. Badlands National Park; 2. Sharps Corner; 3. Wounded Knee; 4. Whiteclay, Nebraska; 5. Rapid City, South Dakota. *Source*: Mora Orensanz.

The history of the TVCDC, based just nineteen miles north of Wounded Knee near the hamlet of Sharps Corner, is inseparable from the 1890 massacre. The legacy of Wounded Knee was reinvigorated in 1973, when two hundred supporters of the American Indian Movement, led by Lakota activist Russell Means, occupied the area for seventy-one days to decry current tribal leadership, demand renegotiation of treaties, and advocate for better treatment of Native peoples. TVCDC's current Executive Director, Tatewin Means, Russell Means's daughter, has inherited her father's struggle to improve the lives of Lakota people.

Pine Ridge faces interrelated climatic and political threats. In July 2018, a storm with baseball-sized hailstones severely damaged five hundred homes. Agents from the Federal Emergency Management Agency assessed the destruction but refused to help, arguing that the "total financial damage hadn't been high enough to warrant assistance." The reason was both ironic and devastating to the low-income people impacted: their homes were simply worth too little.[2] About half of Reservation residents own their homes, well below the average for South Dakota. Property values remain extraordinarily low, averaging just $35,000. More than one third of citizens lack health insurance, and two thirds of those with insurance receive coverage through Medicaid.[3]

In a place so often stigmatized by its deficits, the rise of the TVCDC charts a different path, using Indigenous practices to build community agency. TVCDC marshals women-led and youth-focused leadership to assert greater agency over resources and livelihoods, including by instigating climate adaptation informed by traditional Lakota values (figure 10.3).

In 2006, Oglala community members—many of whom are deeply involved with the yearly Thunder Valley Sun Dance gatherings—initiated discussions about launching a nonprofit community development organization. TVCDC was formally established in March 2007. Oglala activist Nick Tilsen served as founding executive director. Tilsen says the organization "came out of a movement" dedicated to "reconnecting young people to culture and spirituality and identity," emphasizing "commitment to place." To Tilsen, TVCDC exists to build "an ecosystem of opportunity" to enable "our community to prosper" through varied activities, including sustainable housing, language revitalization, food sovereignty, workforce development, youth leadership, and social enterprise (figure 10.3).[4] TVCDC

10.3 Many TVCDC initiatives foreground women's leadership and educational opportunities for tribal youth. *Source*: Mora Orensanz.

aimed to address the "root causes of perpetual poverty" by building a community centered on Lakota culture and the traditional *tiospaye* structure of extended families.[5] Instead of accepting the narratives of damage and deficit by which others define the community, Tilsen charged TVCDC with "doing economic development work that is based on regeneration, based on resilience," work that "is as much about healing the human spirit as it is about green buildings." TVCDC envisioned a "net-zero energy community" at "Ground Zero for poverty in America."[6] Tilsen, whose own parents first met in 1973 "in the middle of the revolution at Wounded Knee,"[7] led TVCDC for a dozen years, building a national reputation.[8]

From 2010 to 2013, Tilsen and TVCDC launched a planning process as phase 1 of realizing this bold vision. The group bought thirty-four acres of Reservation land and developed the Thunder Valley Regenerative Community Plan. In 2014, TVCDC construction activity commenced, including both subsidized single-family homes for sale and an apartment building. The TVCDC constructed a series of buildings to demonstrate sustainable agriculture and energy practices and inaugurated a community center in 2019. Behind the community building, they added a playground in 2020. In 2022, TVCDC purchased forty-eight acres just across the state line in Whiteclay, Nebraska, to launch a holistic healing community.[9]

TVCDC's emphasis on "healing" underscores the group's central aspiration. DeCora Hawk, Thunder Valley's longtime Director of Community
Engagement, spoke of the need to build a "regenerative community," one
that can "change the narrative in our area," both for tribal members and
for those who stigmatize them.[10]

THUNDER VALLEY CDC AND EQUITABLE RESILIENCE

Equitable resilience on the Pine Ridge Reservation takes many forms. The
TVCDC champions regenerative environmental practices to cope with climate threats while enhancing governance, security, and livelihoods. Indian[11]
Reservations raise particular complexities with respect to sovereignty and
self-governance. So, this is the dimension of equitable resilience where we
begin. Nonetheless, in this case, as in others that we profile, governance is
inseparable from improving security, livelihoods, and environments.

MODELING INDIGENOUS GOVERNANCE, WITHIN AND BEYOND
TRIBAL GOVERNMENT

Governance of Thunder Valley remains knotty and contested. Although it
is an independent organization, TVCDC aligns its work with the aims of
Oglala Lakota Tribal government. Under Tilsen's direction, TVCDC coordinated the Tribe's regional planning, funded by the federal Sustainable
Communities Regional Planning Program. Architect Christina Hoxie, previously of BNIM, the Kansas City firm that has worked with the Tribe and
TVCDC, marveled that Tilsen brought together approximately "fifty different local organizations, including the Tribal Council" in the regional
planning effort.[12] Completed in 2012, and credited to nearly three hundred named individuals, the plan is written partly in Lakota and partly in
English.[13] It is entitled Oyate Omniciyé, translated as "The Circle Meetings
of the People." Explaining the "deeper meanings" of the title, the plan
states, "First, 'Oyate' does not just refer to humans, but can include all living beings. Secondly, calling for an 'Omniciyé' is not to be taken lightly.
This word signifies that very important things are to be considered, and in
the way of the Lakota, the ultimate goal is to seek consensus for all who
wished to remain in the conversation."[14] As a group of elders launched the

conversations that led to the plan, they were asked whether the Lakota language had a word for "sustainability." "Oyate Omniciyé" was their closest approximation, rooted in the meetings of headmen from various Lakota *tiospayes* and larger bands that have long gathered to discuss important issues facing the future of the people.[15]

To enhance tribal collective efficacy, the Oyate Omniciyé: Oglala Lakota Plan advances twelve core initiatives, including creation of "model communities," small-scale developments that deliver "more than housing." The Oyate Omniciyé Plan featured TVCDC's proposed site plan as a "case study" to validate such a model.[16]

RESERVATIONS ABOUT RESERVATION GOVERNANCE

The Oyate Omniciyé Plan proposes a wholesale rethinking of tribal governance, noting the ongoing cultural disconnect between decentralized traditional practices and the centralized tribal government system (with its one-house legislature and popularly elected single president) imposed by the Indian Reorganization Act (IRA) in 1934. Additional tensions remain with various US government agencies, including the Bureau of Indian Affairs (BIA), the Indian Health Service, and the tribal Housing Authority.[17] *The State of the Native Nations*, a 2008 study of efforts at "self-determination," laments that the IRA-imposed governance system for Pine Ridge ignored "notably successful" Lakota traditions that provided "parliamentary-type structures in which leaders gathered in council selected multiple executives to carry out administrative functions and an independent society resolved disputes and provided for law and order." In the early twentieth century, the Tribe spent a decade arguing for such a system, but the resultant IRA terms proved to be "a poor match with Lakota standards of legitimacy and authority."[18] The Oyate Omniciyé contends that this history explains why "the IRA government at Pine Ridge is subject to turmoil and experiences great difficulty in exercising stable, sovereign authority or in winning the allegiance of the community."[19] The plan aspires to "increase self-governance" but also pulls some punches. Its authors worry that pushing too hard against the BIA and other federal agencies could trigger retrenchment, yielding "widespread job loss through cutting of services and erosion of sovereignty."[20]

Tatewin Means, Tilsen's successor as TVCDC executive director, is well situated to understand the Tribe's institutional complexity. Her education began in Pine Ridge at the Little Wound School, but she left "the safe, loving environment of the rez where people look like me" for middle school and high school in Rapid City. That "culture shock" included her "first experience with direct and overt racism." Her formal education culminated in a surely unique combination: an undergraduate degree from Stanford in environmental engineering with a minor in comparative studies in race and ethnicity, a law degree from the University of Minnesota with a concentration in human rights law; and a master's in Lakota leadership and management from Oglala Lakota College. Following law school, she accepted an appointment as the youngest-ever attorney general for the Tribe. While in that post, she encountered "the almost insurmountable barriers that we face when we try to work inside of a broken system . . . [that] was designed to oppress people that look like me."[21]

Before becoming TVCDC's executive director in 2019, Means spent two years on the board and served as outside counsel. Her connection with TVCDC runs even deeper: she had long Sun Danced with the Spiritual Circle that included the organization's founders. When offered the opportunity to succeed Tilsen in charting the organization's next steps, she brought both immersion in Pine Ridge and skepticism of the Reservation's multilayered governance.[22]

PATHWAYS TO SELF-SOVEREIGNTY FROM THUNDER VALLEY

To Tatewin Means, concepts such as governance, sovereignty, self-determination, and freedom are personal matters, rooted in the quest for what she calls "self-sovereignty." In this, she draws upon insights from her late father, Russell (1939–2012). "My father used to always speak about freedom," she recalls. "He wanted freedom for his people; he identified himself as a freedom fighter." When she was younger, she thought this meant separating from the US government and being "our own Nation." But Russell Means wanted more. He wanted Lakota people "to have the freedom to be responsible," to be "free from those chains of colonization . . . that have bound us for centuries as Indigenous people." Tatewin Means interprets her father's vision as a commitment to "*self*-sovereignty," saying, "If we ourselves are not sovereign, if we are not self-liberated, then collectively we have no movement toward liberation."[23]

Building from this concept of self-sovereignty, Means says that governance at Thunder Valley begins with a "ripple effect that extends to our immediate family, to our extended family, to our community." She says, "When you live in a pervasive state of poverty," there's an "overwhelming sense of hopelessness." In this context of protracted despair, Means argues that "that word 'resiliency'" reminds people that "something bad is around the corner," echoing critiques of resilience as inviting neoliberal exploitation. In response, Means and TVCDC are committed to "changing the narrative in our communities about who we are." In their framing, changing the narrative and restoring hope among Pine Ridge residents starts with healing from historical and ongoing traumas. She continues: "We can build the most beautiful communities, the most beautiful homes, community centers, but it will not have the effect we want, if we don't . . . change the mindset of our community, to be liberated to think and act once again as Lakota people." For the staff and beneficiaries of TVCDC, making every decision requires reflection: "Is this a colonized mindset? [or] Is it a liberated Lakota perspective?"[24]

Kimberly Pelkofsky, an architect who became TVCDC's Director of Design and Planning in 2018, has been a rare non-Native key player in the organization. She arrived after fifteen years of assisting other marginalized communities on community design and construction projects, seeking to "support their vision for what they want to achieve." Acutely aware of her complex role as both insider and outsider, Pelkofsky celebrates that "our development" is "a place where we're trying to decolonize architecture and planning, trying to challenge Eurocentric conceptions of spatial organization." She wants the Regenerative Community Plan to support "Lakota identity" and reflect "who they are as a people and how they organize themselves." To Pelkofsky, TVCDC's work responds to the harms perpetrated by "generations of policies," by nurturing a "holistic system" that can support a "community of healing."[25]

Pelkofsky emphasizes that TVCDC's development reflects contemporary best practices in design while drawing on "traditional patterns of development that connect people to who they are as Lakota." When the first design consultant proposed linear rows of houses that "almost looked like army barracks," the TVCDC team reacted viscerally, making clear, "No. This isn't us." To Pelkofsky, this disconnect underscored the extent to which the "community engagement component is really important in everything we do." Instead of barracks, the revised plan arranged twenty-one homes into three circles of seven structures. The symbolically freighted number 7 references the Oglala band of the Seven Council Fires of the Lakota Nation, seven sacred rites, and seven Lakota virtues. Circles permeate Lakota traditions, including the Sun Dance Circle or Sacred Hoop. DeCora Hawk lives in one of the first seven houses in the community, having returned to the Reservation after leaving to study law. As she looks out upon the growing community that has been both her workplace and her home, she comments that TVCDC is "more than an organization; it's a lifeway."[26]

The entrance to each home symbolically faces east toward the sunrise. Each cluster of homes features a central communal space, a circle within the circle (figure 10.4). Once the seventh house in each circle is occupied, residents come together to decide whether to use the shared space for a community garden, a kids' pool, an equipment shed, a space to run around, or something else entirely. For TVCDC and the designers, the dialogue spurred by these spaces is as important as the result.[27]

Efforts to marry contemporary needs with traditional values extend to other parts of the plan. In addition to the home circles, the architectural language of the community building utilizes a "stone base, like a teepee." Like the overall site plan, early design ideas for the playground also began inauspiciously but evolved through community engagement. Architect Pelkofsky recalls a meeting with designers soon after her arrival in Thunder Valley. Charged with creating "a playground based on Lakota teaching," they responded with superficial symbols—"it was like 'medicine wheel, circle pattern—there's your culture.'" Fortunately, the cultural engagement on the playground has markedly improved, with ideas for a massive mural and other elements featuring constellations, foods, stories about animals, language panels, and multiple opportunities for interaction, including a storytelling area featuring audio of elders.[28]

Even the process of naming streets reflects the assertion of personal and collective identity as part of the larger pursuit of self-sovereignty. In the land of Mount Rushmore and a national park whose very name, Badlands, casts aspersions on its inhabitants, the Thunder Valley community

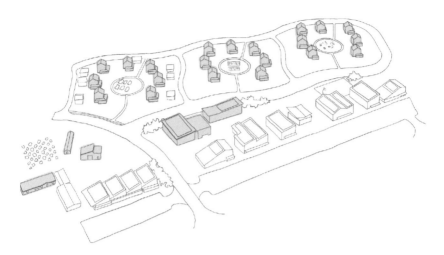

10.4 Site plan of the Thunder Valley community, with completed buildings shown in gray. In addition to the circles of single-family homes, there are apartments, a playground, and community and agriculture facilities. *Source*: Mora Orensanz.

names its roads after chiefs, such as Thatháŋka Íyotaka (Sitting Bull) and Thašúŋke Witkó (Crazy Horse)—pointedly erecting those signs only in the Lakota language. They do so even though (or because) 90 percent of tribal members speak little or no Lakota and only 3 percent are fluent, largely elders.[29] Underscoring the independence of this effort, the team also refused to display the names or logos of outside funders on benches or sidewalks throughout the community.[30]

SHARING RESOURCES: RELATIONS WITH PINE RIDGE

TVCDC has struggled to negotiate the needs and demands from both Pine Ridge Reservation leaders and its own community. Because TVCDC purchased the land from the tribal government, their development is both within the Reservation and separate from it.

The TVCDC rents its community center to the Tribe for meetings, weddings, and events such as the Pine Ridge High School prom and a summer youth camp. Half of the building functions as a kind of bunkhouse, where groups such as dancers or musicians participating in powwows can rent bunks. While it may seem odd for the Tribe to rent facilities from Thunder Valley on their own reservation, the community center provides a heavily discounted, technologically equipped space that is preferable to the alternatives. Still, this rental relationship has caused tensions in the past, and TVCDC leaders "hope to be able to offer facilities free of charge at some point."[31] As with other aspects of Thunder Valley's efforts, tensions surrounding the new community center demonstrate both the value of new facilities to beneficiaries and the need and insecurity that remains in the larger Lakota Nation.

SECURING LAND, SECURING PEOPLE

The search for equitably resilient forms of security at Thunder Valley and across the Pine Ridge Reservation remains a work in progress, both in terms of residential security of tenure and as measured by personal safety and well-being.

LAND AND EXTRACTION ON THE RESERVATION

Many Americans assume that Indian reservations are bounded territories on which Native communities have political sovereignty. Native sovereignty is undermined (sometimes by literal mining) by the fact that the US federal government holds most lands in trust. While this trust theoretically imposes a moral and fiduciary responsibility, it also means that reservation lands can be used by the allottee but not sold, at least not without considerable complexity. Because Native allottees cannot use their land as collateral, they cannot access conventional loans to develop housing, businesses, or infrastructure. Instead, Native land "owners" often feel they have no choice but to lease out their property to outsiders to attain economic security.[32]

At Pine Ridge, land ownership remains mired in complexity rooted in historical inequities. The General Allotment Act of 1877 (better known as the Dawes Act) divided Native lands that had previously been communally managed into individually owned parcels of 40–160 acres. After granting these allotments, most of the remaining land—the majority of the Reservation—was deemed "surplus" and opened to non-Native settlement. Pine Ridge harbors multiple types of land ownership, distributed across an irregular checkerboard that intersperses trust lands (titles held by the federal government), restricted fee lands (on which tribes or individual members hold legal title but uses and transfers are restricted), and fee simple lands (where the owner, often federal or state governments, holds title and can freely transfer). Additional regulations depend on whether the land is owned by tribes, individual members, or non-Indians, creating a maze of conflicting legal jurisdictions. This fragmentation in land tenure causes a range of problems: obstructing the assembly of large parcels for farming, ranching, or development; and restricting freedom to access sacred sites.[33]

Imposed land tenure regimes have long facilitated wealth and resource extraction. Much of the fertile agricultural land within the Reservation is owned by outsiders. One study found that less than one third of nearly $33 million in agricultural receipts from Pine Ridge went to tribal members, and "just 20 people control nearly 46 percent of reservation lands on Pine Ridge through leasing."[34] The Native Lands Advocacy Project developed a Lost Agriculture Revenue Database (the LARD acronym plays on the Lakota phrase *Wasi'chu*, which means "takes the fat," referring to settler resource extraction). The project revealed that, nationally, non-Native farmers and ranchers collected an average of 86 percent of the agriculture revenue from Native lands between 2012 and 2017.[35] This history of outsider exploitation of Native lands contributed to TVCDC's choice to build their community on fee simple land that could be purchased without protracted government and tribal negotiation.

CONTRADICTIONS OF OWNERSHIP

In his autobiography, Russell Means questions the very concept of owning land, saying, "one must understand that to any Indian, ownership of land is a foreign concept. The earth is our Grandmother, who provides us with everything we need to survive. How can you *own* your grandmother? How can you *sell* her?"[36] His daughter, trained as a lawyer and deeply immersed in traditional practices, has inherited this core contradiction as she leads TVCDC. She recognizes that "this notion of individual land ownership and accumulation of assets" resides at the "heart of existing property law," whereas "our ideology as local people is directly opposite." She wants tribal laws to change "so that they're not just a facsimile of Western thought and law." Despite this commitment, the initial build-out of the Thunder Valley community has emphasized ownership of single-family homes. Aside from the shared central circle, each surrounding household is allotted about an eighth of an acre for itself.[37]

In hindsight, Means thinks that the homeownership model (developed before her tenure at TVCDC) may have "jumped ten steps ahead of where our community is." In her view, building a "regenerative community" need not assume "everyone should strive for" individual homeownership—as in, "'Here's this goal, so *get* there.'" Instead, she argues, "We don't even like that system, right? This idea of wealth accumulation in this very Western sense is not *us*. That's not liberation."[38]

As an exercise linking land tenure and communal governance, TVCDC seeks to counter the common perception that home ownership is a purely individualistic endeavor. Rather than emphasizing individual accumulation, Thunder Valley aims to build *community* wealth, which is not associated with "how much you have in your bank account or how much profit you can make in a business." When Means describes TVCDC's desire to build an "alternative economy," she recognizes that "returning to" more communal forms of wealth is "beyond our imagination" and "can be really intimidating." With respect to housing and homeownership, for instance, she wants to "redefine what it means to take care of the home," such that residents see themselves as "a caretaker of a home" and "not use that word 'ownership.'" Still, the contradictions between individual and community wealth building persist. As Kimberly Pelkofsky points out, the goal of building "intergenerational wealth through equity in a home" can combat "intergenerational poverty . . . but also conflict[s] with traditional society."[39]

TVCDC's apartment building, with twelve units ranging from one to three bedrooms, has "a long waiting list" of Indigenous households. Half of the apartments are subsidized for households with very low incomes (50–60 percent of area median income). Others are offered at market rate. TVCDC recognizes there is an enormous need for housing across the Reservation. This need includes their own staff. Most of the current residents are Thunder Valley employees, with still more on the waiting list. While surely understandable, this situation could create the perception that TVCDC's efforts are self-serving. At base, however, the organization remains committed to serving people "from here, who have connections here," rather than serving "a whole bunch of outsiders crossing our gates to live in the

Reservation." Still, Pelkofsky acknowledges that some apartments may go to non-Natives, including teachers who work in Reservation schools.[40]

Oglala Lakota anthropologist Richie Meyers, who ran graduate programs at Oglala Lakota College from 2017 to 2022, is impressed that TVCDC has tried to develop an "alternative" system to tribal government. Nonetheless, he remarks that some cast Thunder Valley as an extension of "nepotism and cronyism" or an arm of the "nonprofit industrial complex." Others, he points out, see their land purchase as an attempt to avoid neighbors in nearby Sharps Corner, while doing nothing to resolve the conflict between fee-simple ownership and trust land.[41]

SECURING HEALTHY FUTURES

Thunder Valley CDC leaders recognize that while there is much to celebrate, as the Oyate Omniciyé Plan put it, some "things must change." Plagued by intergenerational poverty rooted in historical dispossession and other forms of vulnerabilization, Reservation residents face intense problems such as high rates of addiction and suicide. Pine Ridge has seen recent crises in youth suicide, including one extreme period that prompted leaders to declare a "state of emergency" in 2015.[42]

Alcohol sales are prohibited on the Reservation. Tribal leaders fought for decades to ban alcohol imports onto the Reservation from across the state line in Whiteclay, Nebraska. Although Whiteclay has a population of only about a dozen residents, until recently, the town hosted four stores that collectively sold between 4.3 and 4.9 million cans of beer annually, mostly to Indians who would then resell them on the Reservation. After more than a century of advocacy, lawsuits, and presidential executive orders, the Tribe succeeded in forcing the closure of these stores in 2017.[43]

Predatory alcohol sales are a matter of public health and safety for Reservation communities. On the Reservation, one in four babies suffer from fetal alcohol exposure.[44] Fatal car crashes caused by drunk drivers occur at ten times the rate elsewhere in South Dakota. Tatewin Means estimates that "98% of the crimes committed here are committed under the influence of alcohol." In response to these problems, TVCDC's work aims to immerse youth in tribal traditions to regain their "sacredness as a people," including respect for women as the "focal point of our Nation's strength." To Means, this labor entails counteracting patterns by which "colonization created lateral oppression within our communities," where Native people are too often "the first to tear each other down."[45] In this deeply personal approach to countering violence, the quest for security is inseparable from the realm of livelihoods, another part of a holistic approach to equitable resilience.

REGENERATING LIVELIHOODS ACROSS PINE RIDGE

For TVCDC, building a regenerative community also requires counter-ing the despair that accompanies protracted unemployment and poverty. To date, TVCDC is estimated to have created between sixty-five and a hundred jobs across several different initiatives. This includes Thunder Valley Farms, a cooperative with five hundred chickens laying up to three hundred dozen eggs each week (figure 10.5). The farm also has a geother-mal greenhouse for growing produce sold at stores and farmers markets on and off the Reservation. Farm profits support daycare facilities in the community. TVCDC also emphasizes youth-focused training, education, and empowerment, and their development efforts incorporate work-force development in building trades. Possible future initiatives include a youth shelter and on-site retail facilities.[46]

To counter the long history of wealth extraction, TVCDC aims to influ-ence larger economic and employment practices on and off the Reserva-tion. Since the Pine Ridge Reservation currently imports nearly all of its food, Tatewin Means emphasizes that the goal of "food sovereignty" is about more than what happens on TVCDC's thirty-four acres. It is about transforming systems and mindsets. Instead of the present situation where cattle ranchers lease tribal land, and "100 percent" of what they raise "goes off Reservation," she imagines changing ranching and renegotiating the Lakota relationship with the US Department of Agriculture's commodity foods programs to build a network of growers and producers. Means envi-sions a Reservation-wide food system where "everything is locally grown, produced and stays here."[47] Whether building a more sustainable local food system or training young people for jobs in environmentally progres-sive construction, TVCDC's vision for improved livelihoods is linked to their environmental stewardship aims.

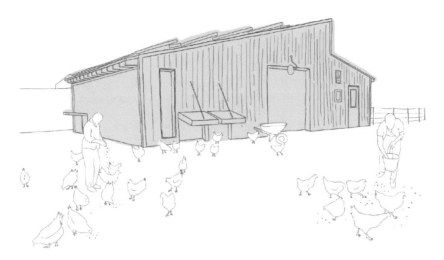

10.5 TVCDC has undertaken a range of agricultural and food sovereignty initiatives. *Source*: Mora Orensanz.

ENVIRONMENT: GREENING THUNDER VALLEY

TVCDC's environmentally focused efforts range from food and agriculture education to waste reduction and green building (figure 10.6). The group's farm is both a source of livelihood and a venue for environmental education through which staff teach about soil health and the value of local plants such as choke cherries, wild plums, and buffalo berries. Starting in 2020, TVCDC built the first large recycling collection point on the Reservation. The recycling initiative emerged from a TVCDC survey of community environmental priorities. DeCora Hawk, whose team led the survey, reports "a huge increase in usage" where previously "people were just landfilling everything." Pelkofsky notes that tribal leaders appreciated the waste diversion and recycling initiatives, given that their landfill, which was planned to accommodate twenty-five years' worth of trash, filled after less than five years. She goes on to express the hope that the success of the recycling program will prompt the Lakota Tribe to "move ahead with their own recycling initiative."[48]

Early in TVCDC's development project, they constructed a demonstration building using straw bales to create superinsulated walls, with the intention of using such practices throughout the project. After discovering that farmers in the region did not produce bales of the right size, however,

10.6 Thunder Valley's work includes environmental education. *Source:* Mora Orensanz.

they moved away from straw-bale construction, instead emphasizing solar-power generation and other methods of achieving their aspiration of "net zero energy use." Other environmental priorities for TVCDC's development include wastewater reuse, soil remediation, and landscape-based stormwater management. Executive Director Means remains committed to the project's dual aims: implementing "climate resilient, sustainable, green buildings" while also pursuing "consistent and persistent subsidy" to make the housing affordable to community members. Fortunately, TVCDC does not operate alone. Pine Ridge benefits from many other complementary efforts, including the tribally owned Lakota Solar Enterprises and the Red Cloud Renewable Energy Center's Wiconi project.[49]

Water and sanitation present another set of complex environmental challenges for Pine Ridge. TVCDC's houses are dependent on septic

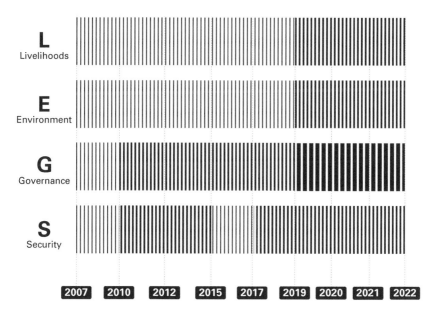

L Livelihoods

E Environment

G Governance

S Security

2007 2010 2012 2015 2017 2019 2020 2021 2022

10.7 TVCDC's first fifteen years have included both the development of a small community and a larger commitment to enhanced communal self-governance. *Source*: Smriti Bhaya. 2007: Thunder Valley CDC formed; 2010–2013: Project planning, thirty-four acres purchased; 2012: Oyate Omniciyé Oglala Lakota Plan completed; 2015: Youth suicide crisis on Pine Ridge; 2017: Legal action forces closure of liquor stores in Whiteclay; 2019: Construction of Thunder Valley homes begins, community center opens, Nick Tilsen succeeded by Tatewin Means; 2020: Playground phase I completed; 2021: Community outreach revisioning begins; 2022: Forty-eight-acre property in Whiteclay, NE, acquired for a holistic healing community.

systems, and the group would need an increased water allocation to expand their development. Given the long-standing acute shortage of housing on the Pine Ridge Reservation, water constraints limit the creation of new communities or enlargement of existing ones. Increased drought pressures add another layer of uncertainty.

CONCLUSION: ACHIEVEMENTS AND CHALLENGES

TVCDC has made great strides in its efforts at community engagement, cultural regeneration, and equitably resilient governance. Although their immediate footprint is small, TVCDC's impacts ripple outward, from the spiritual development of individuals to the edges of Pine Ridge and beyond.

TVCDC's leaders would readily admit that their efforts do not fully address the Reservation's significant challenges, rooted in centuries of marginalization and exploitation. The twenty-one houses and twelve apartments built in their recent development fulfill just a tiny fraction of the Reservation's unmet housing needs, estimated at 1,600 low-income families.[50]

More fundamentally, TVCDC's success involves shifting attitudes. Led by two successive visionary directors, each with deep community roots, the organization has emphasized hope and spiritual development. Rather than despairing over deficits or railing against detractors, Thunder Valley's principles and practices have brought equity to resilience through commitment to youth and to cultural regeneration in Pine Ridge. More than a small innovative enclave, Thunder Valley's invocation of "self-sovereignty" invites both introspection and outreach, inspiring an ever-widening governance coalition (figure 10.7).

It is both heartening and appropriate that the latest TVCDC venture is in Whiteclay, Nebraska. Instead of the scourge of parasitic liquor stores, TVCDC seeks to "change the narrative and the nature of the relationship of that community."[51] In January 2022, the organization announced the purchase of forty-eight acres in Whiteclay to develop a holistic healing community. Recognizing that many Pine Ridge residents were not yet ready for the other Thunder Valley homes and apartments, the organization plans to construct "transitional housing or permanent supportive housing," with much-needed services brought together "in one centralized place."[52] Tatewin Means views this as a "tremendous" opportunity to help "our relatives who are on the periphery, those that are forgotten or invisible or having a harder time accessing resources."[53] Ultimately, she hopes to build such a healing center in each of Pine Ridge's nine districts. The achievements and aspirations of the TVCDC, a multipronged approach to equitable resilience, have provided an inspirational start.

Case 11

COMMUNITY-GENERATED PUBLIC SPACES IN NAIROBI'S KIBERA SETTLEMENTS

OVERVIEW

Since 2006, the Kounkuey Design Initiative has worked with residents and community-based organizations (CBOs) across Nairobi's Kibera settlement to co-design and manage a network of multiuse "Productive Public Spaces." The first eleven of these projects include community centers, schools, sanitation facilities, as well as interventions promoting flood resilience and environmental restoration.

KDI's approach includes extensive and innovative engagement of community expertise to strengthen local governance. This account is based on more than twenty interviews with KDI staff, local partners, other NGO staff, and academic researchers, as well as field visits and reviews of academic literature and popular and social media in English and Kiswahili. Collectively, the eleven projects engage all four of our equitable resilience components: governance, environment, livelihoods, and security.

INTRODUCTION

COLONIALISM AND URBAN GROWTH
Residents of Kibera, one of Africa's largest informal settlements, face enormous challenges rooted in a history of colonialism and racial segregation. At

11.1 Opening ceremony at Kibera Public Space Project 2 in Nairobi. *Source:* Amos Wander, Kounkuey Design Initiative.

the end of the nineteenth century, Nairobi grew as a British colonial transportation nexus between Mombasa and Kisumu on the Kenya–Uganda Railway. The site now occupied by the fast-growing Kenyan capital attracted colonial settlers with its cool climate, sparse inhabitation, and ample water from the surrounding Nairobi and Mbagathi Rivers (figure 11.2). Plans put forth in 1898 and 1899 established Nairobi as "a railway town for Europeans with mixed European and Asian trading posts" serving the growing population of Indian immigrants and British colonial settlers. Plans allocated the desirable high ground to the north and west of the railway to Europeans, with separate settlements for Indians and a small area designated for Africans in the flood-prone area then known as the Eastlands.[1] Colonial-era regulations, anticipating the extreme apartheid policies of South Africa, required Africans to have a pass to leave "native reserves" at the city edges, with males older than fifteen years of age required to wear hated *kipande* cards around their necks or face imprisonment under vagrancy ordinances.[2]

The area that now includes Kibera was once a four-thousand-acre military exercise ground for the King's African Rifles (KAR), Britain's East

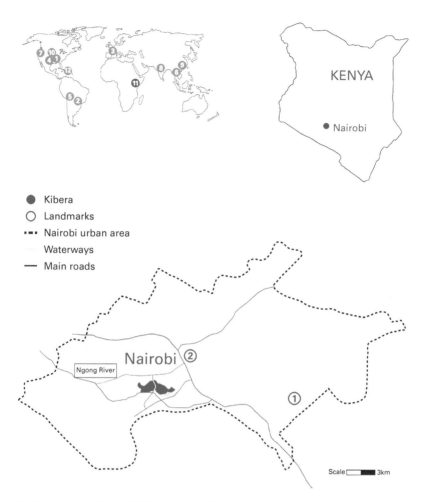

11.2 Location of the KDI Productive Public Space projects in Kibera, Nairobi, Kenya. *Source*: Mora Orensanz, with Elly Wanyoni. 1. Jomo Kenyatta International Airport; 2. Nairobi Central Business District.

African colonial forces, including many Sudanese soldiers brought to Kenya to guard the new railway. KAR veterans began settling in Kibera around 1904. These Sudanese soldiers were regarded as "detribalized Natives" who held no rights to land in Kenya's "Native reserves." Objecting to this, they argued for more than half a century that their service to the British Crown warranted full tenure rights in Kibera (whose name comes from *kibra*, the Sudanic Arabic word for "forest"). The colonial regime resisted tenure regularization and gradually took back the majority of the land, including the portion converted into the Royal Nairobi Golf Club in the 1930s. British policy vacillated between "malicious neglect in an attempt to force the Sudanese out by making Kibera unlivable" and futile efforts to relocate the Sudanese elsewhere, making periodic attempts to regulate the land while waiting until the remaining veterans "eventually died off." As colonial rule gave way to an independent Kenya in 1963, the new regime paid little heed to long-standing disputes with Sudanese settlers, now considered to be non-Kenyans, despite attempts by some to recast themselves as "Nubian" Kenyans. By the time of independence, Kibera had become home to people from a wide variety of ethnicities. No matter the ethnicity of its residents, central governments—whether British colonial or independent Kenya—consistently treated Kibera as a problem.[3]

As Nairobi developed, first as a colonial outpost and then as Kenya's national capital, the early patterns of privilege and precarity persisted, with lower-income residents most vulnerable to lowland flooding. Although the city was developed first for the British and their South Asian workforce, urbanizing Africans became a majority before independence, even as they struggled to find secure shelter. Six decades later, following accelerating rural-to-urban migration, Nairobi has more than 130 informal settlements, home to more than 60 percent of the city's population. Surviving largely through the informal economy, these people constitute the capital's "marginal majority."[4]

KIBERA TODAY

Today's Kibera settlement, occupying about an eighth of the original military ground, has inherited the consequence of more than a century of discriminatory land use and misuse. Kibera, four miles south of the city center, is riddled with flood-prone waterways and starkly bound by a rail line, the Ngong River, the still-extant Royal Nairobi Golf Club, and the reservoir behind the Nairobi Dam—a planned recreational area now fouled by pollution. The eighteen contiguous, densely inhabited informal settlements that comprise Kibera house at least a quarter of a million of the city's most disadvantaged residents. Kibera is not only the "largest slum in Kenya" but may also be among the most populous in Africa.[5] Although the land underlying Kibera is government owned, the settlement lacks adequate government services.

An estimated 15 percent of residents are descended from the settlement's Nubian–Kenyan founders. While most shack owners are Kikuyu (Kenya's majority ethnic group), informal renters come from many other groups, including Luo, Luhya, and Kamba. Ethnic and political tensions, threaded through with socioeconomic inequality, often roil Kibera.[6] In 2017, after more than a decade of legal struggle, the Kenyan government issued a land title to the Nubian community trust encompassing 288 acres—more than half of Kibera. In offering this more secure form of tenure, the government promised to make the land "a model city" by investing in new development. Thus far, this promise has remained difficult for the Nubian community to operationalize, prompting renewed tensions with other groups.[7]

Meanwhile, however, Kibera's residents continue to suffer from inadequate infrastructure and services. In 2020, a UN Habitat team reported that residents had unreliable access to piped water and that "70% of waste collection sites" were "in dilapidated conditions and under no one's management."[8] Each latrine typically serves hundreds of people, and when they fill, "they drain directly into the watercourses that run through the settlement, or are emptied into them,"[9] exacerbating flooding and public health threats.

The UN Habitat team also stressed that Kibera faces an "acute shortage of public space."[10] Kibera's estimated 250,000 residents live in an area of about five hundred acres—just over half the size of New York's Central Park. Less than 2 percent of the territory is dedicated to public space. Importantly, more than one third of that meager allocation of public space has been created or upgraded through the Kibera Public Space Project (KPSP), facilitated by KDI.[11]

KPSP AND EQUITABLE RESILIENCE

At first glance, a modest increase in public space may seem insignificant, given the scale of socioeconomic and environmental hazards facing Kibera's residents. Such dismissal underestimates the ongoing achievements of the community and KDI and misreads the value of enhanced public space. KDI's multisite KPSP matters because it provides mechanisms for Kibera's various publics to reclaim territory. In equitable resilience terms, KDI has used a shared governance approach to launch a network of projects led by CBOs. In turn, many of these projects provide ongoing sources of livelihood, instill new forms of security in areas once prone to crime, and make considerable inroads in mitigating environmental hazards.

KDI's commitment to shared governance manifests through participatory site selection, design, planning, financing, and management. This methodology advances stability, solidarity, and recognition that are at the heart of improved security. Today, there are more than six hundred active CBOs working in Kibera,[12] but KDI's KPSP stands out for its impact, longevity, and true community partnerships.

The design–build process developed by KDI is fundamentally about designing and building trust. Some aspects of the process were present in KDI's earliest Kibera ventures, while other strategies emerged over time through learning from early projects.[13] In its earliest years, KDI proposed projects to communities, but more recently the impetus has come from Kibera CBOs. These CBOs propose sites, program them, identify hazards and needs, co-design interventions, help construct them, and then control, operate, and maintain the sites.

KPSP's first eleven Productive Public Space (PPS) projects (figure 11.3) came about through similar processes although they vary in their programs and partners. The PPS projects have created a range of public spaces, including playgrounds, community meeting spaces, and new bridges (figure 11.4). In addition to construction jobs, the projects enhance livelihood opportunities by accommodating a range of ventures, including markets, small business kiosks, savings and loan programs, childcare facilities, schools, health clinics, women's baking and craft cooperatives, a composting business, a motorcycle taxi service, and access to wireless internet. The projects include environmental enhancements, including flood mitigation and erosion control, planting, community toilets, water provision, washing and laundry spaces, and outdoor lighting. Project partners also range widely, from new CBOs formed explicitly through the KPSP process to manage new spaces (e.g., New Nairobi Dam Community [PPS 1 and 6])[14] to groups with long track records in the area (e.g., Slumcare [PPS 4]). PPS projects deliver benefits across the four dimensions of equitable resilience: livelihoods, environment, governance, and security. The projects that address

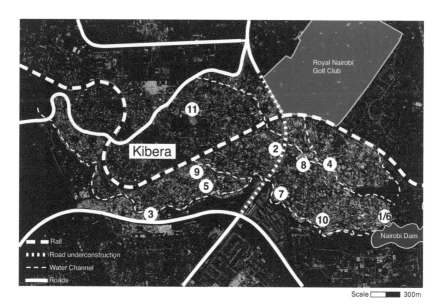

11.3 Location of KDI's first eleven Productive Public Space projects. 1. PPS 1; 2. PPS 2; 3. PPS 3; 4. PPS 4; 5. PPS 5; 6. PPS 6; 7. PPS 7; 8. PPS 8; 9. PPS 9; 10. PPS 10; 11. PPS 11. *Source*: Smriti Bhaya.

mounting flood vulnerability are especially relevant to the pursuit of equitable resilience, characterized by what KDI terms "Community Responsive Adaptation to Flooding."[15]

GOVERNANCE: COMMUNITY DRIVEN, NOT JUST COMMUNITY SERVING

KDI's robust community engagement process reveals how resident-led upgrading can be sustainable and effective. Every project involves either working with an existing community group or forming a group to co-design the project as well as to maintain and operate new facilities after completion. Community groups and residents help to design each project through an iterative series of workshops. At the same time, residents learn management skills and develop programs and businesses to bring life to the sites.

Self-governance remains the cornerstone of KDI's practice, expressed through all eleven of the public-space projects. When KDI launched PPS 1 and 2 in 2006, the initial process included the creation of a new

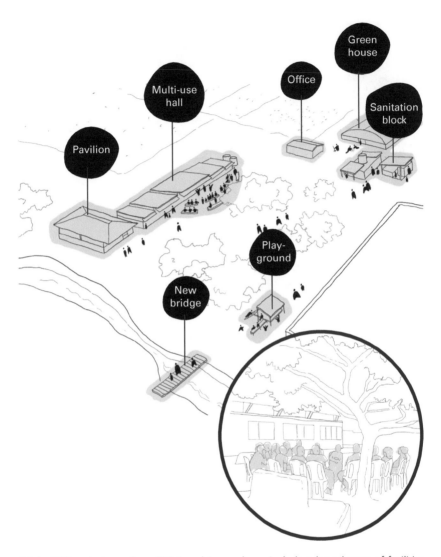

11.4 PPS projects, such as PPS 1 and 6 seen here, include a broad array of facilities linked to the priorities of local CBOs. *Source*: Mora Orensanz.

government-recognized CBO. George Nengo, the chairman of the New Nairobi Dam Community (NNDC) group that was created to facilitate PPS 1, has nothing but praise for how KDI "brought us together." He is adamant that all key ideas came from the community and are focused on advancing community benefits not KDI's interests.[16] Jacqueline Nduku, NNDC's treasurer and the site's waste-management project manager, agrees and observes that the CBO gained even more community buy-in after 2017 once various on-site businesses became profitable.[17]

As KDI's approach matured, so did their commitment to shared governance. In more recent projects, KDI begins by seeking outside investment, including grants. With resources secured, they advertise requests for proposals, describing the funding and how it can be used, inviting CBOs with project and site ideas to apply. Independent CBOs, not just KDI, identify available sites with willing partners. KDI reviews applications, conducts interviews, and selects partners based on transparent evaluation criteria. To guard against running afoul of government plans in project areas, KDI reviews proposals with authorities, in effect seeking tacit approval. KDI then finalizes the selection of CBO partners and signs a Memorandum of Understanding with the "structure owner"—the person or entity holding informal claims to the site. Although the government formally owns most Kibera land, the transparent MOU process aims to ensure buy-in from other empowered parties with unofficial authority over sites. Next, KDI signs another MOU with the CBO partner, outlining the percentage of construction costs the group will contribute, the number of days per week that the CBO's members will volunteer, and KDI's contributions to the process. Next, KDI and the CBO jointly convene community workshops for project visioning, programming, and design. To ensure mutual buy-in, the workshops include area chiefs, elders, youth, men, and women as well as the structure owners and CBO members.

The shared governance process also contributes to livelihood enhancement, since anyone who attends the workshops is eligible for paid work during the building phase. During the programming phase, the KDI–CBO partnership prioritizes the livelihood, environment, and security goals for the project. KDI sources project materials from local businesses, hires local skilled craftspeople, chooses laborers from among workshop participants, and negotiates with either the city utilities or informal cartels for provision of water and electricity. The CBO and KDI then build the project together, holding one another accountable to their MOU agreements. Once construction is complete, the CBO runs and maintains the site, with KDI providing financial and other assistance. One year after completion or as soon as feasible, the CBO takes over full operation and maintenance, and KDI shifts to a purely advisory role. Using this well-honed process, each of the eleven projects has commenced with clear expectations related to self-governance and co-design, tapping community expertise and encouraging robust community ownership of completed projects.

The self-governance aspect of KDI's process has evolved based on learning from early projects in four key ways: (1) partnering with existing CBOs instead of creating new CBOs, (2) using requests for proposals (RFPs) to select CBOs and sites, (3) informally vetting projects with government officials, and (4) formalizing the MOU process.

Partnering with existing CBOs selected through RFPs puts the power of project definition and ownership in the hands of impacted communities from the start. Instead of approaching the community with a site and project idea and building a CBO around it, KDI now invites community groups that already have internal organizational structures and ideas to bring forward a proposal. The old method risked picking historically contested sites, poor locations, or places already targeted for government intervention, as painfully happened when the community center, school, and health center built for PPS 3 were destroyed after the site's owner encouraged the government to move forward on a sewage infrastructure project. Starting with PPS 5, CBO applicants needed to suggest a site, based on their local knowledge of relevant factors, including the willingness of structure owners to participate. Wilson Sageka, KDI's Nairobi office manager from 2011 to 2018, observes: "[we now] let the CBO or the community group" take the lead "because they know that place probably better than we do."[18]

KDI's governance model also evolved to incorporate informal vetting with local officials as well as more formalized MOUs with both the CBO and structure owner. This provides clarity about the division of responsibilities, expected project deliverables, benefits to the structure owner, and timetables for site handover, ensuring mutual accountability.

PUBLIC SPACE AS MEANS TO PROMOTE SELF-EMPOWERMENT

At base, KDI's KPSP seeks to build community capacity to solve problems. The KPSP process does this by gradually transitioning site operation and maintenance to community partners while still offering support as needed. Prisca Okila, KDI's Community Coordinator, observes that the earliest projects, where KDI helped form project partners, did so for a similar reason. KDI CEO Chelina Odbert and her colleagues wanted to help new organizations gain government certification so that they could apply for government and philanthropic funds. For subsequent partnerships with existing CBOs, KDI exercises careful due diligence. As part of the RFP response, CBOs complete a detailed questionnaire about their organizational history, mission, and vision. KDI staff researches potential CBO partners' projects and financial resources and visits CBOs for interviews because, Okila stresses, "we trust what we see on the ground." Due diligence also helps KDI determine if "the space has issues," including contestation over site control. KDI chooses groups that exhibit transparency and internal consensus because "we need a group that is very well organized."[19]

To build capacity among potential partners, KDI has developed leadership training programs. In recent years, community director Regina Opondo points out that KDI "rolled out a program called the Network Challenge, which is a sort of a peer learning exercise, where we get the groups to learn from each other's best practice." They have applied this peer-to-peer learning model to business operations, record keeping and financial management, and membership recruitment and engagement.[20] All of the CBOs start their own business ventures in order to pay for site maintenance and operation. Enterprises range from PPS 4's plastic recycling, childcare programs, and taxi start-up to the independently run schools on the sites of PPS 8 and 9 to the women's cooperatives and the wireless mesh networks at several sites. According to KDI's Okila, a lifelong Kibera resident, KDI expects its partners to follow their own leaders, visions, and constitutions, pursuing their activities independently while still permitting some oversight from KDI.[21]

COMMUNITY CONTROL AND STEWARDSHIP

Every part of KDI's process is designed to be community driven, with the exception of financing. KDI asks partner CBOs to contribute a nominal amount, usually about 5 percent of construction costs, to ensure group investment. CBO partners also must agree to contribute one day a week of volunteer labor during construction. Every person we interviewed lauded KDI for the deep community engagement and participation in their process. Landscape architect Franklin Kirimi estimates that between fifty and two hundred community members have been involved in each site design and that the core management team is "always about between twenty and forty people."[22] Okila emphasizes that KDI aims to engage the full range of community members: "We engage children, engage women, men, local administration, and also village elders." In workshops over the course of up to six months, "We co-design together with them because they are the one who knows that space, they are the one who knows their needs." KDI encourages its community partners to "give us direction on how they want that project to look like, what is the specific things they want in that area," including everything from basic programming and site planning to material and color choices. Once construction begins, KDI and the CBO identify skilled people from the community.

Kirimi recognizes that relying on technical skills is insufficient: "I could build a working infrastructure, a working design. But . . . if I don't involve the community, there is no ownership, so there is no maintenance." Moreover, successful engagement means using "materials that the community knows [it] can access" and using "technology that the community has learned how to use." By doing this, "they can continue to build and repair these buildings," and "as time goes by, they can build their own interventions." As Kirimi says, if KDI's process and designs are successful, community partners "can easily adapt to coming changes where KDI may or may not be around."[23]

KDI's approach, rooted in community control and self-stewardship, distinguishes KDI from most other NGOs operating in Kibera. Wilson Sageka, a longtime Kibera resident, outlines the more typical experience in NGO-saturated Kibera: "You'd wake up in the morning, [and] there's a project, or a program being run very next to your house, and you don't know about it until it starts. And probably when you ask about it, you'll be told, 'Oh, this project has been brought by that NGO;' the big word there is like, *brought*." KDI operates differently. Sageka continues: "[When a] community participates in choosing a project . . . this project goes into addressing their most felt needs" and they "finally get to own the project and run it on their own."[24] Water and sanitation scholar/practitioner Sarah Lebu, who grew up in Nairobi and worked with NGOs in the city's informal settlements, concurs that most NGOs come with fixed, imported concepts: "'Oh, we've done our lit review. And we know that this slum committee has this problem. This is a framework that has worked elsewhere in the world; it will work here: this is the money [so] do it.'" They miss community priorities: "You're coming and saying, 'We think your problem is water and sanitation,' but [they're thinking] 'I can't put food on my table.' So, 'I'm prioritizing food, you're prioritizing water and sanitation.'" By contrast, KDI prioritizes "the community's motivations" and keeps them "the center of the entire process."[25] Given this, once KDI transitions to an advisory role, it is possible to leave the project with community control. KDI's exit is predicated on ensuring that the community leadership operates with transparency, accountability, and financial stability; they cease to provide funding after a year but continue to monitor, evaluate, and support partners in other ways.[26]

PATHWAYS TO RECOGNITION

Since PPS projects do not typically seek or receive formal government approvals, it is more difficult for them to influence larger scales of governance. Projects are vulnerable to conflicting government priorities, as demonstrated by the demolition of facilities from PPS 3. On several other occasions, the government has intervened on portions of sites (including at PPS 5 and 10) for other infrastructure projects. Further displacements remain possible. That said, several interviewees maintain that the government would now be less likely to take land that hosts a treasured KDI project, such as a community center.[27] In the case of PPS 5, KDI's CBO partner successfully petitioned the government not to demolish the entire project. Such efforts demonstrate how PPS projects can contribute to informal solidarity and de facto security of tenure, even in cases without formal recognition.

LIVELIHOODS: SUPPORTING KIBERA RESIDENTS WITHIN AND BEYOND KPSP SITES

Irrespective of whether the government recognizes KDI-initiated public spaces, these spaces contribute substantially to the livelihoods of many Kibera residents. Job creation and new livelihood-supporting facilities

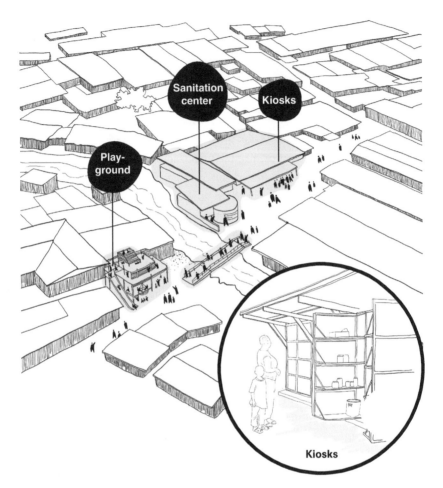

11.5 KDI's PPS projects offer both amenities and sites for local business development, as seen in PPS 2. *Source*: Mora Orensanz.

feature centrally on every KPSP site (figure 11.5). Such businesses serve a dual purpose. First, they provide CBOs income to maintain and operate the sites. Second, they generate personal income for both CBO members and people from the larger community. When asked, KDI leaders measure success less in terms of financial self-sufficiency than on how well the sites are governed. In most cases, the best-governed sites are also the ones with the best financial performance.

VIABLE LIVELIHOODS FOR RESIDENTS

PPS projects facilitate livelihoods in three crucial ways. First, KDI hires residents from the settlement as full-time community coordinators, office managers, and accountants, making them a key part of the organization's staff. Based on observations of their career trajectories, KDI's staff seem to fare better than staff of other local NGOs, where pay for Kibera residents is often substantially lower than other workers.[28]

KDI provides a second kind of livelihood support by hiring community residents for project construction work. Jack Clause, design director of KDI's Nairobi Office, says, "The vast majority of our labor force comes from the very close neighborhood." KDI's "community-based construction process" tracks who attends community design workshops to help identify potential workers and volunteers. Compensation for paid work is at prevailing local rates for day labor, which are quite modest. "Unskilled laborers" lacking construction experience receive 400 shillings (about $4) per day, while skilled tradespeople receive three to four times more. Some PPS interventions, including those with bridges, require engineering expertise hired from outside the community. Project construction only lasts a few weeks or months. As such, local employment practices can build local buy-in, teach new skills, and provide short-term cash infusions but do not create long-term livelihood opportunities.[29] One way that KDI counteracts the shortcomings of project-based hiring is by using and training people to work with locally sourced materials, such as compressed earth blocks, thereby supporting livelihoods in local supply chains.[30]

A third way that KDI is enhancing livelihoods is through the deployment of new mesh wireless internet networks proposed for PPS sites 4, 8, 10, and 11. Assisted by the Civic Data Design Lab at MIT and championed by Kenya's Tunapanda Institute, this "community-based internet and data hub" seeks to "shift the power in who designs, implements, owns, and maintains infrastructure in resource-poor communities and present new approaches by building local partnerships to co-design the network and its uses."[31] Some residents learn how to install, maintain, and use neighborhood Wi-Fi, while others use improved digital access for educational and professional purposes.

ACCOMMODATING LIVELIHOODS

PPS projects support livelihoods through several mechanisms, including community service provision (e.g., water and toilets), savings and loan programs (most notably in PPS 1 and 7), rental kiosks for small businesses (PPS 2, 5, 7, 10, and 11), and community halls, which are used for markets, meetings, and production. Regina Opondo explains the process of generating new enterprises, saying that CBOs set up a "committee that runs the local business." The savings and loan programs operate as a form of "table banking" through which groups of people make small contributions at meetings, which are banked and made available with very low interest to members who need loans for businesses or short-term emergency support.[32]

COPING WITH ENVIRONMENTAL RISKS IN KIBERA

Flooding remains a major environmental risk across Kibera, displacing residents and threatening their health. Many KDI projects are located along Kibera's rivers and tributaries, where some of the settlements' poorest people are also concentrated. When KDI's public-space projects address flooding, they do so by making those particular spaces "flood adaptable"[33] rather than by mitigating flooding in surrounding areas. Some projects incorporate rain gardens, retaining walls for erosion control, and sanitation blocks that divert waste from the rivers (figure 11.6). Even so, KPSP efforts cover a tiny proportion of Kibera and do not provide a settlement-wide solutions to Kibera's flooding, erosion, or water pollution challenges.

BUILDING ENVIRONMENTAL LITERACY

Starting with PPS 10, KDI markedly increased its focus on providing environmental information through its Community-Responsive Adaptation to Flooding projects. As Kirimi details, these information-focused projects engage residents to map flood risks and "create more awareness" about the need to "stop building our houses where the flooding occurs." This participatory mapping, combining community insight with scientific data, also helps discern where best to site community facilities to avoid floodwaters. New maps reveal spaces that may be unusable for a couple of weeks each year due to flooding but are otherwise available for compatible uses.[34] KDI

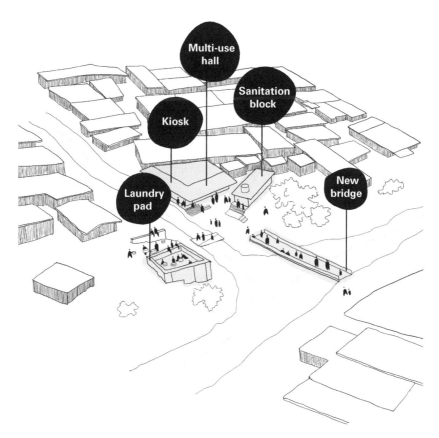

11.6   PPS projects offer new sanitation facilities and localized protection from flooding, including upgraded bridges, as shown in PPS 5. *Source*: Mora Orensanz.

has also co-developed early warning systems with Weather Mtaani and other groups that provide signposts to alert residents to risks such as torrential rain. At PPS 10, for instance, a post is mounted near the river to display a green or red flag, depending on projected flood conditions.[35] At PPS 1 and 6, environmental awareness is linked to livelihood efforts in enterprises that harvest invasive, waterway-clogging hyacinth to provide material for a weaving business.[36]

IMPROVING ECOLOGICAL FUNCTION

Even though KPSP efforts cannot prevent flooding, they demonstrate techniques for reducing adverse health consequences, including deploying sanitation blocks to divert waste and reduce pollution in nearby waterways. While their scale is limited, KDI's use of green infrastructure (e.g., rain gardens) to manage infiltration demonstrates promising strategies. Greater progress in addressing the dire environmental conditions in and around Kibera would require more systemic government infrastructure provision and sensitivity to existing ecological assets.

SECURING KIBERA: PROGRESS AND LIMITATIONS

Compared to their attention to environmental, livelihoods, and self-governance aims, KDI's Kibera projects have been less focused on improving security from displacement and violence. Nonetheless, KPSP efforts have made progress in countering both types of threats. Converting once-hazardous sites into well-loved community spaces and much-needed facilities does not directly deter eviction, although it may help. Physical safety, although not an overarching aim of KDI, has been a project-specific goal when brought forward by partner CBOs and residents. Various KPSP sites have enhanced safety through such strategies as improving lighting, increasing activity in the public realm, and designing safer structures.

INDIVIDUAL SAFETY THROUGH COMMUNITY ENGAGEMENT

Increased engagement through KDI's process can build trust and community investment in the safety of public space. As former office manager Wilson Sageka notes, it is not just that having more activated and attractive spaces deters crime; collaborative design processes also ensure that these "spaces now have the backing of the community," making it "very unlikely that people get tempted to come steal or commit crime in these spaces." Active stewardship drives security. Our interviews suggest that residents do not judge any of the sites to be completely safe, although all have seen improvements. In PPS 4, Slumcare and the women's group and youth organization it nurtured gave special attention to security issues (figure 11.7). Prisca Okila notes that chiefs—backed by residents and the CBOs—have now located some policemen in that area, reducing crime. Still, a leader of the Usalama Bridge Youth Reform Group is not convinced that the "reform" has gone far enough: PPS 4 still suffers from vandalism,

11.7 Residents credit PPS 4 for improving security in this part of Kibera. *Source*: Mora Orensanz.

including tampering with the pipes leading to their tank and theft of security lights.[37] Asha Abdi, chairlady of the Ndovu Development Group, which offered baby care, has similarly mixed feelings about the success of the project. She observes that the group is now just a dozen elderly women, down from a peak of sixty members. Their once-successful daycare facility, she laments, became financially unsustainable, in part because of muggings and thefts. Nonetheless, she praises the overall achievement at PPS 4, viewing "the benefits as clear as day and night," from a "filthy dirty place" of "drug addicts and petty crimes" into an active place where "the community can come and enjoy constant activity."[38]

Ultimately, the limitations on KDI's ongoing work in Kibera inhere in the larger problems of inadequate land-use control, tenure insecurity, and extreme crowding. Any effort to improve conditions risks displacement and invites contestation. It is not just that the government technically owns the land; it is also that many non-state actors, including informal landlords and cartels, also make claims. Despite careful attention to identifying and working with the so-called structure owners, KDI and their CBO partners still operate in contested landscapes of self-appointed stakeholders.[39]

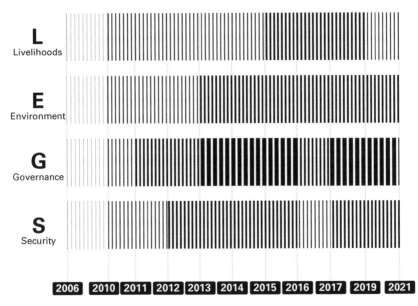

11.8 Through constructing a network of Productive Public Spaces, KDI markedly enhances local capacities for self-governance while also improving environmental conditions, livelihoods, and security. *Source:* Smriti Bhaya. 2006: KDI established and KPSP launched; 2010: PPS 1 opens, community hub; 2011: PPS 2, sanitation center and flood management; 2012: PPS 3, community center, school, clinic; 2013: PPS 4, incubator spaces, sanitation; 2014: PPS 6, sanitation, flood prevention; 2015: PPS 5 and 7, business kiosks; 2016: New Road Link #12 project launched, cuts through Kibera; 2017: Kibera's Nubians secure land rights; 2019: PPS 10, business kiosks, flood adaption; 2021: PPS 11, business kiosks, flood adaption, Wi-Fi.

KDI has closely monitored and documented the government's latest new intervention: a major four-lane north–south highway (known as "Missing Link 12") that slashes a two-hundred-foot-wide swath through Kibera, bisecting the newly won Nubian land claim and breaking the promise of secure tenure. Clearance began in July 2018, destroying at least five schools that served two thousand children, as well as several churches, clinics, and hundreds of homes, leaving tens of thousands homeless. Some in the community decried the mass evictions and worried over escalating rents near the new road. Others welcomed the improved transportation access and associated opportunities for youth employment. KDI's PPS 7 is immediately adjacent to the roadway. The project obtained formal city approval and offers sanitation, laundry facilities, and business kiosks. The neighboring infrastructure investment has yielded opportunities for greater integration and enhanced lighting for pedestrians but may also create new pressure to develop higher-end land uses.[40]

CONCLUSION: EXPANDING FROM VITAL PUBLIC SPACES

KDI's first eleven KPSP projects in Kibera have made great strides in improving residents' self-efficacy, enabling ongoing management of the public spaces, initiatives, and facilities (figure 11.8). One analysis of KPSP's contributions to "transformative adaptation" observed that "KDI acts as a networking link between stakeholders, helping government and communities to engage with one another in the hope to improve relationships, share ideas and plans, and obtain support from both parties, towards more sustainable and equitable solutions."[41] Despite these considerable successes, challenges remain. Given the small scale of KDI's projects and the extended time frame needed to build local trust, KPSP interventions cannot meaningfully address many of the challenges facing Kibera residents, including the sometimes-conflicting needs of land tenure security and government infrastructure provision. There is, therefore, an ongoing need to build upon the insights from KDI's deep engagement with local residents and organizations to forge linkages with the municipal and national institutions (such as Nairobi city agencies and Kenya Railways) that are responsible for major infrastructure projects and investments affecting Kibera. As a model governance process, the KPSP framework charts a path to equitable resilience with a key insight: instead of bringing projects to a community, KDI brings together communities to dream and create projects together.

Case 12

CAÑO MARTÍN PEÑA: BUILDING CHANNELS FOR EQUITABLE GOVERNANCE IN SAN JUAN

OVERVIEW

The multidecade effort to transform el Caño Martín Peña, a water channel in San Juan, Puerto Rico, surrounded by informal settlements, stands as a multifaceted achievement in pursuit of equitable resilience. Local community organizations gained legal power and forged vital partnerships with government agencies to ensure that drainage and flood control interventions did not displace residents from their centrally located neighborhoods. Starting in 2002, these efforts linked eight neighborhoods, an innovative community land trust (CLT), and a government corporation known as ENLACE. Together, these groups have ensured that improving environmental conditions and reducing flood risk would not also wash away the thousands of low-income residents. The ongoing saga of el Caño Martín Peña reveals the complexity of constructing more equitable forms of community governance, coupled with enhanced environmental protection, improved security, and more modest progress in improving the livelihoods of disadvantaged residents. This account is based on visits in San Juan, interviews with a range of stakeholders, observers, and residents, and review of both academic and popular accounts.

12.1 Aerial view of constricted Caño Martín Peña, with wealthier parts of San Juan behind. *Source*: El Fideicomiso de la Tierra.

INTRODUCTION: CHANNELING DEVELOPMENT

El Caño Martín Peña (el Caño) is a tidal estuary that runs nearly four miles from the mangroves of San Juan Bay (the location of the city's main port) to San José Lagoon (adjacent to the airport; figure 12.2). In the early twentieth century, water and transport flowed freely through this passage. The channel averaged two hundred feet in width and was six to eight feet deep. Beginning in the 1920s and 1930s, the areas on either side of el Caño saw mass in-migration driven by its proximity to San Juan's central economic districts. In 1927, out of a misguided attempt to control malaria-transmitting mosquitoes, Puerto Rico's legislature authorized the sale of these areas with the stipulation that they be drained and filled, leading to a surge in both formal and informal development. Rural families converged on San Juan, seeking alternatives to agricultural livelihoods in an industrializing economy. Two deadly hurricanes—San Felipe II (1928) and San Cipriano II (1932)—decimated rural coffee and sugar production and accelerated urban migration. The el Caño settlements gave residents access to opportunities in the city's downtown, but inhabitants faced

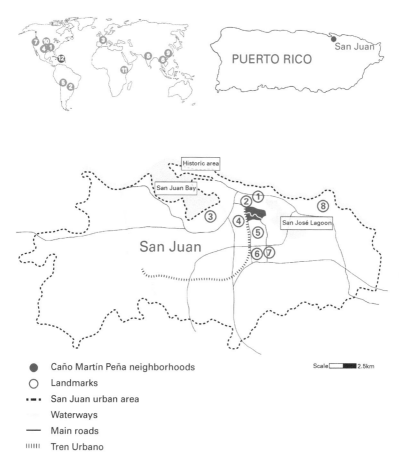

12.2 El Caño Martín Peña runs between San Juan Bay and San José Lagoon near the center of San Juan, Puerto Rico. *Source*: Mora Orensanz. 1. Sagrado Corazón University; 2. Tren Urban Station Sagrado Corazón; 3. Port; 4. Plazas Las Américas shopping mall, Hato Rey; 5. Polytechnic University of Puerto Rico; 6. Hospital; 7. University of Puerto Rico; 8. Airport.

many challenges, as the area remained disconnected from the sanitary sewer system and other critical infrastructure. Despite such limitations, by the 1970s, the prospects of affordable centrality had drawn an estimated 86,000 residents to the burgeoning communities on either side of el Caño. Uncontrolled filling and settlement, along with dumping by outsiders, led to deterioration of the estuarine ecosystem.[1] By the early twenty-first century, some sections of the once-broad waterway were nearly completely obstructed.[2] From the mid-1970s onward, the population of the barrios along el Caño declined due to a succession of redevelopment efforts and deteriorating conditions. A short film released in late 2012, *Mala Agua*, viscerally documents both the dangers of polluted water and the determination of community members to improve conditions.[3]

Climate change further threatens el Caño's ecosystem and its human residents—as previewed by two Category 5 hurricanes, Irma and Maria, which struck in 2017. In addition to more intense tropical storms, torrential rainfall will bring more flooding, particularly near obstructed stormwater drains. Meanwhile, projections of sea level rise predict unprecedented flooding and increased salinity in the estuary, threatening many species.[4] For residents of the surrounding neighborhoods, coping with el Caño's deteriorating conditions has required attention not only to environmental concerns but also to the complexities of politics, urban design, and social relations. Viewing the situation in el Caño through the lens of urban political ecology, however, means understanding these communities not as permanently vulnerable but rather as *vulnerabilized* by unjust processes and practices that residents now seek to alter.[5]

The ongoing transformation of el Caño into an exemplar of equitable resilience reflects a dramatic change in planning practice. Following decades of top-down government interventions that assumed formal redevelopment programs and modern infrastructure would improve conditions for all, recent efforts have taken a radically different form. Residents have galvanized local resistance to implement new forms of productive partnerships in the face of threats from government and private-sector interests bent on removing el Caño's low-income communities. The ongoing struggle has intertwined these two competing channels of development and resistance.

The twenty-first-century transformation of el Caño is an extension of a half century of organized projects, handled with increasing sensitivity to social and ecological needs. In 1966, as part of the US government's "recommended action," the San Juan Metropolitan Area Community Renewal Program proposed "total clearance" of nearly all of the barrios along on the northwest, northeast, and southeast portions of el Caño, and the Model Neighborhood from 1967 prioritized clearance on the western

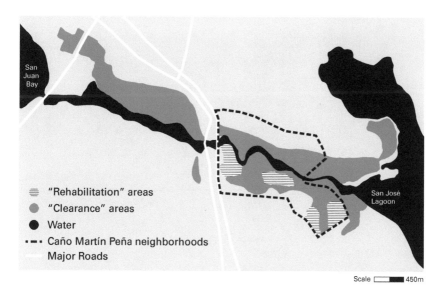

12.3 A 1966 Community Renewal Plan called for replacing informal settlements along el Caño Martín Peña. Dark gray areas were targeted for "clearance," while small hatched areas on the southeast of the channel were proposed for "rehabilitation." The subsequent Model Neighborhood Plan (1967) prioritized clearing only the western half. The dotted line indicates the approximate boundaries of the post-2000 interventions. *Source*: Mora Orensanz, based on the Community Renewal Program for San Juan Metropolitan Area.

side (figure 12.3).[6] Implementation of such plans led to mass eviction, often relocating residents to new multistory public housing with very different living conditions and reduced access to employment. Channel-side redevelopment continued into the 1980s with the clearing of the Tokío neighborhood in Hato Rey by eminent domain. With low-income residents removed, engineers widened the entire western section of el Caño. Planners replaced the cleared communities with housing for middle- and upper-income families, parks, government facilities, commercial areas, a sports complex, and the José Miguel Agrelo Puerto Rico Coliseum.

By the 1990s, federal officials declared that drainage conditions along the eastern end of the channel had also become intolerable. Since Puerto Rico is an unincorporated territory of the US, responsibility for channel dredging fell to the US Army Corps of Engineers. USACE officials viewed the problem as a technical challenge: how to widen and deepen the channel for more efficient drainage. To the tens of thousands of people

12.4 Eight neighborhoods flank the channel's eastern end, including flood-prone informal settlements. *Source*: Mora Orensanz. 1. Barrio Obrero; 2. Barrio Obrero Marina; 3. Buena Vista Santurce; 4. Parada 27; 5. Las Monjas; 6. Buena Vista Hato Rey; 7. Israel-Bitumul; 8. Cantera. A. Community center and Tren Urbano station access; B. G-8 offices.

who lived in the eight neighborhoods near el Caño (figure 12.4), however, the channel-widening proposal triggered renewed concerns about displacement.

Residents from one of those neighborhoods, Cantera, took the early lead in seeking investment without displacement. They successfully lobbied for action following the devastation caused by Hurricane Hugo in 1989. With support from experts at the University of Puerto Rico, they established the Cantera Neighborhood Council, which was instrumental in developing the 1994 Comprehensive Plan for the Development of the Cantera Peninsula. The plan represented the first time that government agencies, the private sector, and low-income residents worked together on a planning effort in Puerto Rico.[7] Inspired by the Cantera community and still outraged by the earlier removals in el Caño's western areas, el Caño's residents undertook broader actions.

Residents knew that mitigating hazards had long been a pretext for mass evictions. A project proposed in 2000, called The Gardens of el Caño, triggered new fears. As community leader Juan Caraballo Pagán recalls, this proposal, which would have erased existing communities to build hotels and high-end housing, underscored for el Caño residents the value of their land to outside developers. Mindful of mounting development pressure, residents sought partnerships to protect the community from further deterioration and displacement.

Between 2002 and 2004, residents of the channel's eastern settlements mobilized, creating the Group of Eight Communities Adjacent to the Caño Martín Peña, Inc., known as the G-8.[8] With other partners and allies, the G-8 drafted a plan for the district, based on hundreds of community meetings and outreach activities.[9] To give this plan official legitimacy, the G-8 championed Law 489, passed by the Government of Puerto Rico in September 2004. Law 489 enabled the Comprehensive Development and Land Use Plan for the Special Planning District of Caño Martín Peña, known as the District's Plan. This, in turn, created both a community land trust, the Fideicomiso de la Tierra (Fideicomiso), and a government organization to implement the plan, the ENLACE Project Corporation, chartered to operate for twenty-five years and then be dissolved.

With ENLACE led for fifteen years by executive director Lyvia Rodríguez Del Valle, the joint work with the Fideicomiso and the G-8 earned national and international acclaim, winning the 2009 American Planning Association's Paul Davidoff Award for Social Change and Diversity, the 2011 Environmental Protection Agency's National Achievements in Environmental Justice Award, and a United Nations World Habitat Award in 2016.[10] ENLACE's structure (requiring that residents and community leaders occupy six out of thirteen board positions) guaranteed significant roles for community decision making.[11] Acting on that community focus, the Fideicomiso aimed to rehouse thousands of residents from the areas close to the channel into nearby neighborhoods rather than relocate them to distant sites (figure 12.5). In this, it extended commitments first forged in Cantera. Alejandro Cotté Morales, who had spent eight years leading community development for the Cantera Project, served as Director of Citizen Participation and Social Development for the Fideicomiso and ENLACE.

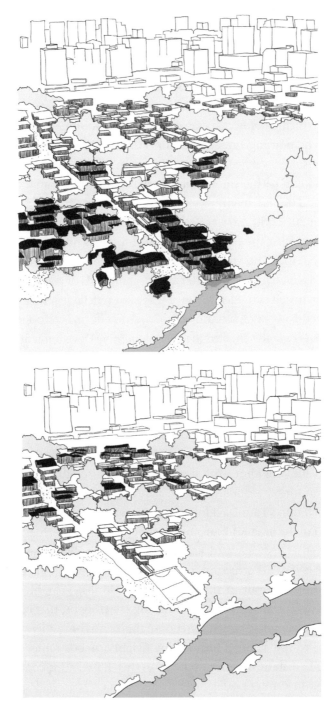

12.5 Channel-side redevelopment proposes to remove precarious housing along the channel, introduce new recreational spaces, and rehouse displaced residents in nearby neighborhoods. *Source*: Mora Orensanz.

EL CAÑO AND EQUITABLE RESILIENCE

The ongoing work to protect, preserve, and enhance the communities surrounding el Caño gets to the heart of equitable resilience. Too often, the adaptability or "resilience" of low-income people living in hazard-prone places becomes an excuse for not improving conditions. In such contexts, making resilience more *equitable* entails both solidarity and struggle. The el Caño effort copes with the environmental exigencies caused by climate change and rapid urbanization in ways that explicitly retain access to a desirable location, even after upgrading. Its aspirations—as well as the successes achieved so far—demonstrate progress in all four of our key dimensions: environmental protection through reduced pollution and flooding, enhanced livelihoods through proximity to jobs and amenities, security against displacement, and, undergirding it all, a sustained commitment to community-led governance. Even though US government agencies initiated the transformation with environmental engineering objectives, the work around el Caño entails far more than technical intervention imposed from outside. Rather, the efforts of ENLACE, G-8, and the Fideicomiso recognize that el Caño is a socio-technical and socio-ecological system. Although it began as an effort to make a waterway more fast moving and efficient, the work at el Caño has become a community process that is necessarily slow moving in order to gain trust. While infrastructure progress and funding ultimately depend in large part on the US Army Corps of Engineers, which has been exceedingly sluggish, much of the value of the project inheres in the longer term engineering a different kind of core army, a community of empowered residents bound in solidarity, skillfully wielding formal and informal power.

GOVERNANCE: NEW CHANNELS FOR EMPOWERED PARTNERSHIP

Governance in el Caño matters at multiple scales. At the broadest level, this place and its people face the challenges of all Puerto Ricans, subjected since 1493 to settler colonial dispossession. Since the island was seized by the US in 1898 after the Spanish–American War, Puerto Rico has held a complex and contradictory status. Its inhabitants have been citizens since 1917, but as an unincorporated territory rather than a state, citizenship has not come with full representation in the federal government. In recent years, the Commonwealth of Puerto Rico has endured extended economic hardship, including the imposition of fiscal austerity and delays in the allocation of federal funds—all while being buffeted by disasters linked to climate change. Taken together, this toxic combination yields what Danielle Rivera has termed "disaster colonialism." Any discussion of governance in el Caño must be seen in light of this longer history and the tensions over land policy and land ownership it has engendered. Across the US, the centrality of land value and property taxes as a source of revenue for municipalities makes retaining affordability on prime real estate challenging.[12] This tension is even more acute in Puerto Rico, given the island's glaring inequality and its history of financial instability.

As a component of equitable resilience, governance entails more than mere "participation." In Puerto Rico, the expectation of more than token participation grew stronger in 2001 when Law 1 passed under the administration of Governor Sila Calderón (who had previously seen the progress of citizen-led planning in Cantera as mayor of San Juan). Law 1 established the Office for Socioeconomic and Community Development, with the aim of ending poverty in hundreds of marginalized "special communities." It required that municipalities "modify their intervention approach," replacing the model of "a paternalist state" with "the active involvement" of residents who would "assume the direction of their own development process."[13] Close observers of el Caño concur that Law 1 "enabled the participatory approach that was used in the ENLACE Project."[14]

Between 2002 and 2004, community leaders estimate that "more than 700 participatory planning, action, and reflection activities" took place across el Caño neighborhoods. By 2007, once the Puerto Rico Planning Board had adopted the District's Plan, low-income residents could justifiably regard themselves as a vital part of the governance structure (figure 12.6). More than fifteen years later, participation in community governance remains robust. As ENLACE's Estrella Santiago Pérez states, "This project is not about voting once. It's continuous engagement." Lawyers and planners associated with ENLACE and the Fideicomiso estimate that there are "approximately 120 community leaders active within the G-8," most of them women, and nearly half of them under the age of twenty-five.[15] These young leaders are

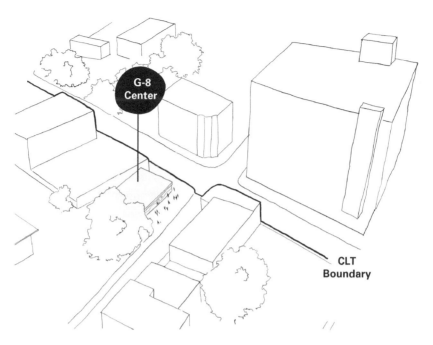

12.6 Community-based governance takes multiple forms in el Caño, anchored by the G-8 and the community land trust. *Source*: Mora Orensanz.

vital, especially since ENLACE itself is meant to be phased out after twenty-five years in 2029.

This deeply rooted commitment to community control, linked to legal authority granted by government institutions, extends to the governance structure of the Caño CLT. The Fideicomiso features an eleven-member board that includes four members who live on CLT-owned land and two additional el Caño residents chosen by the G-8. Two other board members are nonresidents chosen for their specialized skills. The remaining three board members represent government organizations: one from ENLACE, one from the San Juan Municipality selected by the mayor, and one chosen by Puerto Rico's governor.[16] Line Algoed, who served on the World Habitat evaluation committee that gave its prize to el Caño, argues that, more than the technicalities of the CLT's collective land tenure model, it is "the social processes and networks that are built in order to have this collective tenure" that matter most.

These robust social networks were clearly visible in the community's response to Hurricane Maria.[17] Lyvia Rodriguez Del Valle, then ENLACE's executive director, testified to the US House Subcommittee on Environment in November 2017 about el Caño's rapid mobilization: "During the first month after María, we recruited close to 500 external volunteers, removed 122 truckloads of vegetative material, cleaned the flooded homes of the elderly and sick, made referrals of families in need to community health centers, provided over 800 [roof] tarps, assisted over 400 families with FEMA individual assistance applications, worked to ensure over 500 blue roof applications were processed, distributed around 5,000 hot meals, 1,500 canned food bags, over 800 mosquito nets, 4,000 mosquito repellants, as well as cash to 150 families, and coordinated cultural activities." All of this, she stresses, was possible because of "15 years of community organizing and partnership building in Puerto Rico and abroad."[18]

For residents clinging to the well-situated but flood-prone ground around el Caño, community organization emerged slowly. "When we first came in as a community organization," G-8 president Lucy Cruz Rivera observes, "Residents thought we were a political party . . . We had to gain their trust and show them that we're here to represent them. We represent *them*, not the current or the future mayor. Now, we go out and people say, 'They're from G-8; they're the only ones that are helping me.'"[19] Community activist José Caraballo Pagán attests to the critical role of residents' experiential knowledge in shaping decisions: "We invite other types of knowledge . . . but we always elevate our community knowledge as the most important."[20] Since 2019, ENLACE itself has been led by a community resident, Mario Núñez Mercado, who has lived in Las Monjas for six decades.[21]

The overall project of widening the channel and constructing associated infrastructure entails moving about a thousand households. ENLACE and allied organizations have focused on ensuring that relocation does not harm disadvantaged families. As ENLACE's manager of housing, Jennifer Rivera-De Jesus emphasizes, "Our rules also consider the emotional value of a home, rather than simply seeing it as a transaction." As of mid-2023, the Fideicomiso estimated that about half of the households living in the area to be dredged had been relocated, but hundreds more remained.[22] Proposed relocation of a household starts with a visit from the relocation committee to determine the residents' needs. The committee is composed of other residents who have themselves previously relocated. Only after the relocation committee's assessment does ENLACE conduct an independent assessment of the house's value, make an offer to purchase, and assist with relocation. ENLACE executive director Núñez Mercado estimates that about two thirds of relocated households elect to stay in the district, while others follow children who have moved away or "return to their towns of origin."[23] ENLACE offers support throughout the process, "including mental health services to mitigate the emotional stress relocation can cause."[24] ENLACE's Estrella Santiago-Pérez stresses the customized approach of the Relocation Committee: "In many cases, they know the families personally and can better assist in relocating them. The committee is critical because they can ensure that the families' rights are being protected and that the process is thoughtful and respectful. It is also more appropriate for this help to come from someone who has lived through it first-hand."[25] This personal engagement is what enables the communities along el Caño to make progress in self-governance—the sense of collective efficacy that emerges from enhanced self-worth. As community leader José Caraballo Pagán phrases it, "Ultimately, opening people's consciousness and returning them their dignity is the biggest success."[26] Once a channel-side house is acquired, ENLACE and its partners erect a sign to remind everyone of the collective enterprise and deter anyone from rebuilding in this space: "El Caño es Tuyo ¡Protégelo!" (The Channel is Yours. Protect It!).[27]

The district plan includes constructing 420 new homes and rehabilitating eighty others. Fideicomiso and ENLACE have worked with residents to generate five housing designs that meet their needs.[28] Architect Carlos Muñiz notes that single-family homes with a backyard are culturally engrained for many residents because this is what people could build when they initially settled the precarious landscape along el Caño. "People feel empowered knowing that they themselves built these spaces," he observes.[29] Planner Deepak Lamba-Nieves, research director at Puerto Rico's Center for a New Economy, underscores the need for professionals to respect resident preferences: "For members of these communities, having their own house and plot is seen as an economic and political tool to advance socioeconomically. Land provides them the possibility of self-sufficiency. There are long traditions and historical processes that we need to consider when trying to demand this notion of living in high rises instead of single-family housing."[30] While cognizant of the preference for single-family homes, given high construction costs, ENLACE and its partners are exploring alternative multifamily and row-house configurations that might be suitable. Lamba-Nieves suggests finding ways "to re-parcel the district and create smaller and narrower plots where people can have their own house and still have high density."[31]

ENLACE housing manager Rivera-De Jesus notes that the Fideicomiso is working with partners to develop such housing, but she acknowledges that many residents harbor bad associations with multifamily housing. She recognizes that they are working with "communities who have members that were taken out of their previous communities by the government and relocated to public housing without their voices being heard." She tries to demonstrate to them that "it's not the same as how it's been done before—you can have multifamily [housing] with some privacy and a balcony and other cultural preferences. There are cases where multifamily is actually the best option."[32] Still, for many in el Caño, the single-family home is more than an architectural type; it is a symbol and a mechanism for household control.

While community engagement has been essential to design and planning efforts in el Caño, sustaining these efforts has not always been easy, especially given funding delays for channel dredging. Delays have led to skepticism among some residents about ENLACE's value and effectiveness. Some residents decry the Fideicomiso's community land ownership as a form of creeping socialism.[33] Some government officials have pounced: Governor Luis Fortuño (in office 2009–2013) and San Juan Mayor Jorge Santini (in office 2001–2013) argued for reviving a neoliberal development model for the eastern portion of el Caño, much as was deployed in the western portions. Opponents criticized the very concept of the Fideicomiso and sought to retake the land. Lawyer María Hernández Torrales, who fought the government's expropriation of Fideicomiso property, praises several legal safeguards implemented in 2013: "If the city wants to expropriate the land again, they need to pay an exorbitant amount of money that of course the government does not have. Also, now it would be seen as a political shame."[34] Ultimately, planner Lamba-Nieves convincingly contends, el Caño's nascent community institutions have prevailed and even gained strength from the persistent fights against skeptics.[35]

SECURITY: NEW CHANNELS FOR COLLECTIVE LAND TENURE AND AFFORDABILITY

Security of tenure has been a central aim as the G-8, ENLACE, and the Fideicomiso have developed their innovative governance model in pursuit of development without displacement. For generations, most el Caño residents lacked land titles or obtained them at bargain rates through clientelist political practices. With proposed infrastructure and environmental upgrades, residents feared displacement and gentrification. Such pressures have become so rampant across many parts of San Juan that they inspired a 2022 music-video-cum-anti-displacement documentary by reggaetón superstar Bad Bunny.[36]

In asserting their right to stay put, el Caño's residents initially insisted that they wanted individual titles to land so that they could pass property on to their heirs, access utilities and other public services, improve their credit, and obtain mortgages. Inspired in part by a visit from a member of Boston's celebrated Dudley Street Neighborhood Initiative, however, residents of el Caño began to consider a CLT as an alternative model for securing land collectively and affordably. Still, the Fideicomiso did not emerge easily or quickly. Only after many workshops and community assemblies did residents conclude that this form of collective ownership—despite being unprecedented in Puerto Rico—could be the best solution for gaining security of tenure to ward off gentrification.[37] During the establishment of the CLT, planner Lamba-Nieves extols the importance of "several great lawyers and community members" who jointly developed a way to codify inheritance rights in the Fideicomiso structure, enabling participants to view housing as a source of intergenerational stability. "It's a safe and secure place," he observes, "when you know family generations before you made it their home too."[38] Mariolga Juliá Pacheco, first hired at the Fideicomiso in 2014 and now director of ENLACE's Office of Citizen Participation, reiterates that "people's biggest concern was 'How do I leave something for the next generation?'"[39]

When Law 489 established the Fideicomiso in 2004, it placed two hundred acres of formerly public land in permanent legal community ownership. That initial territory has been slowly augmented by the acquisition
of privately owned houses, bringing the CLT's total land holdings to 272
acres (110 hectares), covering almost two thirds of the Caño Martín Peña
Special Planning District. Unlike many other CLTs that begin with vacant
land or vacant buildings, the Fideicomiso works in a densely occupied territory. The CLT seeks to accommodate residents relocated from flood-prone
areas. Individual households on Fideicomiso land can receive a durable
surface rights deed for their parcel, even as the underlying land remains in
community ownership. CLT rules permit residents to bequeath their home
to their heirs but give the Fideicomiso the right of first refusal if residents
want to sell their house or surface rights. Enabling residents to own surface
use rights is a key part of the Fideicomiso's model, since those rights are
valued at one quarter of the overall value of the land, enabling residents
to build wealth and secure loans. Pacheco describes the dual advantages
of the CLT model as "balanc[ing] between individual rights and aspirations and collective protections and safeguards."[40] The Fideicomiso also
uses resale formulas designed to limit profits and retain affordability. Most
important for long-term security of tenure, the Fideicomiso is required to
manage its land for the common benefit of the component communities
and cannot sell the land.[41]

Despite the clear appeal of the Fideicomiso's approach to advancing
community equity, in several senses of that term, the process of transfer-
ring land titles has met several challenges. Nearly two decades after work
began to reclaim the eastern parts of el Caño, the population of the area has
decreased by about half. Of the approximately eleven thousand people who
live in the G-8 neighborhoods, only about 1,500 households are members of
the Fideicomiso. Of those, only about 147 had fully executed surface deeds
as of mid-2023.[42] Importantly, Line Algoed points out, many residents are
actually "not interested in having an individual title or a collective title,"
arguing that they have resided along el Caño for decades without such recognition and do not want to start "paying tax and services [fees] to a state
that has never supported them." Fortunately, most of those households
relocated for the channel widening have gained both new houses and deeds
to surface rights. Even so, the gentrification-resisting intent of ENLACE and

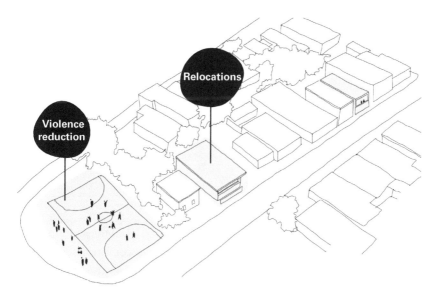

12.7 Youth programs in public spaces address violence and encourage community solidarity. *Source*: Mora Orensanz.

the Fideicomiso falls well short of a guarantee of secure tenure for all residents. Despite these limitations, the Fideicomiso's work represents a landmark achievement, demonstrating the power of a customized CLT model in facilitating relocation from hazardous areas while retaining secure community tenure.[43]

Much of work in el Caño is devoted to minimizing the risk of displacement, but some parts of the initiative also enhance security from other types of violence (figure 12.7). ENLACE's Pacheco points to a youth violence prevention program that has conducted workshops on gender violence, drug-related violence, substance abuse, citizen participation, and citizen responsibility. She observes that the fruits of this work are "qualitative results that you cannot necessarily quantify" but "become clear when you see [youth] grow up and see all they have accomplished."[44] In addition to the anti-violence work, many in the el Caño communities have supported other forms of community solidarity, including among the large Dominican community in the neighborhoods.[45]

Improvements in new public spaces can also enhance security. Carlos Muñiz emphasizes that the channel side is not merely being cleared; it is also being reconfigured to benefit the community, exemplified by a temporary public-gathering platform deployed near the water in Israel Bitumul. Even before the dredging is finished and the public amenities are completed, such projects can help residents become accustomed to using these spaces. In Muñiz's conceptualization, creating spaces "for people to interact with and come in contact with el Caño" and one another "provide[s] a sense of security."[46]

ENVIRONMENT: NEW CHANNELS FOR PERSONAL AND COMMUNITY WELL-BEING

Innovative governance to enhance security for el Caño residents only matters if it yields safe and desirable neighborhood conditions. Efforts to improve environmental well-being in el Caño communities have taken many forms. Reducing vulnerability to flooding is the primary reason for moving residents out of channel-side locations. The prospect of relocation is much more acceptable for residents because it has been paired with both sustained community environmental education and positive alternatives that enable residents to remain in the neighborhood. Estrella Santiago Pérez, ENLACE's manager of environmental affairs, reports that the groups have held workshops on water quality and developed school-based gardens to "to promote certain aspects of food sovereignty and food security." These gardens serve multiple functions, from teaching about water and plant cycles to supplying food for school lunches (figure 12.8).[47]

Beyond environmental education, work by ENLACE and its partners directly supports environmental protection and ecosystem vitality. Given mounting flooding and storm threats, the urgency of seeking safer ground has become obvious for many in the communities. The groups mobilized after Hurricane Irma and Hurricane Maria to address the immediate needs of residents. Maria left at least seventy-five families homeless along el Caño, damaging more than a thousand roofs and flooding 70 percent of the communities. A roof repair program, Techos para el Caño (Roofs for el Caño), repaired 113 roofs in the first two years. Another post-Maria program, Mi Casa Resiliente (My Resilient House), built homes to demonstrate affordable storm-resistant features, although it struggled to expand.[48] ENLACE leaders have also worked with local architects to develop socially

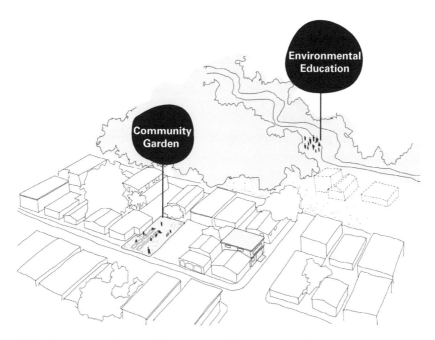

12.8 Environment-focused efforts along el Caño have included community gardens and educational initiatives to test water quality. *Source:* Mora Orensanz.

and environmentally responsive house designs that could accommodate changing family structures. As Carlos Muñiz, manager of ENLACE's Urbanism and Infrastructure Program, puts it, incorporating flexibility meant "the family didn't have to invest in changing their house as their needs changed, too. The kitchen and living room can be easily adjusted to build another room."[49]

Recognizing that six of the adjacent el Caño communities lie either mostly or entirely within the hundred-year floodplain, the communities have worked with several different design teams to envision flood tolerant channel-side open spaces. ENLACE leaders take special care to monitor and protect the areas that have been cleared along the channel. This has sometimes entailed installing temporary uses, such as pavilions, in part to forestall the return of informal occupation of vacated spaces.[50] In 2021, ENLACE commissioned Marvel Architects to design ten new parks on vacant land along the channel, including at least one park in each of the G-8 neighborhoods, each one reflecting community priorities.[51]

In addition to these parks, in March 2017, the EPA funded workshops that yielded proposals for an extensive water plaza immediately adjacent to the channel in the Buena Vista Santurce neighborhood. The water plaza would link across the Paseo Buena Vista Santurce to an inland urban plaza, yielding a single integrated stormwater management strategy with above- and below-ground components. A system of nearby rain gardens and stormwater pods at various levels is intended to store and filter 56,200 cubic feet of water before returning cleaned water to the channel through below-ground pipes. The plaza, which is inspired by the Benthemplein Water Square in Rotterdam, would provide space for sports and theater during dry periods.[52] Moving forward, ENLACE contracted with the landscape firm Olin Studio for a full master plan for the G-8 communities, prioritizing flood mitigation but also seeking to build "community capital," including workforce and entrepreneurial opportunities, according to Olin partner Richard Roark.[53] Equitable resilience depends on such connections between environmental protection and livelihood generation.

LIVELIHOODS: NEW CHANNELS FOR ECONOMIC ADVANCEMENT

In terms of livelihoods, the work on el Caño matters for equitable resilience because it sustains low-income residents' access to areas with abundant amenities and opportunities. The G-8 neighborhoods are walkable to a variety of job opportunities in the adjacent downtown financial district. The construction of the eleven-mile-long Tren Urbano passenger railway line between 1996 and 2004 significantly improved access to these neighborhoods.[54] Longtime resident leader Juan Caraballo Pagán, from Barrio Obrero Marina, points out that "from a planning perspective, we have it all" because we are "close to everything," including "a major train station, banks, hospitals, and universities" (figure 12.9).[55]

It is one thing to protect proximity to economic resources in nearby job-rich neighborhoods, but quite another to ensure that development within el Caño communities themselves will accommodate and sustain residents' livelihoods. In some cases, ENLACE's plans for new housing explicitly treat houses as both homes and workplaces. Architect Muñiz has advocated for designs with floodable ground floors that could be available for livelihood generation during dry times. Other opportunities to incorporate space for

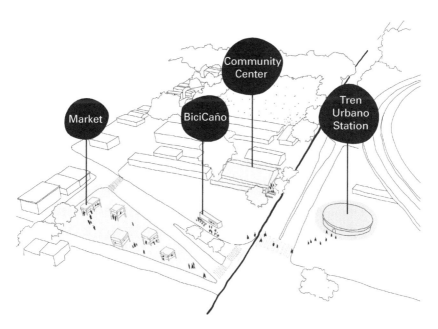

12.9   Redevelopment along el Caño has benefited from proximity to Tren Urbano and has included new business ventures. *Source*: Mora Orensanz.

economically productive activities can emerge from flexible design, including homes where the kitchen and living room space "has its own door so it can be accessed without entering the house, which could then double as an office space."[56] In addition to income generation from homes, residents have developed microenterprises aimed at attracting tourism, such as Bici-Caño (resident-led bike tours around the area).

When it comes to protecting or enhancing livelihoods, significant challenges remain. Santiago Pérez observes that US federal regulations often miss "Puerto Rico's cultural context." For example, when ENLACE tries to help residents relocate home-based facilities for raising chickens or growing vegetables for market, she says "federal regulation does not take them into account," even though they are "very important income sources for our residents."[57]

If the ongoing work along el Caño has a shortcoming thus far, it is in this aspect of livelihood enhancement. In the assessment of Deepak Lamba-Nieves, a member of ENLACE's advisory board since 2005, "Economic development projects haven't advanced as quickly as environmental justice

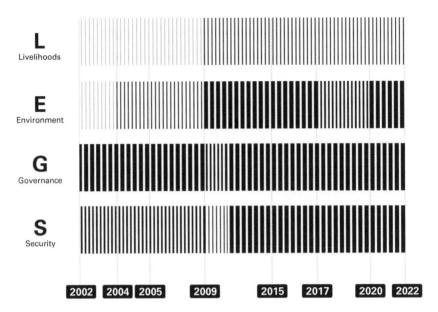

12.10   Although slow to develop, the efforts at el Caño have brought major achievements in communal governance, security of tenure, and environmental protection. *Source*: Smriti Bhaya. 2002: Community engagement; 2005: G-8, ENLACE, CLT formed; 2009: Threat of losing land; 2015: First surface deeds; 2017: Hurricane Maria; 2020: Relocations and infrastructure; 2022: USACE allocates $168 million for el Caño.

projects." He comments, "Expectations about high job creation have not been met—in great part because Puerto Rico has been mired in an economic crisis for almost two decades." That said, he points out, if one looks at less quantifiable measures and includes the realms of skill building and confidence building, the calculus is rather different. "Projects that teach people how to read and write, to look and apply for jobs, that empower women leaders, [and] that raise awareness about violence," he observes, can jointly "advance a new vision about and contribute to economic and social development," even though "traditional metrics" may miss such gains.[58]

CONCLUSION: TOWARD HOLISTIC EQUITABLE RESILIENCE

The Fideicomiso in el Caño has gained worldwide recognition. Although it is just the second CLT to have been established in an informal settlement (following an experiment in Kenya), it has not been the last. The institution-building and the community-engagement processes carried out along el Caño have triggered significant interest elsewhere, including exchanges with favela upgrading efforts in Brazil. This broad reach is not lost on the locals. Lamba-Nieves is delighted to see that "el Caño's CLT is part of a global conversation. It's exchanging knowledge, tools, and capabilities with communities across the world regarding their community development, collective land ownership, and justice."[59] Ultimately, all three words in the name of a CLT matter: community, land, and trust—and the latter is far more than a legal term.

As a sustained undertaking in equitable resilience, the work to support the neighborhoods that flank el Caño Martín Peña is neither perfect nor completed (figure 12.10), but it demonstrates that low-income people can both lead, and benefit from, holistic efforts to adapt to climate change threats.

In July 2022, the USACE signed an agreement to start construction on el Caño's $168 million Ecosystem Restoration Project, bringing new promise.[60] When the dredging broke ground in January 2023, a grateful Governor Pedro Pierluisi credited the activism of G-8 community members: "If it weren't for your efforts. If it weren't for your persistence, your perseverance, your militancy, we wouldn't be here today. You take the credit; neither the federal government nor the government of Puerto Rico does."[61] In a city with stark inequality, el Caño's example shows that environmental gains can be accompanied by radical commitment to improved livelihoods, enhanced security, and empowered self-governance. These successes are not easy to replicate, but they show the transformative potential of pursuing the equitably resilient city.

LEARNING FROM THREE STRUGGLES FOR EQUITABLY RESILIENT GOVERNANCE

The three cases we grouped together in this section highlight the challenges of pursuing equitably resilient governance. Together they underscore the centrality of community co-design. Thunder Valley CDC—operating on the sparsely populated Pine Ridge Reservation—and KDI's KPSP—dispersed across one of Africa's densest informal settlements—have radically different contexts. The work along el Caño Martín Peña takes place at a middle density, functioning holistically across several adjacent neighborhoods. These three attempts to address socio-ecological challenges in diversely situated settlements also differ in their relationship to existing neighborhoods. Whereas Thunder Valley is building a new community on open land, KDI and the el Caño coalition are mitigating environmental risks while infusing new life into rare open space in crowded places lacking basic utilities, services, and amenities.

These examples from Pine Ridge, Kibera, and San Juan share a commitment to enabling community self-governance. KDI initially co-created neighborhood organizations, thereby inventing their own partners. Although this approach had its advantages, KDI soon pivoted to a more radical form of collaboration. They reversed the directionality of their community engagement process, inviting existing community groups to propose projects to KDI. The Thunder Valley and el Caño interventions are, arguably, even more community driven, since both are primarily controlled by groups founded and run by community members themselves.

All three places face threats from climate change, and all suffered acutely during the COVID-19 pandemic. These challenges, whether chronic or sudden, underscore the need for deep community engagement in setting priorities, drawing on practices ranging from participatory budgeting and deliberative democracy to collective ownership. The spiritually grounded work of Thunder Valley's Lakota champions makes clear that local approaches to equitable resilience can draw upon much older practices of self-governance and relationship with earth systems. Thunder Valley's progress in one of the most impoverished places in the US is also a reminder of ongoing struggles for Indigenous self-determination and land recovery in settler colonial societies from the US and Canada to Brazil and New Zealand. Other Indigenous groups have deployed different tactics in cases where they have regained high-value urban land. For instance, members of the Squamish Nation are planning high-density urban development in a project called *Senákw* on recently recovered lands in Vancouver. A trio of First Nations groups are collaborating on even larger projects elsewhere in the region.[1]

The projects advanced by Thunder Valley CDC, KDI, and el Caño's CLT all reveal shared roots in the tangled history of settler colonialism. All are community-based responses to generations of discrimination by regimes that subjugated, segregated, and silenced disempowered people. Each group seeks to reclaim land for community well-being and safer occupation. In each case, these communities are engaged in long-term efforts to claim fuller forms of ownership. The forms of shared ownership used in each case intertwine individual and collective claims, revealing the centrality of governance to aspirations for equitable resilience.

These three examples—whether a single core project, a convening of neighborhood organizations, or a scattered network of small public spaces—underline the necessity to work with larger scales of government (table IV.2). El Caño's CLT and ENLACE were both founded with legislative authority, and the CLT's initial land holdings came from the territory's government. US government funds, however slow to materialize, are essential for infrastructure upgrading. Similarly, Thunder Valley's leaders are renegotiating land rights and economic development policies with federal

Table IV.2

Multiscalar analysis of governance cases

	Thunder Valley CDC	KDI Kibera Public Space Project	El Caño Martín Peña CLT
Individual/ Household Self-Efficacy	Emphasis on "self-sovereignty" and healing	Leadership training	Formal workshops and youth training
Project Control	Circles of houses surrounding communal space	Community-initiated design–build projects to build trust	Community design for green spaces and replacement housing
City/ Region Influence	Tensions with tribal government and federal agencies	Increased recognition from Nairobi authorities	Legislation founded ENLACE and supported CLT; networking with global CLTs; tensions with federal authorities remain

agencies, including the BIA. All of KDI's activities in Kibera face threats from Nairobi city and Kenyan national authorities that assert ownership over Kibera's land and plan infrastructure projects with little regard for Kibera residents. All three of these efforts impart a commitment to improving environmental conditions and socioeconomic livelihoods through shared governance. That said, these cases show that progress toward equitable resilience, even in places with sincere allegiance to distributed governance, proceeds in fits and starts. These three final cases—with their incremental progress and inevitable setbacks—demonstrate remarkable achievements that stand firmly on all four "LEGS" of equitable resilience.

CONCLUSION: AXIOMS OF EQUITABLE RESILIENCE

What have we learned from these twelve places and their people, struggling to build solidarity in the face of mounting climate crises? What do these stories reveal about the role of environmental, security, livelihoods, and governance conditions in adapting to climate change threats across scales? What can they tell us about progress toward "the equitably resilient city"? To assess, we return to the four words that form our title. After revisiting the terms "equitable," "resilient," "city," and "the," we propose ten axioms of equitable resilience, drawing together insights from the solidarities and struggles of the dozen cases detailed in the previous chapters.

Even though most of the featured case studies focus on the first two decades of the twenty-first century, it is clear that *equitable* intervention in settlements must consider longer time frames. A robust approach to equity must go beyond a discrete project or set of projects in a particular place; it also matters how interventions relate to broader historical and ongoing patterns of inequity. Equitable transformation of settlements is impossible without trust between impacted residents and institutions seeking to intervene. Building such trust requires untangling the roots of suspicion by recognizing emplaced histories. Settler colonial practices, racialized segregation and subjugation, ethnic rivalries, discriminatory land and housing policies, and unacknowledged past traumas lurk in many contested landscapes. These deep-seated injuries contribute to environmental vulnerabilization, inadequate livelihoods, insecure tenure, and disempowering governance regimes. In many of the impressive initiatives that we profile, acknowledging and seeking to repair past inequities has been central to pursuing equitably resilient futures. Even so, the case studies also clearly demonstrate that the path to equity is never instantaneous and is rarely marked by continuous steady improvement. Seeking justice entails struggle, with both moments of triumph and periods of backlash, retrenchment, and disappointment. Productive struggle requires enduring solidarities, joining people together in community and linking communities to outside sources of stability, insight, and support.

The twelve cases that we profile also make clear that pursuing *resilient* futures requires more than reducing vulnerabilities to climate change and other environmental threats. If transformations in housing, infrastructure, or land-use patterns make settlements environmentally safer but do not attend to other aspects of equitable resilience, disempowered groups will not benefit. This is suggested by growing fears of green gentrification, including among participants in the Gentilly Resilience District and Paris OASIS projects. If interventions improve security from displacement but do not sustain or enhance livelihoods, residents may not be able to afford to stay, as illustrated by mounting costs of living for residents in the Paraisópolis Condomínios and the canal-side Baan Mankong communities in Bangkok. Finally, if a project or program improves environmental, security, and livelihood conditions without enhancing local self-governance, interventions are unlikely to reflect the values of residents or to be sustained, as demonstrated by the struggles of migrant painter workers in Dafen, who were economically marginalized when local governments pivoted to different models of revenue generation.

Our discussion of these four dimensions of equitable resilience is not meant to suggest that every intervention must address all four dimensions in equal measure or that they must proceed in the same sequence in which we present them—from environment to security to livelihoods to governance. Differing contextual factors and community priorities will lead different initiatives to emphasize some dimensions of resilience more than others over time. In fact, as we discuss below, there can be tensions and trade-offs among environment, livelihoods, security, and governance goals.

The twelve cases also provide insight into what it means to invoke the equitably resilient *city*. Our examples range from small interventions in rural places to upgrading in dense urban neighborhoods. They vary from single sites to national programs with hundreds or thousands of sites. Collectively, these examples show the necessity of multiscalar analysis and action. Many critiques point out that calls for resilience are often used to shift the burdens of climate adaptation and disaster recovery onto individuals and households, furthering neoliberal goals of privatization of risk and state retrenchment. There is doubtless validity to this critique. Our case studies make clear that states and other empowered institutions have essential roles to play in advancing equitable resilience, even when they are not the central actors.

The cases that we profile also highlight the limits of project-based approaches when not coupled with broader systemic change. Intervention in a single Shenzhen urban village such as Dafen cannot be separated from national household registration policies. Creating a "resilience district" in Gentilly matters little if New Orleans's larger infrastructural systems fail or if longtime residents are displaced due to gentrification. Site-specific planning and design interventions such as those that we profile cannot wholly redress the radically uneven vulnerabilities faced by long-disempowered groups in cities. As we explain throughout the book, in some cases, flashy built environment renovations undertaken in the name of urban resilience can actually disadvantage already-marginalized groups or distract from deeper system injustices. Systemic change in a range of domains, from insurance and taxation to housing and social service provision, is essential to equitable adaptation. Nonetheless, site-specific built environment interventions can be crucial to advancing equitable resilience, especially when they go beyond narrow technical changes to include environmental well-being, livelihood enhancement, personal and property security, and empowered self-governance across scales. Investigating exemplary site-specific interventions can reveal tangible, desirable, and realizable steps that improve the lives of disadvantaged people while contributing to necessary structural change.

Lastly, interrogating our title means asking what these twelve cases tell us about whether there is some definitive set of attributes common to *the* equitably resilient city. Given the multidimensional contested nature of resilience, the range of scales that fall under the term "city," and the historical complexity of what may count as equitable, there is reason to be skeptical of generalization. Comparing the trajectories of the four LEGS dimensions across the twelve cases confirms that initiatives pursuing equitable resilience follow diverse paths and timetables. Still, the common threads linking these cases suggest some shared axioms of equitable resilience.

1. DEEPLY ROOTED INJUSTICES SHAPE POSSIBLE RESILIENT FUTURES

The case studies make clear that while climate adaptation is an inherently future-oriented endeavor, pursuing equitable resilience often demands reckoning with past and present injustices. Deeply rooted injustices shape how disadvantaged people respond to threats and how they relate to governments and other empowered institutions. Generations of callous and exploitative policy and planning have created deep distrust of authorities among many groups, including Black, Latin, and Native communities in the US, as well as immigrants in Paris, migrant laborers in Shenzhen, and residents of informal settlements in India, Kenya, Brazil, and Thailand. Communities that have been marginalized for decades or centuries cannot be expected to abandon their well-earned wariness and fall in line quickly simply because authorities proclaim that there is an urgent need for climate action.

The injustices at the roots of such distrust take many forms. The Lakota of Thunder Valley inherit the wounds of US government and settler colonial land theft, exploitative economic relations, and attempted genocide. Black New Orleanians have borne injuries, from enslavement and racialized slum clearance to radically unequal recovery from Hurricane Katrina. Residents of Paraisópolis, Yerwada, Kibera, and Bangkok's informal settlements have endured decades of political exploitation and broken promises as they have sought secure footholds in their respective cities. Vulnerability remains inseparable from vulnerabilization.

In some of these cases, efforts to pursue equitable resilience have been derailed or delayed by insufficient attention to these deeply rooted injustices. In the Paris OASIS pilot projects, participatory planning and design efforts for schoolyard renovations struggled to engage low-income immigrant residents. Planners of the Gentilly Resilience District expressed similar frustrations, noting that wary residents cannot be expected to take up the burden of maintaining green infrastructure in their neighborhoods. Some efforts have explicitly addressed long-simmering distrust in disadvantaged communities. Thunder Valley CDC frames their work as advancing intergenerational healing to advance self-sovereignty. CODI founded the Baan Mankong program in Thailand with an explicit commitment to reversing decades of paternalistic treatment of poor residents, pledging a new model of upgrading that treats residents as experts in solving their own problems and providing resources for them to do so.

2. PURSUIT OF EQUITABLE RESILIENCE ENTAILS SUSTAINED STRUGGLE

Overcoming generations of well-deserved mistrust to remake built environments and social systems in the name of equitable resilience is difficult work. None of the cases that we profile saw fast, easy, or steady progress. None were without setbacks or delays. In every case, reforming the inherited socio-ecological systems that create uneven vulnerability has required protracted struggle.

The cases show that transformations can be delayed or set back in several ways. Shifting government priorities or personnel can undermine the long, slow work of equitable resilience. For example, turnover in city staff following a mayoral transition in New Orleans led the Gentilly Resilience District to languish for years, even after receiving a major federal grant. Exogenous shocks can derail even well-formulated ideas, making planned initiatives impossible or forcing changes in organizational priorities to address emergency conditions. In the Paris OASIS project, the compound impacts of the COVID-19 lockdowns and a transit strike hobbled participatory design efforts and delayed the public opening of climate-adaptive schoolyards. The pandemic also disrupted Living Cully's work building "environmental wealth" among communities of color in Portland, Oregon. Remaining undaunted, however, Living Cully's emergency rental assistance, food aid, and technology support for low-income residents enabled the coalition to build relationships with community members, strengthening subsequent organizing efforts and providing a powerful example of strategic adaptation to unforeseen crises.

Many of the struggles that we document relate to challenges in creating and sustaining community ownership and self-governance institutions. The cases demonstrate the power of shared ownership and governance, but they also illustrate that these models can be susceptible to internal divisions and power struggles. The case of Comunidad María Auxiliadora is perhaps the clearest instance where internal rifts undermined community progress toward equitable resilience. Factions within the community who resented their women-led governance and prohibitions against individual speculation destroyed a robust but informal community ownership model. Even in this case, though, the protracted struggle yielded significant progress toward equitable resilience.

3. PROGRESS IN ONE DIMENSION OF EQUITABLE RESILIENCE CAN SUPPORT PROGRESS IN OTHER DIMENSIONS

Although we have organized our discussion of equitable resilience and case studies into separate sections on environment, security, livelihoods, and governance, the cases make clear that these four dimensions are far from independent. Rather, they are often mutually supportive. Action to improve conditions in one domain frequently yields progress in other dimensions.

To take one example, building institutions for self-governance and secure tenure can establish a strong foundation from which people can address other challenges, such as adapting to environmental precarity due to climate change. This is clearly the case with the coalition in el Caño where ENLACE, the G-8, and the CLT created structures through which the communities could articulate a vision for equitable upgrading and flood adaptation. Similarly, the ROC USA model is focused on enabling resident ownership to create stable and affordable communities. Few communities pursue resident ownership for environmental or hazard-mitigation reasons. Yet, when residents gain the agency to manage their own communities, many ROCs invest in mitigating hazards and improving environmental conditions.

In other cases, communities come together to address environmental challenges and only then improve other dimensions of security, livelihoods, and governance. When authorities threatened to evict residents of Bang Bua and other canal-side settlements in Bangkok for allegedly worsening water pollution, they organized and asserted an alternative narrative:

they were, in fact, stewards of the canals. This foregrounding of environmental goals then buoyed the communities as they sought support for collective tenure security and upgrading through Baan Mankong. Similarly, KDI's Kibera Public Space Project began with discrete built environment upgrades to mitigate local flooding and other risks. These environment-focused improvements supported livelihoods by accommodating income-generating activities, spawned new self-governing community groups to manage spaces, and improved perceived tenure security, even if more formal secure tenure remains elusive.

4. THE DIMENSIONS OF EQUITABLE RESILIENCE CAN COME INTO CONFLICT

Many of our twelve case studies show how advances in one dimension of equitable resilience support progress on other fronts, but there are also instances in which conflicts between the LEGS dimensions threaten progress. Tensions and trade-offs within our four-part framework take many forms, including well-documented conflicts between improvements in environmental conditions and security of tenure. Environmental improvements for hazards mitigation, urban greening, and decarbonization can increase property values and intensify displacement pressure on low-income people. While some of the cases that we profile, such as Living Cully and Baan Mankong, explicitly attempt to counteract green gentrification by pairing environmental improvements with increased security of tenure, others do not. In the Gentilly Resilience District, planners did not emphasize security against displacement. When the projects commenced, few proponents anticipated that Gentilly would face substantial gentrification pressure. In the Paris OASIS schoolyard renovations, planners similarly did not see housing security as a concern. Nonetheless, in both cases, project participants worried about contributing to displacement. With rising anxieties about inequality and housing insecurity in many cities and increasing awareness of green gentrification threats, designers and planners must consider how environmental improvements can be linked to anti-displacement efforts.

Another common conflict in the cases that we profile emerges when self-governance inhibits the ability of communities to make timely decisions, including those related to environmental improvements and risk mitigation.

Interviewees in several cases, including the ROC USA communities, Baan Mankong, and el Caño Martín Peña, discussed the stubbornly slow work of collective governance. While collective ownership and self-governance can secure affordable settlements and enable self-determination, democratic self-governance at the community scale often exhibits the same divisions, conflicting priorities, and social rifts that characterize any community of neighbors. Some ROC communities are split between resident owners who prioritize environmental improvements and those focused on maintaining affordability. Baan Mankong canal-side reconfigurations can be derailed by holdouts who are skeptical of collective ownership and upgrading for various reasons, including alarm over increased living costs and desires to maintain conditions that benefit some residents, such as informal rental properties. Long-lived self-governing communities, including those in the ROC USA and Baan Mankong networks, often use proven models for minimizing and overcoming challenges that can derail self-governance.

There are no simple rules for navigating potential trade-offs between various dimensions of equitable resilience. Rather, each case requires weighing community values and priorities—a process that benefits from well-supported community self-governance.

5. ACTIONS AT ONE SCALE CAN THREATEN PROGRESS AT OTHER SCALES

Throughout this book, we discuss the intricacies of pursuing equitable resilience across scales, from the individual to the urban region. The scalar complexity of climate change impacts and adaptation are well documented. Neither emissions nor impacts respect bureaucratic boundaries. In response, many researchers call for multiscalar climate and adaptation governance. In the case studies profiled here, we have documented several instances of scalar conflicts.

Interventions to promote equitable resilience that focus on one scale can actually undermine progress or lead to missed opportunities at other scales. These scalar conflicts can be described in two broad categories: missing the forest for the trees and missing the trees for the forest. Interventions focused on small scales can cause problems at larger scales. For example, the house-by-house approach employed in the Yerwada upgrading projects aimed to build improved houses for residents on the same parcel as their previous homes, requiring enormous effort to negotiate the physical and social complexity of irregular high-density sites. This approach allowed residents to maintain their place-based networks and access to amenities and resources, but it also made it impossible to make significant improvements to public spaces and infrastructure that would benefit all residents. If this scalar trade-off constitutes an instance of missing the forest for the trees, one could describe the scalar trade-off encountered in later Baan Mankong upgrading projects as just the opposite. When the military coup government in Thailand radically expanded canal-side Baan Mankong upgrading in Bangkok, initiating a forest of projects, they intended to improve conditions for more communities and to deliver pollution reduction and flood mitigation benefits across the region. Although many of those benefits are being realized, several project participants expressed concerns that the program's expansion and acceleration has come with a major trade-off: de-emphasizing the "trees"—the hardy and deeply rooted resident empowerment that anchored the model's early towering success. Some degree of scalar trade-off may be inevitable as residents and leaders around the world attempt to remake settlements in response to climate change. However, it is essential that those trade-offs be recognized and subjected to vigorous debate to ensure that they do not harm already-disadvantaged people, undermining equitable resilience.

6. SUSTAINING PRODUCTIVE STRUGGLES DEMANDS COALITIONS THAT BOND AND BRIDGE

Addressing scalar conflicts and other challenges of equitable resilience requires coalitions that bring together actors with different constituencies, capacities, and forms of power. Remaking housing, settlements, and infrastructure in the name of equitable resilience requires diverse participants to form stable constellations. These coalitions can change and shift over time as a program or project evolves, but making progress in the struggle for equitable resilience requires sustaining solidarities that bond people and organizations together and bridge across different sectors and types of power.

The coalitions responsible for pushing forward transformations in each case take unique forms, but two broad categories are recognizable: bonding coalitions that join local individuals and organizations, peer-to-peer networks, and global communities of practice; and bridging coalitions, linking grassroots and community efforts to formally empowered institutions.

Several of our case studies build and rely upon horizontal bonding coalitions. ROC USA and Baan Mankong both operate peer-to-peer networks in which residents and leaders from different communities share resources and experiences, build leadership capacity, and commiserate in the difficult work of community self-governance. While those two peer-to-peer networks bring individual people together, Living Cully is itself a coalition of four core member organizations. Living Cully is also a part of many project-specific coalitions joining interested parties around specific goals, from advancing the community-controlled tax-increment finance district to negotiating community benefits agreements for specific redevelopment projects. Living Cully is focused on a specific geography, but other efforts feature global networks of solidarity that bolster organizations advancing equitable resilience. For instance, the leaders of SPARC (responsible for some of the Yerwada upgrading) and CODI (the primary entity responsible for the Baan Mankong program) are linked with other groups committed to improving conditions for residents of informal settlements around the world through networks such as the Asian Coalition for Housing Rights and Slum and Shack Dwellers International. Similarly, large global organizations conjoin and support disparate efforts around the world. Habitat for Humanity, for instance, has played a role in several cases: partnering with eighty families to build houses in Comunidad María Auxiliadora, providing a crucially timed loan for MASHAL as part of the Yerwada upgrading, and serving as a core member of the Living Cully coalition in Portland. While the global circulation of consultants and best practices associated with urban resilience has rightly been critiqued as supporting a self-perpetuating industrial complex, the global networks of solidarity that we observe suggest more hopeful potentials for community-driven struggles for equitable resilience. Advancing such hope entails documenting the "plausibility of the possible" rather than just brooding over the "inevitability of the probable."[1]

Along with the bonding coalitions that enable individuals and organizations to share resources and expertise, bridging coalitions are also important in linking grassroots efforts to formally empowered institutions, including governments. Many of our twelve cases feature coalitions that connect resident-driven grassroots organizations to state actors. The Yerwada upgrading projects join community-embedded organizations such as Pune Mahila Milan with outside technical experts from groups such as SPARC and MASHAL and supports their combined efforts with resources from governments from the municipal corporation to the national level. Similarly, the resettlement and infrastructure upgrades in el Caño Martín Peña would not be possible without both the allied G-8 community members and the state-supported efforts of ENLACE and the Fideicomiso (CLT). The G-8 communities bring experiential knowledge and embeddedness in the social networks of el Caño, while the state-chartered entities endow the project with land, funding, legal authority, and access to political power.

7. GOVERNMENT ACTION IS A NECESSARY BUT INSUFFICIENT CONDITION FOR EQUITABLE RESILIENCE

Government efforts can both enable and inhibit progress toward equitable resilience, but state authorities rarely make lasting contributions without the support of resident and community groups in impacted areas. Government intervention can directly reverse progress made by other actors, as in the case of the sewer infrastructure installations that destroyed one of KDI's public-space interventions in Kibera. The transformation of the canal-side Baan Mankong projects in Bangkok under the coup government shows another way that governments can undermine nuanced equitable resilience interventions. In this case, changes in government practice hijacked the program, altered its administrative structure to emphasize drainage infrastructure over community development, and imposed politically motivated timelines that made deep community engagement impossible. The wrenching shift from the once-rural Dafen village with its collective land tenure to the more corporate entity of the Dafen Oil Painting Village demonstrates yet another shortcoming of government-led equitable resilience: as Shenzhen's municipal government gained greater control and championed new economic development efforts in Dafen, lower-income migrants struggled to retain their affordable place in the settlement.

Government agencies can easily undermine grassroots equitable resilience efforts, and government-led efforts face serious limitations, but state actors can also be essential partners in advancing transformative equity-enhancing adaptation. In the cases of the Paraisópolis Condomínios, Yerwada, and the early phases of Baan Mankong, policy changes across scales of government enabled more human-centered models of upgrading in informal settlements, shifting from slum clearance and peripheral rehousing to integrated urban upgrading that reduced environmental vulnerability while enabling residents to maintain their social and economic networks. Government actors, primarily at the state and local level, also play a critical role in enabling equitable resilience in resident-owned manufactured home parks. State legislation can enable resident purchases and government entities can support infrastructure upgrading and hazard mitigation in ROCs through loans and grants and by enabling ROCs to connect to public infrastructure.

Critics of resilience efforts often frame such projects as shifting the burden of climate adaptation onto already disadvantaged individuals and communities, but our case studies suggest that state actors are often essential to the success of equitable resilience.

8. LASTING SOLIDARITY REQUIRES BOTH INFORMAL COMMUNITY BUILDING AND FORMAL RECOGNITION

One of the primary ways that governments undermine or enable equitable resilience is by withholding or granting recognition of secure land tenure. The cases that we profile suggest that equitable resilience often requires sustained solidarity supported through a combination of formal recognition and informal mutual assistance at the community level.

Official recognition of community land tenure has been essential to progress toward equitable resilience in several of our cases, including Pasadena Trails and other ROC USA communities, the Caño Martín Peña communities, and the canal-side Baan Mankong settlements in Bangkok. Each of these cases, however, required more than official recognition alone. Their stories demonstrate how residents themselves build internal solidarity as a necessary complement to externally granted tenure security. Legal land titles are essential, but so are the bonds formed among parents hanging out in a laundromat while children play, people eating and dancing together during holiday festivals and fairs, and residents supporting one another through mutual aid during COVID-19 and other ordeals. Conversely, while residents of the Paraisópolis Condomínio projects enjoy enhanced security of tenure in their more formalized living arrangements, several expressed a sense of lost solidarity and community agency that came with their moves from auto-constructed favela dwellings to state-provided housing.

While formal security without informal solidarity may make communities less cohesive, communities that build informal solidarity without gaining formal recognition are exposed to both internal and external threats. The founders and early residents of the Comunidad María Auxiliadora developed a strong collective identity and mission that enabled them to build a settlement in the barren hills of peri-urban Cochabamba, Bolivia. Nevertheless, without formal recognition of their collective land tenure, the group's solidarity collapsed in the face of mounting internal divisions and speculative pressure. KDI's public-space interventions in Kibera have enhanced many dimensions of life in the settlements, but the sites themselves, like the settlement at large, do not have any official recognition of their right to remain. Ultimately, for organizations such as KDI, grassroots buy-in can enable significant progress, but truly sustainable success will also require recognition from official authorities.

9. PRODUCTIVE CONTESTATION CAN ADVANCE EQUITABLE RESILIENCE

Boosters often frame resilience in relentlessly positive terms, promoting win-win interventions that purportedly benefit all and are nearly impossible to oppose. The many examples of inequitable resilience that we discuss, especially in the four LEGS framing chapters, demonstrate the inadequacy of such a rosy depoliticized presentation. Unless consciously centered on equity, resilience interventions often hurt disadvantaged residents, both through direct harms such as displacement and risk-shifting maladaptation and through indirect impacts such as green gentrification.

If we resist resilience boosterism, we are equally skeptical of relentless critique. Instead, we traverse the opportunities and instabilities of the actually existing middle ground. This entails close observation of narratives and methods. The twelve cases profiled here make clear that when disadvantaged communities try to advance their own resilience, they often meet resistance from empowered institutions. Building solidarity to overcome resistance to equitable resilience often entails community-building activities from board meetings to barbecues. In some cases, overtly confrontational tactics have proved both necessary and efficacious.

Productive contestation plays a powerful role in several of the case studies that we feature. Living Cully's coalition's structure enables them to confront empowered institutions through protests at city hall and street theater demonstrations at the offices of landlords and developers without implicating coalition members who may normally operate in more formal collaborative ways with those same institutions. Similarly, there are clear links between Cochabamba's Water War and the activism that launched Comunidad María Auxiliadora. The residents' union in Paraisópolis has been instrumental in organizing favela residents to advocate for improved infrastructure and housing, including the Condomínios. When residents marched with coffins through the streets of the adjacent Morumbi neighborhood, they confronted their wealthy neighbors and the city government, forcing them to recognize the daily precarity faced by favela residents subjected to flooding, fires, and landslides. In Bolivia and Brazil, local practices of insurgent citizen resistance paralleled broader national popular movements.

10. EQUITABLE RESILIENCE ENLISTS DESIGN FOR SYSTEMIC CHANGE, NOT JUST PROJECTS

Design professionals and design methods have been central to the adoption of resilience as a central framing concept and normative aim in urban climate change adaptation since the turn of the twenty-first century. This increasing prominence of design has productively expanded possibilities for how settlements cope with environmental stresses, but it has also brought new challenges. Resilience design practices typically seek syntheses and co-benefits, including opportunities for climate adaptation to improve other aspects of urban life. However, these practices have also generated and reinforced fragmented modes of planning and intervention.

Even as urban resilience initiatives aspire to new modes of integrative action, these initiatives are typically manifested through discrete projects. The twelve cases that we profile here illustrate both the power and limitations of "resilience projects." A project or collection of projects can provide opportunities to develop more integrative planning and design processes, drawing actors with diverse expert and experiential knowledge together around shared goals. Projects can test new types of physical intervention, demonstrating possibilities and clarifying challenges for future interventions. Even so, the cases that we profile also show that projects cannot deliver on promises of equitable resilience without attention to broader systemic and institutional changes.

The Gentilly Resilience District aimed to demonstrate landscape-based green infrastructure but struggled to realize its full ambitions because its project-based approach did not focus on making more systemic institutional changes. Among the cases that we profile, those that aspire to make sustained institutional change appear especially promising as vehicles for advancing equitable resilience. For instance, the Baan Mankong, ROC USA, and el Caño Martín Peña cases feature significant achievements toward equitable resilience enabled through resident-driven institutional and governance reforms. These initiatives have all focused on improving particular places, but they have also developed compelling practices of community engagement based on clearly articulated participatory principles. Living Cully has also developed strategies for moving beyond project-by-project approaches, including the creation of the Clean Energy Community Benefits Fund and the nascent community-driven TIF district. Alternatively, using a more inward-facing strategy of deep reflection, the work of the Thunder Valley CDC has simultaneously embraced both the design of a small community and a more systemic approach to Indigenous cultural and spiritual regeneration. In other words, in equitable resilience terms, the most effective initiatives seem to be those that attempt to "scale up" *principles*, not just replicate projects.

Design matters, not just because designers offer innovative interventions in built environments; design matters because the process of implementing a design thoughtfully and equitably also entails designing institutional reforms, including new forms of collective ownership and self-governance. Design processes that include institutional innovations such as collective savings groups, cooperative management regimes, and community land trusts build new forms of social solidarity. Such structures enable previously destabilized residents to imagine alternative collective futures. While such institutional initiatives may not have immediate physical manifestations, they can lay the groundwork for systemic changes that will spawn many years of built environment projects.

EQUITABLE RESILIENCE: FROM PROJECTS TO PRINCIPLES

With the convergence of increasing urbanization and climate impacts, people and institutions are taking action across scales to adapt settlements to climate change. The ultimate question is not whether settlements will adapt but rather *how* they will adapt: who will benefit, who will be harmed, and who will shape the contours of those adaptation interventions? By drawing on the partial successes from the cases we have presented here, we hope to provide some paths forward as community leaders, professional planners and designers, and residents of urban settlements around the world confront climate crises through solidarities and struggles, seeking new modes of equitable resilience.

NOTES

INTRODUCTION

1. The capitalization of terms referring to racial and ethnic identities is a topic of ongoing discussion and debate. Throughout this book, we capitalize terms such as "Black," "Native," "Indigenous," and "Latinx." We have left "white" uncapitalized, in keeping with evolving conventions in newspapers such as the *New York Times*, https://www.nytimes.com/2020/07/05/insider/capitalized-black.html#:~:text=Times%20policy%20advises%20reporters%20to,but%20both%20will%20remain%20lowercase. Leaving "white" in lower case acknowledges that language, like race, is a social construction. The MacArthur Foundation, among many others, constructs it differently: https://www.macfound.org/press/perspectives/capitalizing-black-and-white-grammatical-justice-and-equity.

2. IPCC Working Group II, *Sixth Assessment Report, Technical Summary* (IPCC, 2021), 66.

3. Jon Coaffee and Peter Lee, *Urban Resilience* (Springer, 2016); Jaimie Hicks Masterson et al., *Planning for Community Resilience* (Island Press, 2014); Rebecca Elliott, *Underwater: Loss, Flood Insurance, and the Moral Economy of Climate Change in the United States* (Columbia University Press, 2021).

4. IPCC, "Summary for Policymakers," in *Climate Change 2021: The Physical Science Basis*, ed. Valérie Masson-Delmotte et al. (Cambridge University Press, 2021).

5. Gavin D. Madakumbura et al., "Anthropogenic Influence on Extreme Precipitation over Global Land Areas Seen in Multiple Observational Datasets," *Nature Communications* 12, no. 1 (2021): 3944.

6. Michael Goss et al., "Climate Change Is Increasing the Likelihood of Extreme Autumn Wildfire Conditions across California," *Environmental Research Letters* 15, no. 9 (2020): 094016.

7. Marc Poumadere et al., "The 2003 Heat Wave in France: Dangerous Climate Change Here and Now," *Risk Analysis: An International Journal* 25, no. 6 (2005): 1483–1494.

8. UNISDR CRED, *Economic Losses, Poverty and Disasters 1998–2017* (Université Catholique de Louvain, 2018), 33.

9. USGCRP, *Fourth National Climate Assessment* (US Global Change Research Program, 2018).

10. A. M. Vicedo-Cabrera et al., "The Burden of Heat-Related Mortality Attributable to Recent Human-Induced Climate Change," *Nature Climate Change* 11, no. 6 (2021): 492–500.

11. Joacim Rocklöv and Robert Dubrow, "Climate Change: An Enduring Challenge for Vector-Borne Disease Prevention and Control," *Nature Immunology* 21, no. 5 (2020): 479–483.

12. Tamma A. Carleton et al., "Valuing the Global Mortality Consequences of Climate Change Accounting for Adaptation Costs and Benefits" (working paper, Working Paper Series, National Bureau of Economic Research, 2020).

13. Mark Pelling, *Adaptation to Climate Change: From Resilience to Transformation* (Routledge, 2011).

14. J. Timmons Roberts, "The International Dimension of Climate Justice and the Need for International Adaptation Funding," *Environmental Justice* 2, no. 4 (2009): 185–190.

15. World Resources Institute, "Greenhouse Gas (GHG) Emissions," Climate Watch, 2021, https://www.climatewatchdata.org/.

16. Kenneth Hewitt, *Interpretations of Calamity from the Viewpoint of Human Ecology* (Allen and Unwin, 1983); Benjamin Wisner et al., *At Risk: Natural Hazards, People's Vulnerability, and Disasters* (Routledge, 2004); Eric Klinenberg, *Heat Wave* (University of Chicago Press, 2002).

17. W. Neil Adger, "Vulnerability," *Global Environmental Change* 16, no. 3 (2006): 268–281.

18. Susan L. Cutter, Bryan J. Boruff, and W. Lynn Shirley, "Social Vulnerability to Environmental Hazards," *Social Science Quarterly* 84, no. 2 (2003): 242–261.

19. Chester W. Hartman and Gregory D. Squires, *There Is No Such Thing as a Natural Disaster* (Taylor and Francis, 2006).

20. Isabelle Anguelovski et al., "Equity Impacts of Urban Land Use Planning for Climate Adaptation," *Journal of Planning Education and Research* 36, no. 3 (2016): 333–348.

21. Benjamin K. Sovacool, Björn-Ola Linnér, and Michael E. Goodsite, "The Political Economy of Climate Adaptation," *Nature Climate Change* 5, no. 7 (2015): 616–618.

22. Melissa Checker, "Eco-Apartheid and Global Greenwaves," *Souls* 10, no. 4 (2008): 390–408.

23. Mike Hodson and Simon Marvin, "'Urban Ecological Security': A New Urban Paradigm?" *International Journal of Urban and Regional Research* 33, no. 1 (2009): 193–215; Mike Hodson and Simon Marvin, "Urbanism in the Anthropocene," *City* 14, no. 3 (2010): 298–313.

24. Jesse M. Keenan, Thomas Hill, and Anurag Gumber, "Climate Gentrification," *Environmental Research Letters* 13, no. 5 (2018); Isabelle Anguelovski et al., "Opinion: Why Green 'Climate Gentrification' Threatens Poor and Vulnerable Populations," *Proceedings of the National Academy of Sciences of the United States of America* 116, no. 52 (2019): 26139–26143.

25. Liz Koslov, "The Case for Retreat," *Public Culture* 28, no. 2 79 (2016): 359–387; Katharine J. Mach and A. R. Siders, "Reframing Strategic, Managed Retreat for Trans-formative Climate Adaptation," *Science* 372, no. 6548 (2021): 1294–1299; Elizabeth Marino, "Adaptation Privilege and Voluntary Buyouts," *Global Environmental Change* 49 (2018): 10–13.

26. Naomi Klein, *The Shock Doctrine: The Rise of Disaster Capitalism* (Picador, 2008); Kevin Fox Gotham and Miriam Greenberg, *Crisis Cities* (Oxford University Press, 2014).

27. Saskia Sassen, *The Global City* (Princeton University Press, 2001).

28. Louis Wirth, "Urbanism as a Way of Life," *American Journal of Sociology* 44, no. 1 (1938): 1–24.

29. Raymond Williams, *The Country and the City*, vol. 423 (Oxford University Press, 1975); Henri Lefebvre, *The Urban Revolution* (University of Minnesota Press, 2003).

30. Richard Burdett et al., *The Endless City* (Phaidon, 2007).

31. Neil Brenner and Christian Schmid, "The 'Urban Age' in Question," *International Journal of Urban and Regional Research* 38, no. 3 (2014): 731–755; Neil Brenner and Christian Schmid, "Planetary Urbanization," in *Urban Constellations*, ed. Matthew Gandy (Jovis, 2011), 10–13.

32. Alex Schafran et al., *The Spatial Contract* (Manchester University Press, 2020).

33. Sara Meerow, Joshua P. Newell, and Melissa Stults, "Defining Urban Resilience: A Review," *Landscape and Urban Planning* 147 (2016): 38–49.

34. Simin Davoudi, "Just Resilience," *City and Community* 17, no. 1 (2018): 3–7.

35. Stephanie Wakefield, "Urban Resilience as Critique," *Political Geography* 79 (2020): 102148.

36. Crawford S. Holling, "Resilience and Stability of Ecological Systems," *Annual Review of Ecology and Systematics* 4, no. 1 (1973): 1–23; Lance H. Gunderson, Craig Reece Allen, and Crawford S. Holling, *Foundations of Ecological Resilience* (Island Press, 2012).

37. W. Neil Adger et al., "Social-Ecological Resilience to Coastal Disasters," *Science* 309, no. 5737 (2005): 1036–1039; Louis Lebel et al., "Governance and the Capacity to Manage Resilience in Regional Social-Ecological Systems," 2006, http://digitalcommons.library.umaine.edu/sms_facpub/52/?utm_source=digitalcomm.

38. Lawrence J. Vale, "The Politics of Resilient Cities," *Building Research and Information* 42, no. 2 (2014): 191–201; Kevin Grove, *Resilience* (Routledge, 2018); Mura Quigley, Neale Blair, and Karen Davison, "Articulating a Social-Ecological Resilience Agenda for Urban Design," *Journal of Urban Design* 23, no. 4 (2018): 581–602.

39. Gilbert F. White, "Natural Hazards Research," in *Natural Hazards: Local, National, Global* (Oxford University Press, 1974), 3–16.

40. Raymond J. Burby, *Cooperating with Nature: Confronting Natural Hazards with Land Use Planning for Sustainable Communities* (Joseph Henry Press, 1998).

41. Ian L. McHarg, *Design with Nature* (American Museum of Natural History, 1969); Anne Whiston Spirn, *The Granite Garden* (Basic Books, 1985).

42. Jon Coaffee et al., *The Everyday Resilience of the City* (Springer, 2009); Jon Coaffee, "Towards Next-Generation Urban Resilience in Planning Practice," *Planning Practice and Research* 28, no. 3 (2013): 323–339.

43. Lawrence J. Vale and Thomas J. Campanella, eds., *The Resilient City* (Oxford University Press, 2005).

44. Joanne Fitzgibbons and Carrie L. Mitchell, "Just Urban Futures? Exploring Equity in '100 Resilient Cities,'" *World Development* 122 (2019): 648–659.

45. Sophie Webber, Helga Leitner, and Eric Sheppard, "Wheeling Out Urban Resilience: Philanthrocapitalism, Marketization, and Local Practice," *Annals of the American Association of Geographers* 111, no. 2 (2021): 343–363.

46. Mehrnaz Ghojeh et al., *Inclusive Planning Playbook* (C40 Cities/World Resources Institute, 2019).

47. UIC Barcelona, "A First Assessment Framework on Public Spaces and Urban Resilience," Master Urban Resilience for Sustainability Transitions (blog), November 8, 2019, https://masterurbanresilience.com/a-first-assessment-framework-on-public-spaces-and-urban-resilience/.

48. Michel Bruneau et al., "A Framework to Quantitatively Assess and Enhance the Seismic Resilience of Communities," *Earthquake Spectra* 19, no. 4 (2003): 733–752.

49. Urban Land Institute, *Ten Principles for Building Resilience* (Urban Land Institute, 2018).

50. Reza Banai, "Pandemic and the Planning of Resilient Cities and Regions," *Cities* 106 (2020): 102929.

51. Linda Shi, "From Progressive Cities to Resilient Cities," *Urban Affairs Review* 57, no. 5 (2020): 1442–1479; Malini Ranganathan and Eve Bratman, "From Urban Resilience to Abolitionist Climate Justice in Washington, DC," *Antipode* 53, no. 1 (2021): 115–137.

52. Sara Meerow and Joshua P. Newell, "Urban Resilience for Whom, What, When, Where, and Why?" *Urban Geography* 40, no. 3 (2019): 309–329.

53. Simin Davoudi, "Resilience: A Bridging Concept or a Dead End?" *Planning Theory and Practice* 13, no. 2 (2012): 299–307.

54. Pelling, *Adaptation to Climate Change*; Vale, "The Politics of Resilient Cities"; Susan Fainstein, "Resilience and Justice," *International Journal of Urban and Regional Research* 39, no. 1 (2015): 157–167.

55. Jeremy Walker and Melinda Cooper, "Genealogies of Resilience from Systems Ecology to the Political Economy of Crisis Adaptation," *Security Dialogue* 42, no. 2 (2011): 143–160.

56. David Chandler, *Resilience: The Governance of Complexity* (Routledge, 2014); Grove, *Resilience.*

57. Michael J. Watts, "Resilient Planet," *Dialogues in Human Geography* 9, no. 2 (2019): 186–189.

58. Chandler, *Resilience.*

59. Al Jazeera English, "In Deep Water: A Way of Life in Peril | Fault Lines," 2010, https://www.youtube.com/watch?v=Q4itfAVq19U.

60. Vlad Mykhnenko, "Resilience: A Right-Wingers' Ploy?" in *Handbook of Neoliberalism* (Routledge, 2016), 218–234.

61. U.S. Army Center for Initial Military Training, Ready and Resilient (R2) Program; https://usacimt.tradoc.army.mil/rr.html.

62. Brad Plumer, "Rick Perry's Plan to Rescue Struggling Coal and Nuclear Plants Is Rejected," *New York Times*, January 8, 2018, sec. Climate, https://www.nytimes.com /2018/01/08/climate/trump-coal-nuclear.html.

63. Jessica Wehrman, "How 'Resilience' Became a Politically Safe Word for 'Climate Change,'" *Roll Call*, September 30, 2019, https://www.rollcall.com/2019/09/30/how -resilience-became-a-politically-safe-word-for-climate-change/.

64. Vale, "The Politics of Resilient Cities."

65. Mick Lennon, Mark Scott, and Eoin O'Neill, "Urban Design and Adapting to Flood Risk," *Journal of Urban Design* 19, no. 5 (2014): 745–758.

66. Fainstein, "Resilience and Justice"; Shi, "From Progressive Cities to Resilient Cities."

67. Torstein Eckhoff, *Justice: Its Determinants in Social Interaction* (Rotterdam University Press, 1974).

68. Wisner et al., *At Risk.*

69. Robert D. Bullard, *Dumping in Dixie* (Routledge, 1990); David Naguib Pellow, *Resisting Global Toxics* (MIT Press, 2007).

70. David Schlosberg and Lisette B. Collins, "From Environmental to Climate Justice," *WIREs Climate Change* 5, no. 3 (2014): 359–374.

71. Sara Meerow, Pani Pajouhesh, and Thaddeus R. Miller, "Social Equity in Urban Resilience Planning," *Local Environment* 24, no. 9 (2019): 793–808.

72. Grove, *Resilience.*

73. Grove, *Resilience*, 7.

74. Stephen Collier, Savannah Cox, and Kevin Grove, "Rebuilding by Design in Post-Sandy New York," *Limn* 5, no. 7 (2016); https://limn.it/articles/rebuilding-by-design -in-post-sandy-new-york/

75. Wakefield, "Urban Resilience as Critique."

76. Aditya Bahadur and Thomas Tanner, "Transformational Resilience Thinking," *Environment and Urbanization* 26, no. 1 (2014): 200–214.

77. Linda Shi and Susanne Moser, "Transformative Climate Adaptation in the United States," *Science* 372, no. 6549 (2021).

78. Vanesa Castán Broto and Enora Robin, "Climate Urbanism as Critical Urban Theory," *Urban Geography* 42, no. 6 (2021): 715–720.

79. Pelling, *Adaptation to Climate Change*.

80. Patricia Romero-Lankao et al., "Urban Transformative Potential in a Changing Climate," *Nature Climate Change* 8, no. 9 (2018): 754–756.

81. Amartya Sen, *Commodities and Capabilities* (Oxford University Press, 1999).

82. Samuel D. Brody et al., "Comprehensive Framework for Coastal Flood Risk Reduction," in *A Blueprint for Coastal Adaptation*, ed. Carolyn Kousky, Billy Fleming, and Alan M. Berger (Island Press, 2021), 2–28.

83. Jon Barnett and Saffron O'Neill, "Maladaptation," *Global Environmental Change* 20, no. 2 (2010): 211–213.

84. Michelle A. Hummel et al., "Economic Evaluation of Sea-Level Rise Adaptation Strongly Influenced by Hydrodynamic Feedbacks," *Proceedings of the National Academy of Sciences of the United States of America* 118, no. 29 (2021): e2025961118.

85. Watts, "Resilient Planet."

86. Niki Frantzeskaki et al., "Nature-Based Solutions Accelerating Urban Sustainability Transitions in Cities," in *Nature-Based Solutions to Climate Change Adaptation in Urban Areas* (Springer, 2017), 65–88.

EQUITABLY RESILIENT ENVIRONMENTS

1. Adger, "Vulnerability."

2. Cutter et al., "Social Vulnerability to Environmental Hazards"; Alice Fothergill and Lori A. Peek, "Poverty and Disasters in the United States: A Review of Recent Sociological Findings," *Natural Hazards* 32, no. 1 (2004): 89–110.

3. Darryn McEvoy, David Mitchell, and Alexei Trundle, "Land Tenure and Urban Climate Resilience in the South Pacific," *Climate and Development* 12, no. 1 (2020): 1–11.

4. Robert C. Bolin and Patricia A. Bolton, *Race, Religion, and Ethnicity in Disaster Recovery* (Institute of Behavioral Science, University of Colorado, 1986); Elaine Enarson, Alice Fothergill, and Lori Peek, "Gender and Disaster," in *Handbook of Disaster Research* (Springer, 2018), 205–223.

5. Anguelovski et al., "Equity Impacts of Urban Land Use Planning."

6. Bullard, *Dumping in Dixie*.

7. Aristotle and R. F. Stalley, *Politics*, trans. Ernest Barker, revised ed. (Oxford University Press, 2009).

8. Vitruvius, *Vitruvius: The Ten Books on Architecture*, trans. Morris Hicky Morgan (Dover Publications, 1960).

9. Zhinguo Ye, "Cities Under Siege: The Flood of 1931 and the Environmental Challenges of Chinese Urban Modernization," in *International Planning History Society Proceedings*, vol. 2 (17th IPHS Conference, History-Urbanism-Resilience, TU Delft Open,

2016), 79–88; Barbara E. Mundy, *The Death of Aztec Tenochtitlan, the Life of Mexico City* (University of Texas Press, 2015).

10. Axel I. Mundigo and Dora P. Crouch, "The City Planning Ordinances of the Laws of the Indies Revisited. Part I," *The Town Planning Review* 48, no. 3 (1977): 247–268.

11. Spiro Kostof, *The City Shaped: Urban Patterns and Meanings through History* (Thames and Hudson, 1991), 209–277.

12. Lawrence J. Vale, *Architecture, Power, and National Identity*, 2nd ed. (Routledge, 2008).

13. Maria Kaika, *City of Flows* (Routledge, 2004); Matthew Gandy, *Concrete and Clay* (MIT Press, 2002).

14. Frederick Law Olmsted, "Public Parks and the Enlargement of Towns," in *The City Reader*, ed. Richard LeGates and Frederick Stout, 5th ed. (Routledge, 2011), 321–327; Theodore S. Eisenman, "Frederick Law Olmsted, Green Infrastructure, and the Evolving City," *Journal of Planning History* 12, no. 4 (2013): 287–311.

15. Patrick Geddes, *Cities in Evolution* (Williams and Norgate, 1915).

16. McHarg, *Design with Nature*.

17. Spirn, *The Granite Garden*; Michael Hough, *Cities and Natural Process* (Routledge, 2004).

18. Kevin Lynch, *Good City Form* (MIT Press, 1984).

19. Charles Waldheim, ed., *The Landscape Urbanism Reader* (Princeton Architectural Press, 2006).

20. Zachary Lamb, "The Politics of Designing with Nature: Reflections from New Orleans and Dhaka," *Socio-Ecological Practice Research* 1 (2019): 227–232.

21. Michael J. Watts, "Now and Then: The Origins of Political Ecology and the Rebirth of Adaptation as a Form of Thought," in *The Routledge Handbook of Political Ecology*, ed. Tom Perreault, Gavin Bridge, and James McCarthy, 1st ed. (Routledge, 2015), 19–50.

22. Elisabeth M. Hamin and Nicole Gurran, "Urban Form and Climate Change," *Habitat International* 33, no. 3 (2009): 238–245.

23. W. Matt Jolly et al., "Climate-Induced Variations in Global Wildfire Danger from 1979 to 2013," *Nature Communications* 6, no. 1 (2015): 7537; Bob Berwyn, "With Lengthening Hurricane Season, Meteorologists Will Ditch Greek Names and Start Forecasts Earlier," Inside Climate News (blog), March 19, 2021, https://insideclimatenews .org/news/19032021/hurricane-season-greek-names/.

24. World Meteorological Organization, *WMO Atlas of Mortality and Economic Losses from Weather, Climate, and Water Extremes (1970–2019)* (World Meteorological Organization, 2021).

25. NOAA US Department of Commerce, *Weather Related Fatality and Injury Statistics* (NOAA's National Weather Service, 2020), https://www.weather.gov/hazstat/; https:// climateandcities.org/about-us/south-asia-heat-health-information-network/.

26. Klinenberg, *Heat Wave*.

27. Zoé Hamstead and Paul Coseo, "Critical Heat Studies," *Journal of Extreme Events* 6, no. 3 and 4 (2019): 2003001.

28. Rob Nixon, *Slow Violence and the Environmentalism of the Poor* (Harvard University Press, 2011).

29. Carleton et al., "Valuing the Global Mortality Consequences"; Susanne Amelie Benz and Jennifer Anne Burney, "Widespread Race and Class Disparities in Surface Urban Heat Extremes Across the United States," *Earth's Future* 9, no. 7 (2021).

30. Daniel P. Johnson, "Satellites Zoom in on Cities' Hottest Neighborhoods to Help Combat the Urban Heat Island Effect," *The Conversation*, June 14, 2022.

31. C. J. Gabbe and Gregory Pierce, "Extreme Heat Vulnerability of Subsidized Housing Residents in California," *Housing Policy Debate* 30, no. 5 (2020): 823–842.

32. Karen Peterson, "Extreme Heat Is Killing People in Arizona's Mobile Homes," *Washington Post*, July 2, 2021, https://www.washingtonpost.com/climate-environment/2021/07/02/arizona-mobile-home-deaths/.

33. Marc Reisner, *Cadillac Desert*, 2nd ed. (Penguin Books, 1993).

34. Jack Healy and Sophie Kasakove, "A Drought So Dire That a Utah Town Pulled the Plug on Growth," *New York Times*, July 20, 2021; Thomas Fuller, "Small Towns Grow Desperate for Water in California," *New York Times*, August 14, 2021.

35. Amy Maxmen, "As Cape Town Water Crisis Deepens, Scientists Prepare for 'Day Zero,'" *Nature* 554, no. 7690 (2018): 13–15.

36. Farhana Sultana and Alex Loftus, eds., *Water Politics* (Routledge, 2019).

37. Karen Bakker, "Neoliberalizing Nature?" *Annals of the Association of American Geographers* 95, no. 3 (2005): 542–565; Malini Ranganathan and Carolina Balazs, "Water Marginalization at the Urban Fringe: Environmental Justice and Urban Political Ecology across the North–South Divide," *Urban Geography* 36, no. 3 (2015): 403–423; Nikhil Anand, *Hydraulic City* (Duke University Press, 2017).

38. John T. Abatzoglou and A. Park Williams, "Impact of Anthropogenic Climate Change on Wildfire across Western US Forests," *Proceedings of the National Academy of Sciences of the United States of America* 113, no. 42 (2016): 11770–11775.

39. Andrew Aurand et al., *Taking Stock: Natural Hazards and Federally Assisted Housing* (Washington, DC: Public and Affordable Housing Research Corporation/National Low Income Housing Coalition, 2021).

40. Timothy W. Collins, "The Political Ecology of Hazard Vulnerability," *Journal of Political Ecology* 15, no. 1 (2008): 21–43.

41. C. J. Gabbe, Gregory Pierce, and Efren Oxlaj, "Subsidized Households and Wildfire Hazards in California," *Environmental Management* 66, no. 5 (2020): 873–883.

42. Jessica Debats Garrison and Travis E. Huxman, "A Tale of Two Suburbias: Turning Up the Heat in Southern California's Flammable Wildland–Urban Interface," *Cities* 104 (2020): 102725; Gregory L. Simon, *Flame and Fortune in the American West* (University of California Press, 2016).

43. United Nations Environment Programme, "Spreading Like Wildfire," UNEP Rapid Response Assessment (Nairobi, 2022).

44. Emily Feng, "In Just 3 Days, An Entire Year's Worth of Rain Has Fallen on Zhengzhou, China," *NPR*, All Things Considered, July 23, 2021, https://www.npr.org/2021/07/23/1019892596/in-just-3-days-an-entire-years-worth-of-rain-has-fallen-on-zhengzhou-china.

45. IPCC, "Summary for Policymakers."

46. John Eligon, Zanele Mji, and Lynsey Chutel, "Housing Crisis Propels High Death Toll in South Africa Floods," *New York Times*, April 20, 2022.

47. Steven T. Moga, *Urban Lowlands* (University of Chicago Press, 2020).

48. Anne Whiston Spirn, "Restoring Mill Creek," *Landscape Research* 30, no. 3 (2005): 395–413.

49. Mihir Zaveri et al., "How the Storm Turned Basement Apartments into Death Traps," *New York Times*, September 2, 2021.

50. Anuradha Mathur and Dilip Da Cunha, *Mississippi Floods* (Yale University Press, 2001); Debjani Bhattacharyya, *Empire and Ecology in the Bengal Delta* (Cambridge University Press, 2018).

51. Daniel Baker, Scott D. Hamshaw, and Kelly A. Hamshaw, "Rapid Flood Exposure Assessment of Vermont Mobile Home Parks Following Tropical Storm Irene," *Natural Hazards Review* 15, no. 1 (2014): 27–37.

52. Stephanie Rosoff and Jessica Yager, "Housing in the US Floodplain," Data Brief (Furman Center, NYU, 2017).

53. Farhana Sultana, "Living in Hazardous Waterscapes: Gendered Vulnerabilities and Experiences of Floods and Disasters," *Environmental Hazards* 9, no. 1 (2010): 43–53.

54. USGCRP, *Fourth National Climate Assessment*, https://nca2018.globalchange.gov and https://nca2018.globalchange.gov/chapter/8.

55. IPCC, "Summary for Policymakers."

56. Jim Morrison, "Sunny-Day Flooding Is About to Become More Than a Nuisance," Wired, August 2, 2021, https://www.wired.com/story/sunny-day-flooding-is-about-to-become-more-than-a-nuisance/.

57. Ellen Plane, Kristina Hill, and Christine May, "A Rapid Assessment Method to Identify Potential Groundwater Flooding Hotspots as Sea Levels Rise in Coastal Cities," *Water* 11, no. 11 (2019): 2228.

58. Mathew E. Hauer, "Migration Induced by Sea-Level Rise Could Reshape the US Population Landscape," *Nature Climate Change* 7, no. 5 (2017): 321–25.

59. Elliott, *Underwater*.

60. Masterson et al., *Planning for Community Resilience*; Brody et al., "Comprehensive Framework for Coastal Flood Risk Reduction."

61. Christine Meisner Rosen, *The Limits of Power* (Cambridge University Press, 2003); Kim Bellware, "Hurricane Andrew Transformed Florida's Building Codes," *Washington Post*, June 30, 2021, https://www.washingtonpost.com/history/2021/06/30/florida-building-codes/.

62. Lisa K. Bates, "Post-Katrina Housing," *The Black Scholar* 36, no. 4 (2006): 13–31.

63. IPCC Working Group II, *Sixth Assessment Report, Technical Summary*, 7.

64. E. Melanie DuPuis and Miriam Greenberg. "The Right to the Resilient City: Progressive Politics and the Green Growth Machine in New York City," *Journal of Environmental Studies and Sciences* 9, no. 3 (2019): 352–363.

65. Li Liu, Ole Fryd, and Shuhan Zhang, "Blue-Green Infrastructure for Sustainable Urban Stormwater Management," *Water* 11, no. 10 (2019): 2024; Stefan Al, "Multi-Functional Urban Design Approaches to Manage Floods: Examples from Dutch Cities," *Journal of Urban Design* 27, no. 2 (2022): 270–278.

66. Ian Burton, Robert William Kates, and Gilbert F. White, "The Human Ecology of Extreme Geophysical Events" (Natural Hazard Working Paper No. 1, NHRP 6701, 1968).

67. Hummel et al., "Economic Evaluation of Sea-Level Rise Adaptation."

68. Anguelovski et al., "Opinion."

69. Charisma Acey, "Rise of the Synthetic City: Eko Atlantic and Practices of Dispossession and Repossession in Nigeria," in *Disassembled Cities: Social and Spatial Strategies to Reassemble Communities*, ed. Elizabeth L Sweet (Routledge, 2018), 51–61.

70. Kian Goh, *Form and Flow: The Spatial Politics of Urban Resilience and Climate Justice* (MIT Press, 2021).

71. Alexandra Pacurar, "A Project for the Next 100 Years: Clippership Wharf," October 15, 2019, https://www.multihousingnews.com/post/a-project-for-the-next-100 -years-clippership-wharf/.

72. Burby, *Cooperating with Nature*.

73. Keenan et al., "Climate Gentrification."

74. Mach and Siders, "Reframing Strategic, Managed Retreat."

75. World Commission on Dams, *Dams and Development: A New Framework for Decision-Making* (Earthscan Publications, 2000); Karilyn Crockett, *People before Highways* (University of Massachusetts Press, 2018); Lawrence J. Vale, *After the Projects* (Oxford University Press, 2019).

76. Christina Finch, Christopher T. Emrich, and Susan L. Cutter, "Disaster Disparities and Differential Recovery in New Orleans," *Population and Environment* 31, no. 4 (2010): 179–202.

77. Koslov, "The Case for Retreat."

78. Marino, "Adaptation Privilege and Voluntary Buyouts."

79. A. R. Siders, Miyuki Hino, and Katharine J. Mach, "The Case for Strategic and Managed Climate Retreat," *Science* 365, no. 6455 (2019): 761–763.

80. Zachary Lamb, "Connecting the Dots: The Origins, Evolutions, and Implications of the Map that Changed Post-Katrina Recovery Planning in New Orleans," in *Louisiana's Response to Extreme Weather*, ed. Shirley Laska (Springer, 2019), 65–91.

81. Nate Millington, "Linear Parks and the Political Ecologies of Permeability," *International Journal of Urban and Regional Research* 42, no. 5 (2018): 864–881.

82. Lamb, "The Politics of Designing with Nature."

83. Leonie Sandercock and Peter Lyssiotis, *Cosmopolis II* (A&C Black, 2003).

84. Benjamin K. Sovacool and Björn-Ola Linnér, "Degraded Seascapes," in *The Political Economy of Climate Change Adaptation* (Springer, 2016), 54–80.

85. Nadja Kabisch et al., eds., *Nature-Based Solutions to Climate Change Adaptation in Urban Areas* (Springer Open, 2017).

86. Amanda Brown-Stevens, Emma Greenbaum, and Resilient by Design Teams, *Resilient by Design*, ed. Zoe Siegel (California Coastal Commission, 2019).

87. Md Feroz Islam et al., "Enhancing Effectiveness of Tidal River Management in Southwest Bangladesh Polders by Improving Sedimentation and Shortening Inundation Time," *Journal of Hydrology* 590 (2020): 125228; Marc Schut, Cees Leeuwis, and Annemarie van Paassen, "Room for the River: Room for Research?" *Science and Public Policy* 37, no. 8 (2010): 611–627; Natalie S. Peyronnin et al., "Optimizing Sediment Diversion Operations," *Water* 9, no. 6 (2017): 368.

88. Luna Khirfan, Niloofar Mohtat, and Megan Peck, "A Systematic Literature Review and Content Analysis Combination to 'Shed Some Light' on Stream Daylighting (Deculverting)," *Water Security* 10 (2020): 100067.

89. T. C. Wild et al., "Deculverting," *Water and Environment Journal* 25, no. 3 (2011): 412–421.

90. Galia Shokry, James J. T. Connolly, and Isabelle Anguelovski, "Understanding Climate Gentrification and Shifting Landscapes of Protection and Vulnerability in Green Resilient Philadelphia," *Urban Climate* 31 (2020): 100539.

91. Hui Li et al., "Sponge City Construction in China," *Water* 9, no. 9 (2017): 594.

92. Christos Zografos et al., "The Everyday Politics of Urban Transformational Adaptation: Struggles for Authority," *Cities* 99 (2020): 102613.

93. Watts, "Resilient Planet."

94. Gregory Bankoff, "Rendering the World Unsafe," *Disasters* 25, no. 1 (2001): 19–35.

95. Anne Whiston Spirn, "Ecological Urbanism," in *Companion to Urban Design*, ed. Tridib Banerjee and Anastasia Loukaitou-Sideris (Routledge, 2011), 614–624.

96. Brenda Brown, Terry Harkness, and Doug Johnston, "Eco-Revelatory Design: Nature Constructed/Nature Revealed." *Landscape Journal* 17, no. 2 (1998): 15–17.

97. Spirn, "Restoring Mill Creek."

98. "Ripple Effect," https://rippleeffectnola.com/.

99. Brown-Stevens et al., *Resilient by Design*.

100. "Women's Action Towards Climate Resilience for Urban Poor in South Asia," https://unfccc.int/climate-action/momentum-for-change/women-for-results/mahila -housing-trust.

101. "ISeeChange—Community Climate and Weather Journal," https://www.isee change.org/.

102. "WOEIP," West Oakland Environmental Indicators Project, https://woeip.org/.

103. "Peta Jakarta · Banjir," https://petajakarta.org/banjir/en/index.html.

104. E. L. Quarantelli, "Disaster Related Social Behavior," 1999, http://udspace.udel.edu/handle/19716/289.

CASE 1: GENTILLY RESILIENCE DISTRICT

1. Richard Campanella, *Draining New Orleans* (Louisiana State University Press, 2023).

2. Data Center, "New Orleans Neighborhood Data," The Data Center, February 24, 2021, https://www.datacenterresearch.org/data-resources/neighborhood-data/district-6.

3. Robert Smith, "A Last New Orleans Neighborhood to Dry Out," NPR, September 21, 2005, sec. Katrina and Beyond, https://www.npr.org/templates/story/story.php?storyId=4857138.

4. Clyde Woods, *Development Drowned and Reborn*, ed. Laura Pulido and Jordan T. Camp (University of Georgia Press, 2017).

5. Lamb, "Connecting the Dots."

6. Andy Horowitz, *Katrina: A History, 1915–2015* (Harvard University Press, 2020); Lamb, "The Politics of Designing with Nature."

7. City of New Orleans, "Reshaping the Urban Delta: National Disaster Resilience Competition, Phase 2 Submission," March 2015, https://www.nola.gov/resilience/resources/ndrc_phase2_neworleans_narrative_graphics/.

8. Han Meyer, Dale Morris, and David Waggonner, *Dutch Dialogues* (SUN Amsterdam, 2009).

9. Waggonner and Ball, "Greater New Orleans Urban Water Plan: Implementation" (Waggonner and Ball Architects, 2013).

10. Lamb, "Connecting the Dots."

11. Marla Nelson, Renia Ehrenfeucht, and Shirley Laska, "Planning, Plans, and People: Professional Expertise, Local Knowledge, and Governmental Action in Post-Hurricane Katrina New Orleans," *Cityscape* 9, no. 3 (2007): 45.

12. Mary Kincaid, "Gentilly Resilience District Microgrid Project Scoping," Presentation slides. City of New Orleans, Project Delivery Unit, New Orleans, LA, 2021.

13. Farrah D. Gafford, "'It Was a Real Village': Community Identity Formation among Black Middle-Class Residents in Pontchartrain Park," *Journal of Urban History* 39, no. 1 (2013): 36–58.

14. Dana Brown Associates, "Homepage," Dana Brown and Associates (blog), June 25, 2022, https://www.danabrownassociates.com/.

15. Spirn, "Ecological Urbanism."

16. Greg LaRose, "Gentilly Resiliency District Goes against Flow of How New Orleans Handles Stormwater," NOLA.com, January 26, 2016, https://www.nola.com/news/politics/article_422e71d2-2db8-544a-8c77-2bceed9bb141.html.

17. Interview with Brenda Breaux, June 2021.

18. Interview with Aron Chang, June 2021.

19. Interview with Mary Kincaid, June 2021.

20. Anonymous interview, October 2021.

21. Interview with John Kleinschmidt, March 2022.

22. Interview with Breaux.

23. Interview with Rami Diaz, March 2022.

24. "Resilience and Sustainability—Gentilly Resilience District," City of New Orleans, January 17, 2022, https://www.nola.gov/resilience-sustainability/gentilly-resilience -district/.

25. Interview with Chang.

26. Anonymous interview, October 2021.

27. Interview with Jeff Hébert, August 2017.

28. Interview with Seth Knudsen, August 2021.

29. New Orleans Redevelopment Authority, "Community Adaptation Program," 2018, https://www.noraworks.org/documents/community-adaptation-program-1-pager /download.

30. Interview with Breaux.

31. Interview with Chang.

32. Interview with Kincaid.

33. Gentilly Messenger, "Gentilly Resilience District Launches Lantern Walk, Online Program, Public Art Project," Gentilly Messenger (blog), September 9, 2020, https:// gentillymessenger.com/gentilly-resilience-district-launches-lantern-walk-online -program-public-art-project/.

34. Interview with Chang.

35. Helga Leitner et al., "Globalizing Urban Resilience," *Urban Geography* 39, no. 8 (2018): 1276–1284.

36. Interview with Chang.

37. Interview with Chang.

38. Interview with Kincaid.

39. Interview with Chang.

40. Anonymous interview, October 2021.

41. Interview with Chang.

42. Anguelovski et al., "Opinion."

CASE 2: PARAISÓPOLIS CONDOMÍNIOS

1. Teresa Caldeira, *City of Walls* (University of California Press, 2001).

2. Janice E. Perlman, *The Myth of Marginality* (University of California Press, 1980) and *Favela* (Oxford University Press, 2010).

3. "Programa Paraisópolis: Regularização Fundiária." City of São Paulo, May 2008 Powerpoint Presentation.

4. "Paraisópolis PAC 1 Post-Occupation Survey." City of São Paulo, October 2016, http://www.habitasampa.inf.br/wp-content/uploads/2017/01/Relato%CC%81rio -Po%CC%81s-Ocupac%CC%A7a%CC%83o-Paraiso%CC%81polis-PAC-1_Outubro _2016_entregue-em-20_10_2016.pdf.

5. S. P. Taschner, "São Paulo e as favelas," *Revista do Programa de Pós-graduação em Arquitetura e Urbanismo da FAUUSP* 19 (2006): 180.

6. City of São Paulo, "História: Como Surgiu Paraisópolis," 2009, https://www .prefeitura.sp.gov.br/cidade/secretarias/subprefeituras/noticias/?p=4385

7. Interview with Joildo Santos, November 2021; Nelson Antonio Alessi, "Formam-se favelas e ganham importância no cenário urbano São Paulo: Heliópolis e Paraisópolis" (master's thesis, Universidade de São Paulo, 2009).

8. SP Cultura, "União do Moradores e do Comércio de Paraisópolis," https://spcultura .prefeitura.sp.gov.br/espaco/88/; "Paraisópolis PAC 1 Post-Occupation Survey"; Maria Carolina Maziviero and Alane Santos da Silva, "O caso do Complexo Paraisópolis em gestões," *Revista Brasileira de Gestão Urbana* 10 (2018): 500–520.

9. "Paraisópolis PAC 1 Post-Occupation Survey"; M. L. Zuquim, "Urbanização de Assentamentos Precários no Município de São Paulo," in *Anais do 2° Encontro da Associação Nacional de Pesquisa e Pós-graduação em Arquitetura e Urbanismo* (ANPARQ, 2012), 7.

10. "Participação da iniciativa privada na construção da cidade" (PMSP-SEMPLA 1992, 21) cited in Luiz Guilherme River de Castro, "Operações Urbanas em São Paulo: inter-esse público ou construção especulativa do lugar." PhD Dissertation, University of São Paulo, 2006. https://www.researchgate.net/profile/Luiz-Castro/publication/224002503 _As_operacoes_urbanas_em_Sao_Paulo_interesse_publico_ou_construcao_especulativa _do_lugar/links/5c9e795292851cf0aea0b802/As-operacoes-urbanas-em-Sao-Paulo -interesse-publico-ou-construcao-especulativa-do-lugar.pdf).

11. Interview with Maria Teresa Diniz, September 2021.

12. City of São Paulo, "História: Como Surgiu Paraisópolis."

13. Elisabete França, "Favelas em São Paulo (1980–2008)," 2009, https://dspace .mackenzie.br/items/22953cc8-cd6c-4caa-a8e4-03308f9a70a7/full.

14. França, "Favelas em São Paulo." There are five types of ZEIS classifications. ZEIS 1 encompasses informal/irregular settlements predominantly inhabited by low-income populations, such as Paraisópolis. ZEIS designations 2–5 encompass "vacant or under-utilized" land or buildings in various other situations.

15. Prefeitura do Município de São Paulo Legislação, Lei No. 13.430 de 13 de Setem-bro 2002, https://www.prefeitura.sp.gov.br/cidade/upload/b9e06_Lei_N_13.430-02 _PDE.pdf

16. Legislação Municipal, Prefeitura de São Paulo, Decreto No. 42.871 de 18 de Fevereiro de 2003, http://legislacao.prefeitura.sp.gov.br/leis/decreto-42871-de-18-de -fevereiro-de-2003/consolidado; França, "Favelas em São Paulo."

17. "Paraisópolis PAC 1 Post-Occupation Survey."

18. França, "Favelas em São Paulo"; "Paraisópolis PAC 1 Post-Occupation Survey."

19. Interviews with residents: Raimundo, January 2022, and Vanessa, March 2022. All of these residents' names are pseudonyms.

20. São Paulo Municipal Housing Secretariat, "Set of Nine Interventions in Paraisópolis Will Benefit 20 Thousand Families," May 9, 2013, https://www.prefeitura.sp.gov.br /cidade/secretarias/habitacao/.

21. Maziviero and Santos da Silva, "O caso do Complexo Paraisópolis em gestões"; interview with Carol Maziviero, May 2021; Raquel Machado Werneck, "As Percepções dos Moradores do Grotão da Favela de Paraisópolis/SP Sobre o Processo de Urbanização" ("Perceptions of Residents of Grotão in the Paraisópolis Favela (SP) about the Urbanization Process") (master's thesis, University of São Paulo, 2018).

22. In addition to these eight Condomínios, the program included others located in the far northwest portion of Jardim Colombo and in the Grotinho sector to the southwest.

23. Interview with Diniz.

24. Although this particular site was directly expropriated, the Transfer of the Right to Build (Transferência do Direito de Construir) is a broader and innovative planning instrument regulated for the specific case of Paraisópolis (Decree 47.272 of May 12, 2006). This allows absentee owners of properties in Paraisópolis to donate properties to the City government and, in exchange, receive additional building rights.

25. Elisabete França quoted in Ana Carolina Carvalho Figueiredo, "Habitação Social e o Programa de Urbanização de Favelas: o conjunto Paraisópolis, em São Paulo," *Revista de Pesquisa em Arquitetura e Urbanismo* 19 (2021): 1–15, Ava Hoffman trans.

26. Divesare, "Elito Arquitetos, Paraisópolis Housing," https://divisare.com/projects /221586-elito-arquitetos-fabio-knoll-paraisopolis-housing. Interviews with residents, however, confirm that transporting heavy items up several flights of stairs can be quite problematic.

27. França quoted in Figueiredo.

28. Condomínio H is now known as Vila Andrade G and operated separately.

29. Interview with Elito.

30. Interview with Diniz.

31. Interviews with Vanessa, March 2022, and Daniel, February 2022.

32. Interview with Joildo Santos, November 2021.

33. Interview with João, July 2022.

34. "Casas Desabam em Paraisópolis, na Zona Sul de São Paulo," Globo.com, June 5, 2016, https://g1.globo.com/sao-paulo/noticia/2016/06/bombeiros-sao-acionados-para -desabamento-em-paraisopolis.html.

35. "Desabamento de Sobrado em Paraisópolis Deixa um Morto e 4 Feridos," Folha de S. Paulo, October 16, 2021, https://www1.folha.uol.com.br/cotidiano/2021/10 /desabamento-de-sobrado-em-paraisopolis-deixa-um-morto-e-4-feridos.shtml.

36. "Incêndio em Paraisópolis Deixa 280 Famílias Desabrigadas, Globo.com, December 10, 2013, https://g1.globo.com/sao-paulo/noticia/2013/12/incendio-em-paraisopolis -deixa-280-familias-desabrigadas.html; "Incêndio Atinge Cerca de 400 Barracos em Paraosópolis," March 1, 2017, Uol.com, https://noticias.uol.com.br/cotidiano/ultimas -noticias/2017/03/01/incendio-atinge-cerca-de-50-barracos-em-paraisopolis.htm; "Novo Incêndio em Paraisópolis Atinge Pelo Menos 20 Moradias," March 10, 2017, *Jornal do Comércio*, https://www.jornaldocomercio.com/_conteudo/2017/03/geral/551071 -novo-incendio-em-paraisopolis-atinge-pelo-menos-20-moradias.html; "Após Incêndio, Moradores de Paraisópos Querem Reconstruier Casas," Agência Brasil, March 3, 2017, https://agenciabrasil.ebc.com.br/geral/noticia/2017-03/apos-incendio-moradores-de -paraisopolis-prometem-reconstruir-casas.

37. Interview with Elito.

38. Daniela Chiarello Fastofski, Marco Aurélio Stumpf González, and Andrea Parisi Kern, "Sustainability Analysis of Housing Developments through the Brazilian Environmental Rating System Selo Casa Azul," *Habitat International* 67 (2017): 47 (table 2).

39. Caixa Econômica Federal, "Casa Azul, Quadio Resumo—Categorias, Critérios e Clasificação, Condomínio E/Condomínio G—Paraisópolis," https://www.caixa.gov.br /Downloads/selo_casa_azul/Ficha_Selo_Paraisopolis.pdf; https://www.caixa.gov.br /sustentabilidade/negocios-sustentaveis/selo-casa-azul-caixa/Paginas/default.aspx; Mantovani quoted in Cidade de São Paulo Comunicação, "Moradias de Paraisópolis Recebem Selo Casa Azul da Caixa Econômica Federal," June 15, 2012, https://www.prefeitura.sp .gov.br/cidade/secretarias/comunicacao/noticias/?p=107027, as translated by Ava Hoffman. Since applications depended on having funding from Caixa Econômica, only the later-phase Condomínios applied for certification, which begins at the initial project design phase. https://www.prefeitura.sp.gov.br/cidade/secretarias/comunicacao/noticias /?p=107027

40. Interview with Elito.

41. Interview with Diniz.

42. Interview with Paula Santoro, September 2021.

43. Interview with Juliana, June 2022.

44. Interviews with Daniel and Márcia.

45. Interviews with Márcia and Vanessa.

46. Ariadne P. Silva, "Casa própria para quê? Um estudo dos condicionantes sociológicos das transações imobiliárias em Conjuntos Habitacionais de Interesse Social" (Paper presented at the 31st conference of the Association of Latin American Sociologists, December 2017).

She found that about 85 percent of her sample of fifty-six households had relocated from elsewhere within Paraisópolis.

47. Interview with Santoro, September 2021.

48. João Fernando Meyer et al., "The Residential Real Estate Market in Paraisópolis: What Has Changed in the Past Ten Years," Seminário UrbFavelas, 2016, https:// www.researchgate.net/profile/Ariadne-Silva-2/publication/338263850_Seminario

_URBFAVELAS_2016_Rio_de_Janeiro_-RJ_-Brasil_MERCADO_IMOBILIARIO
_RESIDENCIAL_EM_PARAISOPOLIS_O_QUE_MUDOU_NOS_ULTIMOS_DEZ_ANOS
/links/5e0b6a8e299bf10bc3855695/Seminario-URBFAVELAS-2016-Rio-de-Janeiro-RJ
-Brasil-MERCADO-IMOBILIARIO-RESIDENCIAL-EM-PARAISOPOLIS-O-QUE-MUDOU
-NOS-ULTIMOS-DEZ-ANOS.pdf.

49. Interview with Santos.

50. Interview with Santos.

51. Interviews with Marcelo, June 2002, and Letícia, June 2022.

52. Interview with Diniz.

53. Interview with Edson Elito, May 2021.

54. Interview with Santoro.

55. Interview with Elito.

56. Interview with Diniz.

57. Interview with Santoro.

58. Interviews with Juliana and Daniel.

59. "Linha de Monotrilho do Aeriporto de Congohas faz 10 Anos Inacabada e com Incertezas," September 11, 2021, Folha de S. Paulo, https://www1.folha.uol.com.br /cotidiano/2021/09/linha-de-monotrilho-do-aeroporto-de-congonhas-faz-10-anos -inacabada-e-com-incertezas.shtml.

60. Interviews with Santos and Santoro.

61. Interviews with Marcelo, Márcia, Daniel, and Lucas, July 2022; Aline, July 2022; and Antônia, July 2022.

62. "Moradores de Paraisópolis voltam a barracos" ("Residents Return to their Shacks"), Jornal da Gazeta, August 29, 2016, https://www.youtube.com/watch?v=vG2h4yEpmTA. The latter estimate seems overstated, given that a study of informal transactions, also conducted in 2016, found that almost three-quarters of residents were still the original beneficiaries; Meyer et al., "The Residential Real Estate Market in Paraisópolis."

63. Interview with Raimundo. An interview with Daniel concurred that a substantial majority of original residents from 2010 remained more than a decade later.

64. Interview with Santoro.

65. Interview with Diniz.

66. Interview with Santoro.

67. Interview with Santos.

68. Márcio Alexandre Barbosa Lima, "Direito à moradia: práticas e legitimidades numa sociedade desigual" ("The Right to Housing: Practices and Legitimacy in an Unequal Society") (PhD diss., Universidade de Brasília, Brazil, 2013), 8.

69. Rafaela Benzi Alves, "Um outro olhar para a Urbanização de Favelas O caso de Paraisópolis em São Paulo" ("A Different Look at Favela Urbanization: The Case of Paraisópolis in São Paulo") (master's thesis, University of Coimbra, 2020), 179; https:// estudogeral.sib.uc.pt/handle/10316/93916.

70. Interview with Raimundo.

71. Alves, "Um outro olhar para a Urbanização de Favelas," 179.

72. Interview with Santos.

73. Keli Gois, "Raio-X da Urbanização de Paraisópolis" ("X-Ray of Paraisópolis' Urbanization"), *Jornal Espaço do Povo* 33 (2014).

CASE 3: PARIS OASIS

1. UNDRR and Centre for Research on the Epidemiology of Disasters, "Human Cost of Disasters: An Overview of the Last 20 Years 2000–2019" (UNDRR, 2020).

2. Daniel Mitchell et al., "Attributing Human Mortality during Extreme Heat Waves to Anthropogenic Climate Change," *Environmental Research Letters* 11, no. 7 (2016): 074006.

3. Klinenberg, *Heat Wave*.

4. Qi Zhao et al., "Global, Regional, and National Burden of Mortality Associated with Non-Optimal Ambient Temperatures from 2000 to 2019," *The Lancet Planetary Health* 5, no. 7 (2021): e415–425.

5. Spirn, *The Granite Garden*.

6. Lorenzo Mentaschi et al., "Global Long-Term Mapping of Surface Temperature Shows Intensified Intra-City Urban Heat Island Extremes," *Global Environmental Change* 72 (2022): 102441.

7. World Cities Cultural Forum, "Percent of Public Green Space: Parks and Gardens," https://worldcitiescultureforum.com/data/?data-screen=percentage_of_public_green_space_parks_gardens

8. Carly D. Ziter et al., "Scale-Dependent Interactions between Tree Canopy Cover and Impervious Surfaces Reduce Daytime Urban Heat during Summer," *Proceedings of the National Academy of Sciences of the United States of America* 116, no. 15 (2019): 7575–7580.

9. Klinenberg, *Heat Wave*.

10. Jeremy S. Hoffman, Vivek Shandas, and Nicholas Pendleton, "The Effects of Historical Housing Policies on Resident Exposure to Intra-Urban Heat: A Study of 108 US Urban Areas," *Climate* 8, no. 1 (2020): 12.

11. Mairie de Paris, "Paris Resilience Strategy: Fluctuat Nec Mergitur," 2018, Paris, City of Paris, 35. https://resilientcitiesnetwork.org/downloadable_resources/Network/Paris-Resilience-Strategy-English.pdf

12. Leah Flax et al., "Greening Schoolyards—An Urban Resilience Perspective," *Cities* 106 (2020): 102890.

13. City of Paris, "Les cours Oasis," March 10, 2023, https://www.paris.fr/pages/les-cours-oasis-7389.

14. Maria Sitzoglou, "The OASIS Schoolyard Project Journal No. 1" (Urban Innovative Actions, European Union, January 2020).

15. Samuel Ferrer et al., "OASIS Schoolyards: Recommendations Booklet for Transforming Schoolyards," 2022, Paris, City of Paris Resilience Mission.

16. Paris Resilience Office, "Transforming 10 Parisian Schoolyards into Local Community Spaces Adapted to Climate Change," 2020, Paris, City of Paris.

17. Sitzoglou, "The OASIS Schoolyard Project Journal No. 1."

18. Ferrer et al., "OASIS Schoolyards: Recommendations Booklet for Transforming Schoolyards," 32.

19. Interview with Laurence Duffort, March 2021.

20. Interview with Duffort.

21. Ferrer et al., "OASIS Schoolyards: Recommendations Booklet for Transforming Schoolyards."

22. Interview with Duffort.

23. Laboratoire Interdisciplinaire d'Evaluation des Politiques Publiques, "Evaluation Report of the Uses of the School Yards: Summary," 2022, Paris, SciencesPo; https://www.sciencespo.fr/liepp/sites/sciencespo.fr.liepp/files/EN_Summary%20Oasis%20Schoolyard%20uses.pdf

24. Interview with Clara Chevrier, April 2021.

25. Ferrer et al., "OASIS Schoolyards: Recommendations Booklet for Transforming Schoolyards."

26. Sitzoglou, "The OASIS Schoolyard Project Journal No. 1."

27. Ferrer et al., "OASIS Schoolyards: Recommendations Booklet for Transforming Schoolyards."

28. Interview with Martin Hendel, March 2021.

29. Laboratoire Interdisciplinaire D'Evaluation des Politiques Publiques, "Evaluation Report on the Participative Process for the Oasis Playgrounds: Summary," 2022. Paris, SciencesPo.

30. Anonymous interview, 2021.

31. Interview with Hendel.

32. Sitzoglou, "The OASIS Schoolyard Project Journal No. 1."

33. Interview with Raphaëlle Thiollier, March 2021.

34. Interview with Thiollier.

35. Interview with Maria Sitzoglou, March 2021.

36. Interview with Hendel.

37. City of Paris, "168 «rues aux écoles» dans Paris," https://www.paris.fr/pages/57-nouvelles-rues-aux-ecoles-dans-paris-8197.

LEARNING FROM THREE STRUGGLES FOR EQUITABLY RESILIENT ENVIRONMENTS

1. Frantzeskaki et al., "Nature-Based Solutions."

EQUITABLY RESILIENT SECURITY

1. Coaffee et al., *The Everyday Resilience of the City*; Vale and Campanella, *The Resilient City*.

2. Laura E. R. Peters, "Beyond Disaster Vulnerabilities: An Empirical Investigation of the Causal Pathways Linking Conflict to Disaster Risks," *International Journal of Disaster Risk Reduction* 55 (2021): 102092.

3. Robyn Molyneaux et al., "Interpersonal Violence and Mental Health Outcomes following Disaster," *BJPsych Open* 6, no. 1 (2020): e1; Enarson et al., "Gender and Disaster"; Alyssa Mari Thurston, Heidi Stöckl, and Meghna Ranganat, "Natural Hazards, Disasters and Violence against Women and Girls: A Global Mixed-Methods Systematic Review," *BMJ Global Health* 6, no. 4 (2021): e004377.

4. Andrew J. Prelog, "Modeling the Relationship between Natural Disasters and Crime in the United States," *Natural Hazards Review* 17, no. 1 (2016).

5. Mindy Thompson Fullilove, *Root Shock* (New Village Press, 2016).

6. Oscar Newman, *Defensible Space* (Macmillan, 1973); Jane Jacobs, *The Death and Life of Great American Cities* (Vintage, 1961).

7. Diane E. Davis, *Urban Resilience in Situations of Chronic Violence* (MIT Center for International Studies, 2012).

8. Jeff Byles, *Rubble: Unearthing the History of Demolition* (Harmony Books, 2005); David Harvey, *Paris, Capital of Modernity* (Routledge, 2005); Lefebvre, *The Urban Revolution*.

9. World Commission on Dams, *Dams and Development*; Rohan D'Souza, *Drowned and Dammed: Colonial Capitalism and Flood Control in Eastern India* (Oxford University Press, 2006).

10. Amita Baviskar, "Between Violence and Desire," *International Social Science Journal* 55, no. 175 (2003): 90.

11. Hernando de Soto, *The Mystery of Capital* (Basic Books, 2000).

12. Andreana Reale and John Handmer, "Land Tenure, Disasters and Vulnerability," *Disasters* 35, no. 1 (2011): 160–182.

13. Geoffrey Payne, "Urban Land Tenure Policy Options: Titles or Rights?" *Habitat International* 25, no. 3 (2001): 415–429.

14. Geoffrey Payne, Alain Durand-Lasserve, and Carole Rakodi, "The Limits of Land Titling and Home Ownership," *Environment and Urbanization* 21, no. 2 (2009): 443–462.

15. Edesio Fernandes, *Regularization of Informal Settlements in Latin America* (Lincoln Institute of Land Policy, 2011).

16. Reale and Handmer, "Land Tenure, Disasters and Vulnerability"; David Satterthwaite, "The Political Underpinnings of Cities' Accumulated Resilience to Climate Change," *Environment and Urbanization* 25, no. 2 (2013): 381–391.

17. Dan Orcherton, David Mitchell, and Darryn McEvoy, "Perceptions of Climate Vulnerability, Tenure Security and Resettlement Priorities," *Australian Geographer* 48, no. 2 (2017): 235–254.

18. McEvoy et al., "Land Tenure and Urban Climate Resilience"; Juan Pablo Sarmiento, Vicente Sandoval, and Meenakshi Jerath, "The Influence of Land Tenure and Dwelling Occupancy on Disaster Risk Reduction," *Progress in Disaster Science* 5 (2020): 100054.

19. Jee Young Lee and Shannon Van Zandt, "Housing Tenure and Social Vulnerability to Disasters: A Review of the Evidence," *Journal of Planning Literature* 34, no. 2 (2019): 156–170.

20. Mark Brennan et al., "A Perfect Storm? Disasters and Evictions," *Housing Policy Debate* 32, no. 1 (2022): 52–83.

21. E. L. Raymond, T. Green, and M. Kaminski, "Preventing Evictions After Disasters," *Housing Policy Debate* 32, no. 1 (2022): 35–51.

22. Fothergill and Peek, "Poverty and Disasters in the United States"; Aditi Mehta, Mark Brennan, and Justin Steil, "Affordable Housing, Disasters, and Social Equity," *Journal of the American Planning Association* 86, no. 1 (2020): 75–88.

23. Finch et al., "Disaster Disparities and Differential Recovery"; Kevin Fox Gotham, "Racialization and Rescaling: Post-Katrina Rebuilding and the Louisiana Road Home Program," *International Journal of Urban and Regional Research* 38, no. 3 (2014): 773–790.

24. Yang Zhang and Walter Gillis Peacock, "Planning for Housing Recovery? Lessons Learned from Hurricane Andrew," *Journal of the American Planning Association* 76, no. 1 (2009): 5–24.

25. Keenan et al., "Climate Gentrification."

26. Anguelovski et al., "Opinion."

27. DuPuis and Greenberg. "The Right to the Resilient City."

28. Savannah Cox, "Inscriptions of Resilience: Bond Ratings and the Government of Climate Risk in Greater Miami, Florida," *Environment and Planning A: Economy and Space* 54, no. 2 (2021): 295–310.

29. David Satterthwaite et al., *Adapting to Climate Change in Urban Areas* (IIED, 2007); Reale and Handmer, "Land Tenure, Disasters and Vulnerability."

30. Klein, *The Shock Doctrine*; Gotham and Greenberg, *Crisis Cities*.

31. Kenneth A. Gould and Tammy L. Lewis, "Green Gentrification and Disaster Capitalism in Barbuda," *NACLA Report on the Americas* 50, no. 2 (2018): 148–153.

32. Justin Hosbey and J. T. Roane, "A Totally Different Form of Living: On the Legacies of Displacement and Marronage as Black Ecologies," *Southern Cultures* 27, no. 1 (2021): 68–73.

33. Woods, *Development Drowned and Reborn*.

34. Lawrence J. Vale, *Purging the Poorest* (University of Chicago Press, 2013).

35. Justin Farrell et al., "Effects of Land Dispossession and Forced Migration on Indigenous Peoples in North America," *Science*, October 29, 2021.

36. Mona Fawaz, "Planning and the Making of a Propertied Landscape," *Planning Theory and Practice* 18, no. 3 (2017): 365–384.

37. Mathur and Da Cunha, *Mississippi Floods.*

38. Bhattacharyya, *Empire and Ecology in the Bengal Delta.*

39. Malini Ranganathan and Anne Bonds, "Racial Regimes of Property: Introduction to the Special Issue," *Environment and Planning D: Society and Space* 40, no. 2 (2022): 197–207.

40. Koslov, "The Case for Retreat"; Robert Freudenberg et al., *Buy-In for Buyouts: The Case for Managed Retreat from Flood Zones* (Lincoln Institute of Land Policy, 2016).

41. Mach and Siders, "Reframing Strategic, Managed Retreat."

42. Marino, "Adaptation Privilege and Voluntary Buyouts."

43. Balakrishnan Balachandran, Robert B. Olshansky, and Laurie A. Johnson, "Planning for Disaster-Induced Relocation of Communities," *Journal of the American Planning Association* 88, no. 3 (2022): 288–304.

44. Orcherton et al., "Perceptions of Climate Vulnerability."

45. Kasia Paprocki, "Threatening Dystopias: Development and Adaptation Regimes in Bangladesh," *Annals of the American Association of Geographers* 108, no. 4 (2018): 955–973.

46. Checker, "Eco-Apartheid"; Hodson and Mavin, "Urban Ecological Security."

47. Dolores Hayden, "Urban Landscape History: The Sense of Place and the Politics of Space," in *Understanding Ordinary Landscapes*, ed. Paul Erling Groth and Todd W. Bressi (Yale University Press, 1997), 111.

48. W. Neil Adger et al., "This Must Be the Place: Underrepresentation of Identity and Meaning in Climate Change Decision-Making," *Global Environmental Politics* 11, no. 2 (2011): 1–25.

49. Patrick Devine-Wright and Tara Quinn, "Dynamics of Place Attachment in a Climate Changed World," in *Place Attachment*, 2nd ed. (Routledge, 2020), 235.

50. Kevin Lynch, *The Image of the City* (MIT Press, 1960).

51. Thomas F. Gieryn, "A Space for Place in Sociology," *Annual Review of Sociology* 26 (2000): 463–496.

52. Nicola Dempsey and Mel Burton, "Defining Place-Keeping," *Urban Forestry and Urban Greening* 11, no. 1 (2012): 11–20.

53. Briavel Holcomb, "Place Marketing," in *Imaging the City*, ed. Lawrence J. Vale and Sam Bass Warner (CUPR Press, 2001), 33–55.

54. Mike Lydon, *Tactical Urbanism* (Island Press, 2015).

55. David Harvey, *Rebel Cities* (Verso, 2013).

56. Marcus Anthony Hunter et al., "Black Placemaking," *Theory, Culture and Society* 33, no. 7–8 (2016): 31–56; Brandi Thompson Summers, *Black in Place* (University of North Carolina Press, 2019).

57. Patrick Devine-Wright, "Think Global, Act Local? The Relevance of Place Attachments and Place Identities in a Climate Changed World," *Global Environmental Change* 23, no. 1 (2013): 61–69.

58. "Gentilly Resilience District Public Art Project," Arts Council of New Orleans, https://new.artsneworleans.org/gentilly-resilience-district-public-art-project/.

59. Payne et al., "The Limits of Land Titling and Home Ownership"; McEvoy et al., "Land Tenure and Urban Climate Resilience."

60. B. Pérez Salazar, "Lecciones de Gobernabilidad Desde El Urbanismo Social de Montaña: Estudio de Caso de la Intervención en la Quebrada Juan Bobo y el Surgimiento del Sector Nuevo Sol de Oriente en Medellin, Colombia," Unpublished working paper, 2010.

61. Payne et al., "The Limits of Land Titling and Home Ownership."

62. Liz Alden Wily, "Challenges for the New Kid on the Block—Collective Property," in *On Common Ground: International Perspectives on the Community Land Trust*, ed. John Emmeus Davis, Line Algoed, and María E. Hernández-Torrales (Terra Nostra Press, 2020), 63.

63. Daniel Aldana Cohen and Mark Paul, "The Case for Social Housing," The Lab by the Appeal, November 2, 2020, https://theappeal.org/the-lab/report/the-case-for-social-housing/.

64. Vale, *After the Projects*.

65. Linda Shi et al., "Promises and Perils of Collective Land Tenure in Promoting Urban Resilience: Learning from China's Urban Villages," *Habitat International* 77 (2018): 1–11.

66. Shi et al., "Promises and Perils of Collective Land Tenure"; Zachary Lamb et al., "Resident-Owned Resilience: Can Cooperative Land Ownership Enable Transformative Climate Adaptation for Manufactured Housing Communities?" *Housing Policy Debate* 33, no. 5 (2022): 1055–1077.

67. John Emmeus Davis, Line Algoed, and María E. Hernández-Torrales, eds., *On Common Ground: International Perspectives on the Community Land Trust* (Terra Nostra Press, 2020); Patricia Basile and Meagan M. Ehlenz, "Examining Responses to Informality in the Global South: A Framework for Community Land Trusts and Informal Settlements," *Habitat International* 96 (2020): 102108.

68. Jessica Grannis, *Community Land = Community Resilience: How Community Land Trusts Can Support Urban Affordable Housing and Climate Initiatives* (Georgetown Climate Center, 2021).

69. Miriam Axel-Lute, "Gaining Scale: How Some Community Land Trusts Are Getting Bigger," Shelterforce (blog), July 13, 2021, https://shelterforce.org/2021/07/13/gaining-scale-how-some-community-land-trusts-are-getting-bigger/; Meagan M. Ehlenz, "Making Home More Affordable: Community Land Trusts Adopting Cooperative Ownership Models to Expand Affordable Housing," *Journal of Community Practice* 26, no. 3 (2018): 283–307.

CASE 4: PASADENA TRAILS

1. Allan D. Wallis, *Wheel Estate: The Rise and Decline of Mobile Homes* (Oxford University Press, 1991). Typically, the term "mobile home" is used to refer to earlier

generations of factory-built housing. The term "manufactured home" typically refers to homes built after the adoption of the Department of Housing and Urban Development's HUD Manufactured Home Construction and Safety Standards in 1976. Because the communities that we profile feature both mobile and manufactured homes, we use the terms interchangeably.

2. Manufactured Housing Institute, "2020 Manufactured Housing Facts: Industry Overview," May 2020. https://www.manufacturedhousing.org/wp-content/uploads /2020/07/2020-MHI-Quick-Facts-updated-05-2020.pdf.

3. US Census Bureau, "Cost and Size Comparison: New Manufactured Homes and New Single Family Site-Built Homes, 2014–2020," Manufactured Housing Survey, June 2021, https://www.census.gov/data/tables/time-series/econ/mhs/annual-data.html.

4. CFPB, *Manufactured-Housing Consumer Finance in the US* (Consumer Financial Protection Bureau, 2014); Zachary Lamb, Linda Shi, and Jason Spicer, "Why Do Planners Overlook Manufactured Housing and Resident-Owned Communities as Sources of Affordable Housing and Climate Transformation?" *Journal of the American Planning Association* 89, no. 1 (2023): 72–79.

5. Esther Sullivan, *Manufactured Insecurity: Mobile Home Parks and Americans' Tenuous Right to Place* (University of California Press, 2018).

6. Tony Petosa et al., *Commercial Real Estate: Manufactured Home Community Financing Handbook* (Wells Fargo, 2020).

7. Lamb et al., "Resident-Owned Resilience."

8. Gabbe and Pierce, "Extreme Heat Vulnerability"; Gabbe et al., "Subsidized Households and Wildfire Hazards"; Gregory Pierce, C. J. Gabbe, and Annabelle Rosser, "Households Living in Manufactured Housing Face Outsized Exposure to Heat and Wildfire Hazards: Evidence from California," *Natural Hazards Review* 23, no. 3 (2022); Gregory Pierce, C. J. Gabbe, and Silvia R. Gonzalez, "Improperly-Zoned, Spatially-Marginalized, and Poorly-Served? An Analysis of Mobile Home Parks in Los Angeles County," *Land Use Policy* 76 (2018): 178–185.

9. Cutter et al., "Social Vulnerability to Environmental Hazards."

10. Lamb et al., "Resident-Owned Resilience."

11. Freddie Mac Multifamily, *Spotlight on Underserved Markets: Manufactured Housing Resident-Owned Communities* (Freddie Mac Multifamily, 2019).

12. Interview with Paul Bradley, January 2021.

13. We use the term "Latinx" here because of the greater inclusiveness of that term. Elsewhere in the chapter, when describing how residents and leaders refer to their own identities, we use the term "Latino" because that is the term used by resident leaders themselves. For more on the debates surrounding the ethnonyms used to classify people from Spanish or Latin American descent, see Ivis García, "Cultural Insights for Planners: Understanding the Terms Hispanic, Latino, and Latinx," *Journal of the American Planning Association* 86, no. 4 (2020): 393–402.

14. Interview with Mary O'Hara, October 2020.

15. Interview with O'Hara.

16. Mike Snyder, "Mobile Home Park's Residents Celebrate Ownership," Chron, November 16, 2009, https://www.chron.com/neighborhood/pasadena-news/article /Mobile-home-park-s-residents-celebrate-ownership-1727304.php.

17. Interview with Estela Garza, June 2022.

18. "Pasadena Trails Resident Owned Community Mobile Home Park in Pasadena, TX | MHVillage," MHVillage, accessed August 4, 2022, https://www.mhvillage.com/parks /24496.

19. Interview with Blanca Vega, June 2022.

20. Interview with Vega.

21. Interview with Garza.

22. Interview with Garza.

23. Interview with Kim Capen, July 2020.

24. Interview with Garza.

25. Interview with Garza.

26. Interview with Garza.

27. "Lakeville Village Ready to Flip the Switch on Solar," ROC USA® (blog), accessed November 2, 2020, https://rocusa.org/news/solar-lakeville-village/.

28. Interview with Victoria Clark, February 2020.

29. Interview with Garza.

30. Meiquing Li, Pavan Yedavalli, Liubing Xie, Sai Balakrishnan, Zachary Lamb, and Karen Chapple. "COVID-19 and the Future of Urban Life." *Berkeley Planning Journal* 32, no. 1 (2022).

31. Interview with Garza.

32. Interview with Garza.

33. Interview with Milena Polak, December 2020.

34. Interview with Garza.

35. "Pasadena Trails Resident Owned Community Mobile Home Park in Pasadena, TX | MHVillage."

36. Interview with Vega.

37. Mike Bullard, "Texas ROC's Forethought Keeps Harvey Flooding at Bay," ROC USA® (blog), August 31, 2017, https://rocusa.org/news/texas-rocs-forethought-keeps -harvey-flooding-bay/.

38. Interview with Gilsrud.

39. Interview with Natividad Seefeld, July 2020.

40. Interview with Thomas Choate, May 2020.

41. Interview with Cahill.

42. Interview with Seefeld.

43. Interview with Garza.

44. Lamb et al., "Resident-Owned Resilience."

45. Interview with Garza.

46. Elsa Noterman, "Beyond Tragedy: Differential Commoning in a Manufactured Housing Cooperative," *Antipode* 48, no. 2 (2016): 433–452.

47. "Pasadena Trails Resident Owned Community Mobile Home Park in Pasadena, TX | MHVillage."

48. Interview with Garza.

49. Interview with Vega.

50. Romero-Lankao et al., "Urban Transformative Potential in a Changing Climate."

CASE 5: COMUNIDAD MARÍA AUXILIADORA

1. Sarah T. Hines, *Water for All* (University of California Press, 2022), 8, 10; Willem Assies, "David Versus Goliath in Cochabamba," *Latin American Perspectives* 30, no. 3 (2003): 18–19.

2. Assies, "David Versus Goliath in Cochabamba," 21–34; Juan Forero, "Who Will Bring Water to the Bolivian Poor? *New York Times*, December 15, 2005; Frontline World (PBS), "Timeline: Cochabamba Water Revolt," June 2002, https://www.pbs.org /frontlineworld/stories/bolivia/timeline.html; Jim Shultz, "The Water War Dispatches in Full," The Democracy Center, 2000–2001, https://www.democracyctr.org/the -water-war-dispatches-in-full; Amonah Achi and Rebecca Kirchheimer, "Innovar Para Alcanzar El Derecho Humano Al Agua de La Zona Sur de Cochabamba," in *Apoyo a La Gestión de Comités de Agua Potable*, ed. Franz Quiroz, Nicolás Faysse, and Raúl Ampuero (Centro AGUA-UMSS, 2006).

3. Hines, *Water for All*, 2, 5, 7; Morales quoted in Hines, 4.

4. Interviews with Rose Mary Irusta, April 2021, Jhenny, April 2021, and Juana, April 2021.

5. Interviews with Manolo Bellott, April 2021, and Irusta.

6. Patrick Bodart, *Buscar un Lugar . . . y Construir una Comunidad: Comunidad María Auxiliadora, Cochabamba* (Periferia, 2014), 19; https://hic2.puntzero.cat/es/documento -buscar-un-lugar-y-construir-una-comunidad/.

7. Angela Lara García, "Planificación y Gestión Autónoma de los Sevicios Básicos en la Zona Sur de Cochabamba (Bolivia)" (master's thesis, Universidad Internacional de Andalucia, 2011), 189; Partnership Actions for Mitigating Systems of Global Change; see Frank Haupt and Ulrike Müller-Böker, "Grounded Research and Practice: PAMS," *Mountain Research and Development* 25, no. 2 (2005): 100–103.

8. Habitat International Coalition members, https://hic-al.org/miembros/.

9. Aguilar Zeballos quoted in Amy Booth, "The Communities of Cochabamba Taking Control of Their Own Water Supply," *The Guardian*, June 9, 2016, https:// www.theguardian.com/global-development-professionals-network/2016/jun/09 /communities-cochabamba-taking-control-water-supply-bolivia.

10. Interview with Bellott.

11. García, "Planificación y Gestión Autónoma de los Sevicios Básicos," appendix 1.

12. Interview with Bellott.

13. Interview with Juana, June 2022.

14. Interview with María Serrano, April 2021.

15. Interview with Ema, April 2021.

16. Interview with Leonor, August 2022; personal communication with Bellott, October 2022.

17. World Habitat Awards, 2008 Finalist, https://world-habitat.org/world-habitat-awards/winners-and-finalists/habitat-para-la-mujer-the-maria-auxiliadora-community/.

18. American Planning Association, *Supporting Innovation: Case Studies from Latin America on Sustainable Housing and Community Development* (APA, 2014), 23–27, https://planning-org-uploaded-media.s3.amazonaws.com/publication/download_pdf/Supporting-Innovation-Case-Studies-from-Latin-America.pdf

19. Interview with Osvaldo Rocha, August 2022.

20. Interview with Rocha. Other interviewees related similar complaints about losing their lots.

21. Interview with Irusta, 2021.

22. Nasya Sara Razavi, "Social Control and Public Water in Cochabamba, Bolivia," (PhD diss., Queen's University, Canada, 2019), 134.

23. Razavi, "Social Control and Public Water," 134.

24. Sugranyes quoted in "Activistas Piden Libarer a Irusta," Ultimas Noticias Bolivia.com, October 5, 2012, http://unoticiasbolivia.blogspot.com/2012/10/activistas-piden-liberar-irusta.html.

25. Razavi, "Social Control and Public Water," 135.

26. Interview with Irusta, 2021.

27. Interview with Juanita.

28. Interview with Irusta, 2021.

29. Interview with Irusta, June 2022.

30. Interview with Ema.

31. Interview with Rocha.

32. Interview with Irusta, June 2022.

33. Interview with Jhenny.

34. Interview with Irusta, June 2022; interviews with Juanita, Margarita, Teo, and Ema, June 2022.

35. Interview with Rocha.

36. Interview with Bellott.

37. Interviews with Irusta, 2021, 2022.

38. Interview with Margarita, August 2022.

39. Interview with Leonor.

40. Hines, *Water for All*, 14.

41. Interview with Leonor.

42. Interview with Bellott.

43. Interview with Ema.

44. Interview with Irusta, 2021.

45. Nuestro Perfil y Lecciones de Nuestros Mejores Prácticas, *Hábitat para la Mujer Comunidad María Auxiliadora* (Comunidad María Auxiliadora, 2014), 4, https://silo .tips/download/esta-es-nuestra-historia-y-estos-nuestros-sueos.

46. World Habitat Awards, "'Habitat para la Mujer'—the Maria Auxiliadora Community," https://world-habitat.org/world-habitat-awards/winners-and-finalists/habitat -para-la-mujer-the-maria-auxiliadora-community/.

47. American Planning Association, *Supporting Innovation*, 27.

48. World Habitat Awards, "'Habitat para la Mujer.'"

49. Interview with Giovana, April 2021.

50. Interview with Jhenny.

51. Interview with Bellott.

52. Interview with Giovana.

53. Interview with Bellott; interviews with Ema and Maria, April 2021.

54. Interview with Giovana.

55. Interview with Irusta, 2021.

56. Interview with Jhenny.

57. Interview with Bellott.

58. Interview with Bellott.

59. James Holston, "Spaces of Insurgent Citizenship," in *Cities and Citizenship*, ed. James Holston (Duke University Press, 1998), 155–175.

60. Hines, *Water for All*, 13–14.

61. Interview with Irusta, 2021.

CASE 6: BAAN MANKONG

1. Doryane Kermel-Torrès, ed., "Chapter 7. Bangkok and the Bangkok Metropolitan Region," in *Atlas of Thailand* (IRD Éditions, 2020), 151–159.

2. "Baan Mankong Urban—CODI," https://en.codi.or.th/baan-mankong-housing /baan-mankong-rural/; Somsook Boonyabancha and Thomas Kerr, "Lessons from CODI on Co-Production," *Environment and Urbanization* 30, no. 2 (2018): 444–460.

3. "Canal-Side Housing—CODI," https://en.codi.or.th/baan-mankong-housing/canal -upgrading-project/.

4. "Canal-Side Housing—CODI."

5. Hayden Shelby, "Commoning or Being Commoned?" *Planning Theory* 21, no. 4 (2022): 333–353.

6. Interview with Somsook Boonyabancha, November 2021.

7. Asian Coalition for Housing Rights, "A Conversation about Upgrading at Bang Bua," 2008, 3, https://pubs.iied.org/sites/default/files/pdfs/migrate/G02376.pdf.

8. CODI, *Baan Mankong at Klong Bang Bua Community Guidebook* (CODI, 2012).

9. CODI, *Baan Mankong at Klong Bang Bua Community Guidebook*, 8.

10. CODI *Baan Mankong at Klong Bang Bua Community Guidebook*.

11. Interview with Bang Bua Community Leaders, December 2021.

12. Nuttavikhom Phanthuwongpakdee, "Living with Floods: Moving Towards Resilient Local-Level Adaptation in Central Thailand" (PhD diss., King's College London and National University of Singapore, 2016).

13. Interview with Nattawut Usavagovitwong, September 2021.

14. Interview with Boonyabancha.

15. "Baan Mankong Urban—CODI."

16. CODI, *Baan Mankong at Klong Bang Bua Community Guidebook*.

17. Interview with Monthon Janjamsai, October 2021.

18. Rittirong Chutapruttikorn, "'We Speak Louder than Before.' A Reflection on Participatory Housing Design for Low-Income People," *Trialog* 102 (2009): 64–68.

19. Shelby, "Empowerment or Responsibility?" 517.

20. Diane Archer, "Baan Mankong Participatory Slum Upgrading in Bangkok, Thailand," *Habitat International* 36, no. 1 (2012): 178–184.

21. CODI, *Baan Mankong at Klong Bang Bua Community Guidebook*.

22. Interview with Prapassom Chuthong, November 2021.

23. Interview with Thanatch Naruepornpong, August 2021.

24. Archer, "Baan Mankong Participatory Slum Upgrading"; Shelby, "Empowerment or Responsibility?"

25. Interview with Naruepornpong.

26. Interview with Naruepornpong.

27. Shelby, "Empowerment or Responsibility?"

28. Giovanna Astolfo and Camillo Boana, eds., *Bangkok: On Transformation and Urbanism*, Cities, Design, and Transformation (The Bartlett Development Planning Unit, University College London, 2016), 76.

29. CODI, *Baan Mankong at Klong Bang Bua Community Guidebook*.

30. Somsook Boonyabancha, "Land for Housing the Poor—by the Poor," *Environment and Urbanization* 21, no. 2 (2009): 324.

31. Interview with Bang Bua Community Leaders.

32. Interview with Naruepornpong.

33. Archer, "Baan Mankong Participatory Slum Upgrading," 178.

34. Shelby, "Empowerment or Responsibility?" 516.

35. Interview with Naruepornpong.

36. CODI, *Baan Mankong at Klong Bang Bua Community Guidebook*.

37. Interview with Naruepornpong.

38. CODI, *Baan Mankong at Klong Bang Bua Community Guidebook*.

39. CODI, *Baan Mankong at Klong Bang Bua Community Guidebook*.

40. Shelby, "Empowerment or Responsibility?" 519.

41. Interview with Bang Bua Community Leaders.

42. Shelby, "Commoning or Being Commoned?" 2.

43. Interview with Naruepornpong.

44. Interview with Bang Bua Community Leaders.

45. Interview with Naruepornpong.

46. Astolfo and Boana, *Bangkok: On Transformation and Urbanism*, 75.

47. Boonyabancha, November 2021.

48. CODI, *Baan Mankong at Klong Bang Bua Community Guidebook*.

49. Boonyabancha and Kerr, "Lessons from CODI on Co-Production," 450.

50. CODI, *Baan Mankong at Klong Bang Bua Community Guidebook*.

51. Interview with Bang Bua Community Leaders.

52. CODI, *Baan Mankong at Klong Bang Bua Community Guidebook*.

53. CODI, *Baan Mankong at Klong Bang Bua Community Guidebook*.

54. Archer, "Baan Mankong Participatory Slum Upgrading," 179.

55. Interview with Naruepornpong.

56. Asian Coalition for Housing Rights, "A Conversation about Upgrading at Bang Bua," 5.

57. Shelby, "Commoning or Being Commoned?"

58. Tanvi Bhatkal and Paula Lucci, *Community-Driven Development in the Slums: Thailand's Experience* (Overseas Development Institute, 2015), 6.

59. Interview with Usavagovitwong.

60. Anonymous interview, July 2022.

61. Interview with Bang Bua Community Leaders.

LEARNING FROM THREE STRUGGLES FOR EQUITABLY RESILIENT SECURITY

1. Noterman, "Beyond Tragedy."

EQUITABLY RESILIENT LIVELIHOODS

1. Alexandria Ocasio-Cortez, "Recognizing the Duty of the Federal Government to Create a Green New Deal," Pub. L. No. H.R. 109 (2019), https://www.congress.gov /bill/116th-congress/house-resolution/109/text; Zachary Lamb and Todd Vachon, "Working for a Just Adaptation," POWER: Infrastructure in America, February 6, 2020, https://power.buellcenter.columbia.edu/essays/working-just-adaptation.

2. Robert Chambers, "Poverty and Livelihoods: Whose Reality Counts?" *Environment and Urbanization* 7, no. 1 (1995): 174.

3. Chambers, "Poverty and Livelihoods," 173.

4. Chambers, "Poverty and Livelihoods," 189.

5. Sen, *Commodities and Capabilities*.

6. Brian King, "Spatialising Livelihoods," *Transactions of the Institute of British Geographers* 36, no. 2 (2011): 297–313.

7. Robert Chambers and Gordon R. Conway, *Sustainable Rural Livelihoods*, DP 296 (Institute of Development Studies, 1991). This also coincided with the United Nations Committee on Economic, Social and Cultural Rights (UNCESCR) assertion of a right to an adequate standard of living as part of a right to adequate housing; United Nations Office of the High Commissioner for Human "CESCR General Comment No. 4: The Right to Adequate Housing," adopted December 13, 1991, https://www.refworld.org /pdfid/47a7079a1.pdf.

8. Stephen More and Nora McNamara, *Sustainable Livelihood Approach: A Critique of Theory and Practice* (Springer, 2013), 10, 20–23; Anthony Bebbington, "Capitals and Capabilities," *World Development* 27, no. 12 (1999): 2022.

9. Gan Xinyue, Chen Yulin, and Bian Lanchun, "From Redevelopment to In Situ Upgrading," *China City Planning Review* 28, no. 4 (2019): 30–41; Ya Ping Wang, Yanglin Wang, and Jiansheng Wu, "Urbanization and Informal Development in China," *International Journal of Urban and Regional Research* 33, no. 4 (2009): 957–973.

10. Martha Alter Chen, with Jenna Harvey, Caroline Wanjiku Kihato, and Caroline Skinner, *Inclusive Public Spaces for Informal Livelihoods* (WIEGO and Cities Alliance, 2018); Alison Brown, ed., *Rebel Streets and the Informal Economy* (Routledge, 2017); Rodrigo Meneses-Reyes and José A. Caballero-Juárez. "The Right to Work on the Street," *Planning Theory* 13, no. 4 (2014): 370–386; Don Mitchell, *The Right to the City* (Guilford, 2003).

11. Nora Ruth Libertun de Duren, "Why There? Developers' Rationale for Building Social Housing in the Urban Periphery in Latin America," *Cities* 72, Part B (2018): 411–420.

12. Robert M. Buckley, Achilles Kallergis, and Laura Wainer, "Addressing the Housing Challenge: Avoiding the Ozymandias Syndrome," *Environment and Urbanization* 28, no. 1 (2016): 119–138; Vanessa Watson, "African Urban Fantasies: Dreams or Nightmares," *Environment and Urbanization* 26, no. 1 (2014): 215–231.

13. Laura Sara Wainer and Lawrence J. Vale, "Wealthier-but-Poorer: The Complex Sociology of Homeownership at Peripheral Housing in Cartagena, Colombia," *Habitat International* 114, no. 3 (2021): 102388.

14. Arturo Contreras, "Reviving Mexico's Modern Ghost Towns," Pie de Página, April 9, 2021, https://piedepagina.mx/reviving-mexicos-modern-ghost-towns/.

15. Nancy Kwak, "Myth #8: Public Housing Is Only for Poor People," in *Public Housing Myths*, ed. Nicholas Dagen Bloom, Fritz Umbach, and Lawrence J. Vale (Cornell University Press, 2015), 175–186; Hong Kong Housing Authority, *An Illustrated Summary of 40 Years of Public Housing Development in Hong Kong, 1953–1993* (Hong Kong Housing Authority, 1993).

16. Eve Blau, *The Architecture of Red Vienna, 1919–1934* (MIT Press, 1999).

17. Walter Gropius, *Scope of Total Architecture*, World Perspectives (Allen and Unwin, 1956); Dolores Hayden, "What Would a Nonsexist City Be Like?" *Ekistics* 52, no. 310 (1985): 99–107.

18. Blau, *The Architecture of Red Vienna*.

19. The Architecture Lobby, "Statement on the Green New Deal," The Architecture Lobby (blog), May 1, 2019, https://architecture-lobby.org/2019/05/01/statement-on -the-green-new-deal/.

20. Kanako Iuchi and Elizabeth Maly, "Residential Relocation Processes in Coastal Areas," in *Coming Home After Disaster: Multiple Dimensions of Housing Recovery*, ed. Alka Sapat and Ann-Margaret Esnard (Routledge, 2017), 211, 212–213, 215, 219, 221, 223.

21. Judith Shaw and Iftekhar Ahmed, *Design and Delivery of Post-Disaster Housing Resettlement Programs*, Report 6 (Monash Asia Institute, 2009); Graeme Bristol, "Surviving the Second Tsunami," in *Rebuilding After Disasters*, ed. Gonzalo Lizarralde, Cassidy Johnson, and Colin Davidson (Routledge, 2010), 133–148; Robert Muggah, *Relocation Failures in Sri Lanka* (Zed Books, 2008), 101–104; Naomi Klein, "Blanking the Beach: 'The Second Tsunami,'" in Klein, *The Shock Doctrine*, 385–405; Sri Lanka Tourist Board quoted in N. Shanmugaratnam, "Challenges of Post-Disaster Development of Coastal Areas in Sri Lanka," Unpublished manuscript. November 2005, 2. https://www.umb .no/statisk/noragric/publications/other_publications/shan_challenges_postdisaster _nov2005.pdf

22. Laurie A. Johnson and Robert B. Olshansky, *After Great Disasters* (Lincoln Institute of Land Policy, 2017); Elizabeth Ferris, "Climate-Induced Resettlement," *SAIS Review of International Affairs* 35, no. 1 (2015): 109–117; Alka Sapat and Ann-Margaret Esnard, eds., *Coming Home After Disaster: Multiple Dimensions of Housing Recovery* (Routledge, 2017); Gonzalo Lizarralde et al., "Meta-Patterns in Post-Disaster Housing Reconstruction and Recovery," in *Coming Home After Disaster: Multiple Dimensions of Housing Recovery*, ed. Alka Sapat and Ann-Margaret Esnard (Routledge, 2017), 232–233.

23. Jennifer Duyne Barenstein, "Post-Disaster Reconstruction: Informal Settlers and the Right to Adequate Housing," in *Coming Home After Disaster: Multiple Dimensions of Housing Recovery*, ed. Alka Sapat and Ann-Margaret Esnard (Routledge, 2017), 257–258.

24. Johnson and Olshansky, *After Great Disasters*, 21, 45–47; Yang Zhang and William Drake, "Planning for Housing Recovery after the 2008 Wenchuan Earthquake in China," in *Coming Home After Disaster: Multiple Dimensions of Housing Recovery*, ed. Alka Sapat and Ann-Margaret Esnard (Routledge, 2017), 198, 204; Daniel Benjamin Abramson and Yu Qi, "Urban–Rural Integration' in the Earthquake Zone," *Pacific Affairs* 84, no. 3 (2011): 514.

25. Aaron Schneider, *Renew Orleans?* (University Of Minnesota Press, 2018).

26. IPCC Working Group II, *Sixth Assessment Report, Technical Summary*, 22.

27. Woods, *Development Drowned and Reborn*.

28. Sultana, "Living in Hazardous Waterscapes."

29. IPCC Working Group II, *Sixth Assessment Report, Technical Summary*.

30. International Labour Organization, *Working on a Warmer Planet: The Effect of Heat Stress on Productivity and Decent Work* (International Labour Organization, 2019), 13.

31. International Labour Organization, "Increase in Heat Stress Predicted to Bring Productivity Loss Equivalent to 80 Million Jobs," July 1, 2019. https://www.ilo.org/global/about-the-ilo/newsroom/news/WCMS_711917/lang--en/index.htm

32. IPCC Working Group II, *Sixth Assessment Report, Technical Summary*, 37.

33. IPCC Working Group II, *Sixth Assessment Report, Technical Summary*, 20.

34. Annah E. Piggott-McKellar et al., "Moving People in a Changing Climate: Lessons from Two Case Studies in Fiji," *Social Sciences* 8, no. 5 (2019): 136, 138–146.

35. W. Courtland Robinson, "Risks and Rights: The Causes, Consequences, and Challenges of Development-Induced Displacement," Brookings Institution, 2003, https://www.brookings.edu/wp-content/uploads/2016/06/didreport.pdf.

36. Michael M. Cernea, "Risks, Safeguards and Reconstruction," *Economic and Political Weekly*, October 7, 2000, 3662.

37. Keston K. Perry, "The New 'Bond-Age,' Climate Crisis and the Case for Climate Reparations," *Geoforum* 126 (2021): 361–371.

38. US Government Accountability Office, "Disaster Recovery: Experiences from Past Disasters offer Insights for Effective Collaboration after Catastrophic Events," Report GAO-09-811, August 31, 2009, https://www.gao.gov/assets/a293534.html.

39. Alice Fothergill, *Heads above Water: Gender, Class, and Family in the Grand Forks Flood* (State University of New York Press, 2004), 24.

40. City of Grand Forks, North Dakota, "Grand Forks, ND: Historic Flood, Remarkable Recovery," 2007, https://www.grandforksgov.com/home/showpublisheddocument/508/635332501257070000.

41. US Government Accountability Office, "Disaster Recovery."

42. FEMA, "20 Years Later: A Resilient Recovery after Catastrophe," April 10, 2017, https://www.fema.gov/blog/20-years-later-resilient-recovery-after-catastrophe.

43. Sandy Gerber, "Disaster Recovery for Low-Income People: Lessons from the Grand Forks Flood," Federal Reserve Bank of Minneapolis, 2006, https://www.minneapolisfed

.org/article/2006/disaster-recovery-for-lowincome-people-lessons-from-the-grand
-forks-flood.

44. Elaine Enarson and Maureen Fordham, "Lines That Divide, Ties That Bind: Race, Class, and Gender in Women's Flood Recovery in the US and UK," *Australian Journal of Emergency Management* 15, no. 4 (2001): 49.

45. Enarson and Fordham, "Lines That Divide," 46, 47, 49, 50.

46. Studies by Larry Burd and by Alana Knudson-Buresh, cited in City of Grand Forks, "Grand Forks Flood Disaster and Recovery Lessons Learned," https://www.icpsr.umich .edu/web/HMCA/studies/3313.

47. Fothergill, *Heads above Water*, 157–172, 212.

48. Enarson and Fordham, "Lines That Divide," 45.

49. Achmad Sukarsono, "Tsunami-Hit Indonesia Coast to Get Buffer Zone," Reuters, February 7, 2005, https://reliefweb.int/report/indonesia/tsunami-hit-indonesia-coast -get-buffer-zone. This initial proposal, from Aceh-based urban planners, allowed fisherfolk to reside in the middle portion of the setback, although not within the first three hundred meters.

50. Lawrence J. Vale, "Toward SDG11 Cities for All: From the Resilient City to the Equitably Resilient City," Lecture at the Bandung Institute of Technology, Bandung Indonesia, August 2020.

51. Joe Henley, "Marooned by Morakot," Climate Home News, December 17, 2021, https://www.climatechangenews.com/2021/12/17/marooned-by-morakot-indigenous -taiwanese-typhoon-survivors-long-to-return-home/.

52. Hsiao-Ming Chang, Chin-Lung Chou, and Cheng-Lung Wu, "Post-Disaster Tribe Relocated Reconstruction and Sustainable Development of Ulaljuc Village," *E3S Web of Conferences* 218 (2020): 02032, https://www.e3s-conferences.org/articles/e3sconf /pdf/2020/78/e3sconf_iseese2020_02032.pdf.

53. Chang et al., "Post-Disaster Tribe Relocated Reconstruction," 4–5.

54. Piggott-McKellar et al., "Moving People in a Changing Climate," 133.

55. Interview with Alejandro Echeverri, March 2022.

56. Interview with Belen Bagnati, Chief of Cabinet of Secretaria de Integración Social y Urbana (SISU), Buenos Aires, January 2000.

57. Assessment of YPF housing comes from site visits and interviews conducted by team members between August 2019 and January 2020. Subsequent interviews suggest that conditions worsened with the onset of the COVID-19 pandemic. See also City of Buenos Aires, *Proyecto de Transformación Urbana del AMBA* (Project of Urban Transformation of the Metropolitan Area of Buenos Aires-Under highway Relocation Project) (Secretaria de Integración Social y Urbana, 2019).

58. Mine Sato, "A Fresh Look at Capacity Development from Insiders' Perspectives: A Case Study of an Urban Redevelopment Project in Medellín, Colombia," (Tokyo: JICA Research Institute, 2013).

59. Lawrence J. Vale, et al. "What Affordable Housing Should Afford: Housing for Resilient Cities." *Cityscape* 16, no. 2 (2014): 21–49.; Florian Steinberg, "Housing Reconstruction and Rehabilitation in Aceh and Nias, Indonesia—Rebuilding Lives," *Habitat International* 31 (2007): 150–166; Ade Syukrizal, Wardah Hafidz, and Gabriela Sauter, *Reconstructing Life after the Tsunami: The Work of Uplink Banda Aceh in Indonesia* (International Institute for Environment and Development, 2009).

60. Kartik Chandraby, "India: Women Act to Make Slums Climate-Resilient, One House at a Time," PreventionWeb, March 16, 2020, https://www.preventionweb.net /news/india-women-act-make-slums-climate-resilient-one-house-time.

CASE 7: LIVING CULLY

1. Isabelle Anguelovski, "From Toxic Sites to Parks as (Green) LULUs? New Challenges of Inequity, Privilege, Gentrification, and Exclusion for Urban Environmental Justice," *Journal of Planning Literature* 31, no. 1 (2016): 23–36; Anguelovski et al., "Opinion."

2. "Living Cully Community Energy Plan," Living Cully, March 2018, 4.

3. Prosper Portland, *A Preliminary Report on The Creation of a Community-Led TIF District in Cully* (Prosper Portland, 2022).

4. Ricardo Banuelos et al., "Not in Cully: Anti-Displacement Strategies for the Cully Neighborhood," Master of Urban and Regional Planning Workshop Projects (Portland State University, 2013).

5. Angela Glover Blackwell, "America Needs a Raise," PolicyLink (blog), January 29, 2014, https://www.policylink.org/sites/default/files/americas-tomorrow-january2920 14.pdf.

6. Interview with Vivian Satterfield, July 2022.

7. Mike Baker and Sergio Olmos, "The Pacific Northwest, Built for Mild Summers, Is Scorching Yet Again," *New York Times*, August 13, 2021, sec. US, https://www.nytimes .com/2021/08/13/us/excessive-heat-warning-seattle-portland.html.

8. Interview with Cameron Harrington, August 2022.

9. Hoffman et al., "The Effects of Historical Housing Policies."

10. Banuelos et al., "Not in Cully."

11. St. Louis Fed, "All-Transaction House Price Index for Portland-Vancouver-Hillsboro, OR-WA (MSA), Federal Reserve Bank of St. Louis, https://fred.stlouisfed.org /series/ATNHPIUS38900Q.

12. Barbara Brown Wilson, *Resilience for All* (Island Press, 2018).

13. "Living Cully Partners," Living Cully, https://www.livingcully.org/about-living -cully/partners/.

14. Noah Enelow and Taylor Hesselgrave, *Verde and Living Cully: A Venture in Place-making* (Ecotrust/Economics for Equity and Environment, 2015), 4.

15. Living Cully, "Memorandum of Understanding among Habitat for Humanity Portland/Metro East, Hacienda Community Development Corporation, Native

American Youth and Family Center, Verde," 2014. https://nayapdx.org/wp-content
/uploads/2013/10/Living-Cully-MOU-executed.pdf

16. Interview with Harrington.

17. Wilson, *Resilience for All*, 143.

18. Noah Enelow et al., *Jobs and Equity in the Urban Forest* (Ecotrust and Policy Link, 2017).

19. Interview with Andria Jacob, August 2022.

20. Enelow and Hesselgrave, *Verde and Living Cully*.

21. Enelow et al., *Jobs and Equity*, 74.

22. Barbara Brown Wilson, "When Green Infrastructure Is an Anti-Poverty Strategy," Next City, July 8, 2019, https://nextcity.org/features/when-green-infrastructure-is-an-anti-poverty-strategy.

23. Interview with Harrington.

24. Interview with Harrington.

25. Interview with Brian Garcia, August 2021.

26. "Living Cully Community Energy Plan"; interview with Satterfield, July 2022.

27. Interview with Jacob.

28. Tony DeFalco and Kari Christensen, *Cully Park Community Health Indicators Project* (Oregon Health Authority, 2014).

29. Enelow and Hesselgrave, *Verde and Living Cully*, 15.

30. Interview with Satterfield.

31. Banuelos et al., "Not in Cully."

32. Banuelos et al., "Not in Cully."

33. DeFalco and Christensen, *Cully Park Community Health Indicators Project*, 22–23.

34. Anguelovski et al., "Opinion."

35. Enelow and Hesselgrave, *Verde and Living Cully*, 10.

36. "Living Cully Community Energy Plan," 4.

37. "Living Cully Community Energy Plan," 4.

38. "Oak Leaf Mobile Home Park," Portland.gov, accessed August 25, 2022, https://www.portland.gov/phb/construction/oak-leaf-mobile-home-park.

39. Interview with Harrington.

40. Interview with Garcia.

41. Interview with Satterfield.

42. Interview with Harrington.

43. Interview with Satterfield.

44. Interview with Harrington.

45. Prosper Portland, *A Preliminary Report*; "Exploring a Community-Led Development District," Living Cully, https://www.livingcully.org/tif/.

46. Prosper Portland, *A Preliminary Report*, 32.

47. Interview with Satterfield.

48. Interview with Harrington.

49. Prosper Portland, *A Preliminary Report*.

50. Prosper Portland, *A Preliminary Report*.

51. Interview with Satterfield.

CASE 8: YERWADA

1. Charles Correa, *A Place in the Shade* (Hatje Cantz, 2012).

2. Ananya Roy, "Why India Cannot Plan Its Cities," *Planning Theory* 8, 1 (2009): 76–87.

3. SPARC, "BSUP Cities 11—NTAG Study of BSUP Projects to Examine Potential for Community Participation by Society for Promotion of Area Resource Centres (SPARC)," 2012. https://sparcindia.org/pdf/studies/BSUP%2011%20City%20Review,%20SPARC,%202012_low%20res.pdf

4. Rajiv Ribeiro, *Designing with Informality: A Case of In-Situ Slum Upgradation in Yerwada* (Goa College of Architecture, 2018).

5. Smita Rawoot, "Governance at the Margins: The Challenge of Implementing Slum Housing Policy in Maharashtra, India" (master's thesis, Massachusetts Institute of Technology, 2015).

6. Interview with Narmada Vetaale, June 2022.

7. SPARC, "BSUP Cities 11."

8. Shelter Associates, *BSUP in Pune: A Critique* (Shelter Associates, 2012).

9. SPARC, "SPARC Annual Review 2010–11," 2011. https://www.sparcindia.org/pdf/sparc_annual_reports/SPARC%20Annual%20Report%202010-11.pdf

10. Interview with Savita Tai, August 2022.

11. Prasanna Desai, "BSUP In Situ Rehabilitation for the Urban Poor under JNNURM at Yerwada, Pune," Unpublished report, 2009, 7.

12. Interview with Vetaale.

13. Rawoot, "Governance at the Margins."

14. Interview with Filipe Balestra, December 2021.

15. SPARC, "BSUP Cities 11."

16. Interview with Vetaale.

17. Desai, "BSUP In Situ Rehabilitation," 3.

18. Rawoot, "Governance at the Margins."

19. Shelter Associates, *BSUP in Pune: A Critique*.

20. Interview with Keya Kunte, November 2021.

21. Rawoot, "Governance at the Margins."

22. Ribeiro, *Designing with Informality*.

23. SPARC, "BSUP Cities 11."

24. Ribeiro, *Designing with Informality*, 59.

25. Informal communication with Savita Tai, August 2022.

26. Rawoot, "Governance at the Margins."

27. Rawoot, "Governance at the Margins."

28. Rawoot, "Governance at the Margins."

29. Interview with Vetaale.

30. Interview with Dhananjay Sadlapure, January 2022.

31. Sheela Patel, "Upgrade, Rehouse or Resettle? An Assessment of the Indian Government's Basic Services for the Urban Poor (BSUP) Programme," *Environment and Urbanization* 25, no. 1 (2013): 177–188.

32. Subhashree Nath, "Slum Upgrading Schemes for Better Liveability: Case of Pune, India" (master's thesis, Bauhaus Universitat Weimar, 2020).

33. Reall, *India Impact Summary Brief: The Impact of REALL's Affordable Housing Investments on Quality of Life in Urban India* (Reall, 2020); Interview with Tai, August 2022.

34. Interview with Vetaale.

35. Interview with Sadlapure.

36. Nath, "Slum Upgrading"; Shelter Associates, "BSUP in Pune."

37. Shelter Associates, "BSUP in Pune," 3.

38. Interview with Tai.

39. Ribeiro, *Designing with Informality*, 63.

40. SPARC, "BSUP Cities 11."

41. Shelter Associates, "BSUP in Pune."

42. Ribeiro, *Designing with Informality*, 82.

CASE 9: DAFEN

1. Mary Ann O'Donnell, "Laying Siege to the Villages," in *Learning from Shenzhen*, ed. Mary Ann O'Donnell, Winnie Wong, and Jonathan Bach (University of Chicago Press, 2017), 107, 110; Shi et al., "Promises and Perils of Collective Land Tenure."

2. O'Donnell, "Laying Siege to the Villages," 109, 110.

3. O'Donnell, "Laying Siege to the Villages," 110.

4. Gan et al., "From Redevelopment to In Situ Upgrading," 30.

5. O'Donnell, "Laying Siege to the Villages," 107, 111, 121; Eduardo Jaramillo, "China's Hukou Reform in 2022," Center for Strategic and International Studies, April 20, 2022, https://www.csis.org/blogs/new-perspectives-asia/chinas-hukou-reform-2022-do-they-mean-it-time-0.

6. Winnie Wong, "Shenzhen's Model Bohemia and the Creative China Dream," in *Learning from Shenzhen*, ed. Mary Ann O'Donnell, Winnie Wong, and Jonathan Bach (University of Chicago Press, 2017), 196, 197.

7. Winnie Won Ying Wong, *Van Gogh on Demand* (University Chicago Press, 2013), 48.

8. Interview with Zhou, May 2022.

9. Wong, *Van Gogh on Demand*, 51–53, 58–59, 64.

10. June Wang and Si-ming Li, "State Territorialization, Neoliberal Governmentality," *Urban Geography* 38, no. 5 (2016): 719. Their work draws upon a Chinese language publication: Youping Wen, *The Rise of Dafen* (Haitian, 2006).

11. See, for example, DafenVillageOnLine.com, established in 2006.

12. Graeme Evans, "Hard-Branding the Cultural City," *International Journal of Urban and Regional Research* 27 (2003): 417–440; Richard Florida, *The Rise of the Creative Class* (Basic Books, 2002).

13. Ren Xiaofeng and Zho Zhiquan, eds., *Dafencun de Teshu Cunmin* (Special Villagers of Dafen) (Lingnan Meishu Chubanshe, 2006).

14. Wong, "Shenzhen's Model Bohemia," 198, 199, 200.

15. Wong, "Shenzhen's Model Bohemia," 202.

16. Wong, "Shenzhen's Model Bohemia," 200.

17. Karita Kan, "Building SoHo in Shenzhen," *Geoforum* 111 (2020), 5.

18. Newspapers quoted in Si-ming Li, Hung-ha Cheng, and Jun Wang, "Making a Cultural Cluster," *Habitat International* 41 (2014): 156.

19. Wong, *Van Gogh on Demand*, 212; "World's Largest Mona Lisa Appears at Shanghai Expo," CCTV.com, April 12, 2010, https://english.cctv.com/20100412/103035 .shtml.

20. Kan, "Building SoHo in Shenzhen," 5, 8.

21. Stephane Hallegatte et al., "Future Flood Losses in Major Coastal Cities." *Nature Climate Change* 3, no. 9 (2013): 802–806.

22. Li et al., "Making a Cultural Cluster"; Wong, *Van Gogh on Demand*, 27.

23. Interview with executive vice chairman of the Dafen Artists' Association who arrived in Dafen in 1993, May 2022; interview with Zhou, May 2022.

24. Wong, *Van Gogh on Demand*, 55, 62–63, 68–69, 78–79. Interviews with painters in 2022 confirm that many see this as a positive alternative to factory conditions.

25. Pei Pei, "Shenzhen Village Moves from Making Replica Paintings to Creating Originals," ChinaDaily.com, April 19, 2021, http://www.chinadaily.com.cn/a/202104/19 /WS607ccb7ba31024ad0bab6518.html. See also "From Imitation to Creation, Chinese Village Paints New Life," ChinaDaily.com, March 3, 2019, http://global.chinadaily .com.cn/a/201903/03/WS5c7b4a31a3106c65c34ec6c0_5.html.

26. Kan, "Building SoHo in Shenzhen," 6; interview with CR Land official (in Chinese), May 2022.

27. Li et al., "Making a Cultural Cluster," 161; Kan, "Building SoHo in Shenzhen," 7; Mr. Zhou discussed his support for deaf painters in interviews in May and June 2022.

28. Interviews with artists, May 2022.

29. Wang and Li, "State Territorialization, Neoliberal Governmentality," 723.

30. Interviews with artists, May and June 2022.

31. Interviews with artists, May and June 2022.

32. Wong, *Van Gogh on Demand*, 105–109; interview with artist who has worked in Dafen since 1993, May 2022.

33. Wang and Li, "State Territorialization, Neoliberal Governmentality," 717–718.

34. Longgang District Housing and Construction Bureau, "Dafen Oil Painting Village Public Rental Housing Administrative Measures and Implementation Plan" (in Chinese), n.d. but likely 2012.

35. Wong, *Van Gogh on Demand*, 220.

36. Longgang District Housing and Construction Bureau, "Dafen Oil Painting Village Public Rental Housing"; more recent rent data from Dafen's official public WeChat platform (in Chinese); interview with artist, May 2022.

37. Wang and Li, "State Territorialization, Neoliberal Governmentality," 723.

38. Interviews with Dafen artists, May and June 2022.

39. Interview with CR Land official.

40. Wang and Li, "State Territorialization, Neoliberal Governmentality," 721.

41. Kan, "Building SoHo in Shenzhen," 7, 8; interviews with artists, May and June 2022.

42. Li et al., "Making a Cultural Cluster," 159, 160.

43. Interview with CR Land official.

44. Kan, "Building SoHo in Shenzhen," 1.

45. Wang and Li, "State Territorialization, Neoliberal Governmentality," 724.

46. Wong, "Shenzhen's Model Bohemia,"209.

47. Gan Xinyue, personal communication, February 2023.

48. Dafen Shui Flood Diversion Project; see https://www.mccht.com/ht/xwzx67/mysr60/p1_817478/index.html (in Chinese).

49. Interviews with residents and Dafen workers, May and June 2022.

50. Canfei He with Lei Yang, "Urban Development and Climate Change in China's Pearl River Delta," *Land Lines*, July 2011, https://www.lincolninst.edu/publications/articles/urban-development-climate-change-chinas-pearl-river-delta; Genevieve Donnellon-May and Guangtao Fu, "Are 'Sponge Cities' the Answer to Shenzhen's Water Scarcity?" *The Diplomat*, January 5, 2022, https://thediplomat.com/2022/01/are-sponge-cities-the-answer-to-shenzhens-water-scarcity/; Faith Ka Shun Chan et al., "'Sponge City' in China," *Land Use Policy* 76 (2018): 772–778.

51. Li et al., "Making a Cultural Cluster," 163.

52. Gan et al., "From Redevelopment to In Situ Upgrading," 39.

EQUITABLY RESILIENT GOVERNANCE

1. Isabelle Anguelovski and JoAnn Carmin, "Something Borrowed, Everything New: Innovation and Institutionalization in Urban Climate Governance," *Current Opinion in Environmental Sustainability* 3, no. 3 (2011): 169–175.

2. IPCC Working Group II, *Sixth Assessment Report, Technical Summary*.

3. Cassim Shepard, *Citymakers* (Monacelli Press, 2017); Kian Goh, Vinit Mukhija, and Anastasia Loukaitou-Sideris, eds., *Just Urban Design* (MIT Press, 2022).

4. Carolina Pacchi, "Epistemological Critiques to the Technocratic Planning Model: The Role of Jane Jacobs, Paul Davidoff, Reyner Banham and Giancarlo De Carlo in the 1960s," *City, Territory and Architecture* 5, no. 1 (2018): 1–8.

5. Jacobs, *The Death and Life of Great American Cities*.

6. Paul Davidoff, "Advocacy and Pluralism in Planning," *Journal of the American Institute of Planners* 31, no. 4 (1965): 331–338.

7. Horst Rittel and Melvin M. Webber, "Dilemmas in a General Theory of Planning," *Policy Sciences* 4, no. 2 (1973): 155–169.

8. Judith E. Innes and David E. Booher, "Consensus Building and Complex Adaptive Systems," *Journal of the American Planning Association* 65, no. 4 (1999): 412–423; Joshua Cohen, "Procedure and Substance in Deliberative Democracy," in *Democracy and Difference*, ed. Seyla Benhabib (Princeton University Press, 1996), 95–121.

9. Archon Fung and Eric Olin Wright, eds., *Deepening Democracy: Institutional Innovations in Empowered Participatory Governance* (Verso, 2003).

10. Doug Kelbaugh, "Design Charrettes," in *Companion to Urban Design*, ed. Tridib Banerjee and Anastasia Loukaitou-Sideris (Routledge, 2011), 317–328.

11. Randolph T. Hester, *Design for Ecological Democracy* (MIT Press, 2010).

12. Jeffrey Hou, "Participation and Beyond," in *Companion to Urban Design*, ed. Tridib Banerjee and Anastasia Loukaitou-Sideris (Routledge, 2011), 329; Wilson, *Resilience for All*.

13. John F. C. Turner and Robert Fichter, *Freedom to Build* (Macmillan, 1972).

14. Manon Mollard, "Revisit Aranya Low-Cost Housing," *The Architectural Review*, August 7, 2019, 114–123.

15. Orangi Pilot Project, "Orangi Pilot Project—Research and Training Institute," http://www.opp.org.pk/.

16. Ariana Zilliacus, "Half a House Builds a Whole Community: Elemental's Controversial Social Housing," *ArchDaily*, October 24, 2016, https://www.archdaily.com/797779/half-a-house-builds-a-whole-community-elementals-controversial-social-housing.

17. Teresa P. R. Caldeira, "Peripheral Urbanization," *Environment and Planning D: Society and Space* 35, no. 1 (2017): 3–20.

18. N. J. Habraken, *Variations: The Systematic Design of Supports* (Laboratory of Architecture and Planning, MIT, 1976).

19. Vinit Mukhija, "The Value of Incremental Development and Design in Affordable Housing," *Cityscape* 16, no. 2 (2014): 11–20.

20. "Experimental Housing at Svartlamon | Nøysom Arkitekter," Archello, Accessed November 21, 2023, https://archello.com/project/experimental-housing-at-svartlamon.

21. Sherry R. Arnstein, "A Ladder of Citizen Participation," *Journal of the American Institute of Planners* 35, no. 4 (1969): 216–224.

22. Shannon Mattern, "Post-It Note City," *Places Journal*, February 11, 2020.

23. David Satterthwaite, "From Professionally Driven to People-Driven Poverty Reduction," *Environment and Urbanization* 13, no. 2 (2001): 135–138.

24. Faranak Miraftab, "Insurgent Planning," *Planning Theory* 8, no. 1 (2009): 32–50; Ananya Roy, "Praxis in the Time of Empire," *Planning Theory* 5, no. 1 (2006): 7–29.

25. Oren Yiftachel, "Critical Theory and 'Gray Space,'" *City* 13, no. 2–3 (2009): 246–263; Ananya Roy. "Urban Informality." In *The Oxford Handbook of Urban Planning*, edited by Randall Crane and Rachel Weber (Oxford University Press, 2012).

26. Miraftab, "Insurgent Planning."

27. Holston, "Spaces of Insurgent Citizenship."

28. Goh, *Form and Flow.*

29. Penn Loh and Boone Shear, "Solidarity Economy and Community Development: Emerging Cases in Three Massachusetts Cities," *Community Development* 46, no. 3 (2015): 244–260.

30. Wily, "Challenges for the New Kid on the Block," 63.

31. Giorgio Agamben, *The State of Exception* (Duke University Press, 2005).

32. Klein, *The Shock Doctrine.*

33. Woods, *Development Drowned and Reborn*; Schneider, *Renew Orleans?*

34. Divya Chandrasekhar, Yang Zhang, and Yu Xiao, "Nontraditional Participation in Disaster Recovery Planning: Cases from China, India, and the United States," *Journal of the American Planning Association* 80, no. 4 (2014): 373–384.

35. Cynthia Rosenzweig, "Cities Lead the Way in Climate-Change Action," *Nature* 467, no. 7318 (2010): 909.

36. Stephen J. Collier, "Neoliberalism and Natural Disaster: Insurance as Political Technology of Catastrophe," *Journal of Cultural Economy* 7, no. 3 (2014): 273–290; Elliott, *Underwater*; Cox, "Inscriptions of Resilience."

37. Erik Swyngedouw, "Apocalypse Forever? Post-Political Populism and the Spectre of Climate Change," *Theory, Culture and Society* 27, no. 2–3 (2010): 213–232.

38. Anguelovski and Carmin, "Something Borrowed, Everything New," 1.

39. IPCC Working Group II, *Sixth Assessment Report, Technical Summary*, 85.

40. Linda Shi, "Promise and Paradox of Metropolitan Regional Climate Adaptation," *Environmental Science and Policy* 92 (2019): 262–274.

41. Harriet Bulkeley and Michele Betsill, "Rethinking Sustainable Cities: Multilevel Governance and the 'Urban' Politics of Climate Change," *Environmental Politics* 14, no. 1 (2005): 42–63.

42. IPCC Working Group II, *Sixth Assessment Report, Technical Summary*, 66, 68.

43. Crawford S. Holling, *Adaptive Environmental Assessment and Management*, International Series on Applied Systems Analysis 3 (International Institute for Applied Systems Analysis, 1978).

44. Judith Layzer, *Natural Experiments* (MIT Press, 2008).

45. Jörn Birkmann et al., "Adaptive Urban Governance," *Sustainability Science* 5, no. 2 (2010): 185–206.

46. Emily Boyd and Sirkku Juhola, "Adaptive Climate Change Governance for Urban Resilience," *Urban Studies* 52, no. 7 (2015): 1234–1264.

47. IPCC Working Group II, *Sixth Assessment Report, Technical Summary*, 82, 66.

48. IPCC Working Group II, *Sixth Assessment Report, Technical Summary*, 66.

49. IPCC Working Group II, *Sixth Assessment Report, Technical Summary*, 59.

50. Shi and Moser, "Transformative Climate Adaptation in the United States," 372.

51. Meerow et al., "Social Equity in Urban Resilience Planning."

52. Goh et al., *Just Urban Design*.

53. IPCC Working Group II, *Sixth Assessment Report, Technical Summary*, 77.

54. IPCC Working Group II, *Sixth Assessment Report, Technical Summary*, 59.

55. Fainstein, "Resilience and Justice," 165.

56. IPCC Working Group II, *Sixth Assessment Report, Technical Summary*, 77.

57. Hannah Reid et al., "Community-Based Adaptation to Climate Change: An Overview," *Participatory Learning and Action* 60, no. 1 (2009): 11–33; E. Lisa F. Schipper et al., *Community-Based Adaptation to Climate Change: Scaling It Up* (Routledge, 2014).

58. Ross Westoby et al., "From Community-Based to Locally Led Adaptation: Evidence from Vanuatu," *Ambio* 49, no. 9 (2020): 1466–1473.

59. Grove, *Resilience*, 16.

60. Henley, "Marooned by Morakot."

61. IPCC Working Group II, *Sixth Assessment Report, Technical Summary*, 86.

62. "LA SAFE: Summary of Strategy Development and Project Selection," 2018. https://s3.amazonaws.com/lasafe/2018/N-04/2018-Summary-Strategy-Development-Project-Selection.pdf

63. WE ACT for Environmental Justice, "#NMCA Northern Manhattan Climate Action: A Draft Plan," November 2016, https://www.weact.org/campaigns/nmca/.

64. Wilson, *Resilience for All.*

65. Jeffrey Moxom, et al., "Cooperation for the Transition to a Green Economy: Global Thematic Research Report." Brussels, Belgium: International Cooperative Alliance, 2021.

66. Lamb et al., "Resident-Owned Resilience."

67. Paul Chatterton, "Towards an Agenda for Post-Carbon Cities: Lessons from Lilac, the UK's First Ecological, Affordable Cohousing Community," *International Journal of Urban and Regional Research* 37, no. 5 (2013): 1654–1674.

68. John Emmeus Davis. "Origins and Evolution of the Community Land Trust in the United States." In *The Community Land Trust Reader*, 1:3–47. Cambridge, MA: Lincoln Institute of Land Policy, 2010.

69. Grannis, *Community Land = Community Resilience.*

70. Roberta M. Feldman and Susan Stall, *The Dignity of Resistance* (Cambridge University Press, 2004); Akira Drake Rodriguez, *Diverging Space for Deviants: The Politics of Atlanta's Public Housing* (University of Georgia Press, 2021).

71. Goh, *Form and Flow.*

72. Robert J. Sampson. "What 'Community' Supplies." In *Urban Problems and Community Development*, edited by Ronald F. Ferguson and William T. Dickens, 241–92. Washington, D.C: Brookings Institution Press, 1999.

73. Laura Grube and Virgil Henry Storr, "The Capacity for Self-Governance and Post-Disaster Resiliency," *The Review of Austrian Economics* 27 (2014): 301–324.

74. Bulkeley and Betsill, "Rethinking Sustainable Cities."

75. IPCC Working Group II, *Sixth Assessment Report, Technical Summary*, 86.

76. Lamb et al., "Resident-Owned Resilience."

77. Gordon C. C. Douglas, "Do-It-Yourself Urban Design: The Social Practice of Informal 'Improvement' through Unauthorized Alteration," *City and Community* 13, no. 1 (2013): 5–25.

78. Wilson, *Resilience for All.*

79. Emily Chamlee-Wright and Virgil Henry Storr, "Social Capital as Collective Narratives and Post-Disaster Community Recovery," *The Sociological Review* 59, no. 2 (2011): 266–282.

80. Marino, "Adaptation Privilege and Voluntary Buyouts."

81. Watts, "Now and Then."

CASE 10: THUNDER VALLEY

1. Partnership with Native Americans, "South Dakota: Pine Ridge Reservation," http://www.nativepartnership.org/site/PageServer?pagename=pwna_native_reservations_pineridge.

2. Saul Elbein, "As Climate Chaos Escalates in Indian Country, Feds Abandon Tribes," Mongabay, July 1, 2019, https://news.mongabay.com/2019/07/as-climate-chaos-escalates-in-indian-country-feds-abandon-tribes/.

3. DataUSA, "Oglala Lakota County, SD: Housing and Living," https://datausa.io /profile/geo/oglala-lakota-county-sd#housing.

4. NDN Collective, "Interview with Nick Tilsen," September 30, 2018, https:// ndncollective.org/podcasts/episode-01-nick-tilsen-talks-building-indigenous-power/.

5. Tilsen quoted in Sarah Sunshine Manning, "Thunder Valley CDC Begins Grass-roots Housing and Community Construction on Pine Ridge," *Indian Country Today*, April 22, 2016.

6. Nick Tilsen, "Building Resilient Communities: A Moral Responsibility," TEDx Rapid City, July 14, 2015, https://www.youtube.com/watch?v=e2Re-KrQNa4.

7. Tilsen quoted in Jim Stasiowski, "Pine Ridge Speakers," *Rapid City Journal*, October 30, 2016.

8. In 2018, Tilsen moved on to found NDN Collective, a national organization devoted to "building Indigenous power" in ways that can "defend, develop and decolonize," https://ndncollective.org.

9. US Department of Agriculture, "USDA Announces $1.97 Million for New Community Center on Pine Ridge Indian Reservation," June 22, 2015, https://www.usda .gov/media/press-releases/2015/06/22/usda-announces-197-million-new-community -center-pine-ridge-indian; Richard Two Bulls, "Thunder Valley to Turn White Clay into Healing Center," *Lakota Times*, January 20, 2022, https://www.lakotatimes.com /articles/thunder-valley-to-turn-white-clay-into-healing-center/.

10. Interview with DeCora Hawk, July 2021.

11. Throughout this chapter and book, we use a range of terms to refer to Native people and institutions, reflecting the pluralism and evolutions in these identities. The term "Indian" remains in common parlance, including in the names of many organizations and government institutions such as the Federal Bureau of Indian Affairs. TVCDC documents and staff frequently use the terms "Lakota people" and "Lakota Nation" to refer to themselves. They sometimes use the term "Tribe" (often capitalized) and often reference Native or Indigenous traditions and identities. Scholar Ned Blackhawk notes that all of these are "problematic" but that each offers "insights into the histories of power and difference." So, he uses "Indigenous," "Native American," and "American Indian" "interchangeably." Ned Blackhawk, *The Rediscovery of America* (Yale University Press, 2023), 449n1.

12. Interview with Christina Hoxie, October 2021.

13. Oyate Omniciyé Consortium and Steering Committee, "Oyate Omniciyé: Oglala Lakota Plan," 2012, 4–5, https://issuu.com/christinahoxie/docs/___oyate_omniciye _final_draft.

14. Oyate Omniciyé Consortium and Steering Committee, "Oyate Omniciyé," 15.

15. Oyate Omniciyé Consortium and Steering Committee, "Oyate Omniciyé," 21, 25.

16. Oyate Omniciyé Consortium and Steering Committee, "Oyate Omniciyé," 104–111, 117.

17. Oyate Omniciyé Consortium and Steering Committee, "Oyate Omniciyé," 65–66.

18. The Harvard Project on American Indian Economic Development, *The State of the Native Nations* (Oxford University Press, 2008), 20. See also Christin Buchholz and

Mark Gustafson, "Local Governance for the Oglala Sioux," Harvard University, Malcolm Wiener Center for Social Policy PRS 00-4, April 4, 2000.

19. Oyate Omniciyé Consortium and Steering Committee, "Oyate Omniciyé," 65–66.

20. Oyate Omniciyé Consortium and Steering Committee, "Oyate Omniciyé," 72.

21. Interview with Tatewin Means, July 2021.

22. Interview with Means.

23. Interview with Means.

24. Interview with Means.

25. Interview with Kimberly Pelkofsky, July 2021.

26. Interviews with Hawk and Pelkofsky.

27. Interview with Pelkofsky.

28. Interview with Pelkofsky.

29. Oyate Omniciyé Consortium and Steering Committee, "Oyate Omniciyé," 81.

30. Interview with Pelkofsky.

31. Interview with Pelkofsky.

32. *Pine Ridge Reservation Allottee Land-Planning Map Book* (Village Earth, 2009), 11, 14, https://ia800101.us.archive.org/4/items/PineRidgeMapBook/PR_Map_Book.pdf.

33. The Harvard Project on American Indian Economic Development, *The State of the Native Nations*, 98; Native Land Information System, "Allotment Timeline on the Pine Ridge Reservation," https://nativeland.info/blog/dashboard/allotment-timeline-on-the -pine-ridge-reservation/.

34. *Pine Ridge Reservation Allottee Land-Planning Map Book*, 3. See also: Indian Land Tenure Foundation, "Land Tenure Issues," https://iltf.org/land-issues/issues/.

35. Native Land Information System, https://nativeland.info/explore-topics/lost-agri culture-revenue-database/.

36. Russell Means with Marvin J Wolf, *Where White Men Fear to Tread* (St. Martin's Press, 1995), 10.

37. Interview with Pelkovsky.

38. Interview with Means.

39. Interview with Pelkofsky; interview with Means.

40. Interview with Pelkofsky.

41. Interview with Richie Meyers, January 2022.

42. Jim Stasiowski, "Youth Suicides Prompt Emergency Declaration on Pine Ridge," *Rapid City Journal*, February 19, 2015; Cecily Hilleary, "Battling Youth Suicide on Pine Ridge Indian Reservation," *Voice of America*, September 8, 2015.

43. *Poor Bear* v. *Nesbitt et al.*, January 29, 2004; 300 F.Supp.2d 904; Carson Walker, "Tribe to Begin Beer Blockade," *Rapid City Journal*, June 27, 2006; Kevin Abourezk, "Oglala Sioux Tribe Sues Whiteclay Stores, Beer Makers, Distributors," *Rapid City*

Journal, February 10, 2012; Zach Pluhacek and Lori Pilger, "Supreme Court Sides with Whiteclay Beer Store Foes," *Chadron Record*, September 29, 2017.

44. Statistics cited in Abourezk.

45. Interview with Means.

46. Interview with Pelkofsky.

47. Interview with Means.

48. Interview with Pelkofsky.

49. Oyate Omniciyé Consortium and Steering Committee, "Oyate Omniciyé," 198, https://sharondavisdesign.com/project/wiconi-community/.

50. Oyate Omniciyé Consortium and Steering Committee, "Oyate Omniciyé," 105.

51. Interview with Pelkofsky, May 2022.

52. Two Bulls, "Thunder Valley to Turn White Clay into Healing Center."

53. Means quoted in Two Bulls.

CASE 11: KOUNKUEY DESIGN INITIATIVE

1. Samuel Owuor and Teresa Mbatia, "Nairobi," in *Capital Cities in Africa*, ed. Simon Bekker and Göran Therborn (HSRC Press, 2012), 120–122.

2. Robert Home, "Colonial Township Laws and Urban Governance in Kenya," *Journal of African Law* 56, no. 2 (2012): 179–180.

3. Timothy Parsons, "'Kibra Is Our Blood': The Sudanese Military Legacy in Nairobi's Kibera Location, 1902–1968," *International Journal of African Historical Studies* 30, no. 1 (1997): 88–89, 96, 101, 103, 105–106, 110, 111, 116, 120–122. See also Samantha Balaton-Chrimes, *Ethnicity, Democracy and Citizenship in Africa* (Routledge, 2016).

4. Lynsey Deanne Farrell, "Hustling NGOs: Coming of Age in Kibera Slum, Nairobi, Kenya," (PhD diss., Boston University, 2015), 37; Owuor and Mbatia, "Nairobi," 132.

5. Joseph Mulligan, "The Nairobi Dam," Re.Think, November 7, 2018, https://rethink.earth/the-nairobi-dam-from-community-resource-to-open-sewer-and-back-again/; Amélie Desgroppes and Sophie Taupin, "Kibera: The Biggest Slum in Africa?" *East African Review* 44 (2011): 23–33.

6. "Kibera Examined," Black Economics, April 1, 2014, https://blackeconomics.co.uk/2014/04/01/kibera-examined/.

7. "President Uhuru Kenyatta Issues title Deed to Kibra Nubians," *Nation*, June 2, 2017, https://nation.africa/kenya/news/Uhuru-issues-title-deed-to-Kibra-Nubians/1056-3953204-9jfwvk/index.html; "After Long Struggle, Kenya's Nubian Minority Secures Land Rights," June 5, 2017, https://reliefweb.int/report/kenya/after-long-struggle-kenyas-nubian-minority-secures-land-rights.

8. UN Habitat, *Informal Settlements' Vulnerability Mapping in Kenya* (UN Habitat, 2020), 21–22; Owuor and Mbatia, "Nairobi," 122.

9. Mulligan, "The Nairobi Dam," 2018.

10. UN Habitat, 20.

11. WRI Ross Center Prize for Cities, "Kibera Public Space Project," https://prize forcities.org/project/kibera-public-space-project.

12. "Kibera Examined."

13. WRI Ross Center Prize for Cities, "Kibera Public Space Project."

14. https://www.facebook.com/NewNairobiDamCommunity/.

15. https://www.kounkuey.org/projects/community_responsive_adaptation_to _flooding.

16. Interview with George Nengo, April 2022.

17. Interview with Jacqueline Nduku, April 2022.

18. Interview with Wilson Sageka, June 2021.

19. Interview with Prisca Okila, April 2021.

20. Interview with Regina Opondo, March 2021.

21. Interview with Okila.

22. Interview with Franklin Kirimi, April 2021.

23. Interview with Kirimi.

24. Interview with Sageka.

25. Interview with Sarah Lebu, May 2021.

26. Interview with Okila.

27. Interviews with Lebu, Sageka, and Lynsey Farrell, May 2021.

28. Interview with Sabina Ayot, May 2021.

29. Farrell, "Hustling NGOs," 4.

30. Maryrose Bredhauer, "Transformative Adaptation in Informal Settlements: The Case of Kounkuey Design Initiative" (King's College London: Urban Africa Risk Knowledge Working Paper #12, November 2016), 29; Interview with Maryrose Bredhauer Threlfall, March 2021.

31. Living Data Hub, "About the Project," https://livingdatahubs.cargo.site/About.

32. Interview with Opondo.

33. Interview with Kirimi.

34. Interview with Kirimi.

35. See https://www.youtube.com/watch?v=0PFEiknD5zo (in Kiswahili) and https:// infonile.org/en/2020/12/kenya-warning-kibera-residents-of-floods-using-flags -whatsapp-and-facebook-forecasts/.

36. https://kounkuey.org/projects/kibera_public_space_project_06

37. Interview with Usalama leader, April 2022.

38. Interview with Asha Abdi, May 2022.

39. Bredhauer, "Transformative Adaptation in Informal Settlements," 38.

40. United Nations Office of the High Commissioner for Human Rights, "Kenya: Stop Forced Evictions from Nairobi's Kibera Settlement, Say UN Rights Experts," July 26, 2018, https://www.ohchr.org/en/press-releases/2018/07/kenya-stop-forced-evictions -nairobi-kibera-settlement-say-un-rights-experts; KDI, "Division or Development? Missing Link #12," https://www.kounkuey.org/uploads/1500015/1557779180504/Blog _post.pdf. For video coverage of evictions, see "Slumscapes | Kibera—The Road That Divides," https://www.youtube.com/watch?v=vSImaiSBY-s; https://www.youtube.com /watch?v=XQoz1SyeleY; and https://www.youtube.com/watch?v=SJHI3K4CQD4.

41. Bredhauer, "Transformative Adaptation in Informal Settlements," 27.

CASE 12: CAÑO MARTÍN PEÑA

1. MIT Urban Risk Lab, "Moving Together," https://movingtogether.mit.edu/puerto -rico; Danielle Zoe Rivera, "Disaster Colonialism: A Commentary on Disasters beyond Singular Events to Structural Violence," *International Journal of Urban and Regional Research* 46, no. 1 (2022): 126–135.

2. US Army Corps of Engineers, "Caño Martín Peña Ecosystem Restoration Project: Final Feasibility Report Delivery Package," February 2016, Appendix H: Final EIS, 3–4; María E. Hernández-Torrales et al., "Seeding the CLT in Latin America and the Caribbean," in *On Common Ground: International Perspectives on the Community Land Trust*, ed. John Emmeus Davis, Line Algoed, and María E. Hernández-Torrales (Terra Nostra Press, 2020), 193.

3. *Mala Agua* (Caño Martín Peña), December 2012, https://www.youtube.com/watch ?v=rPPbFM-Rvok.

4. Medium, "Sapphire Necklace: Resilient Shoreline Communities Along the San Juan Bay Estuary," https://sapphire-necklace.medium.com/sapphire-necklace-resilient -shoreline-communities-along-the-san-juan-bay-estuary-d335c8da0fcb.

5. Lina Algoed and María E. Hernández-Torrales, "The Land Is Ours. Vulnerabilization and Resistance in Informal Settlements in Puerto Rico: Lessons from the Caño Martín Peña Community Land Trust," *Radical Housing Journal* 1, no. 1 (2019): 38.

6. Puerto Rico Urban Renewal and Housing Administration, "Community Renewal Program for San Juan Metropolitan Area," Local Approval Data, Binder No. 1, October 1966.

7. Laura Moscoso, "Cantera or the Neglect of Public Policy to Tackle Poverty," Centro de Periodismo Investigativo, May 20, 2018, https://periodismoinvestigativo.com/2018 /05/cantera-or-the-neglect-of-public-policy-to-tackle-poverty/.

8. The G-8 (Grupo de las Ocho Comunidades Aledañas al Caño Martín Peña, Inc.) encompasses seven neighborhoods included in the Community Land Trust: Barrio Obrero, Barrio Obrero Marina, Buena Vista Santurce, Israel-Bitumul, Buena Vista Hato Rey, Las Monjas, and Parada 27. The eighth neighborhood in the G-8, but not part of the Caño CLT, is the adjacent Cantera Peninsula community; Hernández-Torrales et al., "Seeding the CLT," 208, notes 1 and 6.

9. Algoed and Hernández-Torrales, "The Land Is Ours," 33.

10. World Habitat Organization, "Caño Martín Peña Community Land Trust," World Habitat, May 8, 2018, https://world-habitat.org/world-habitat-awards/winners-and -finalists/cano-martin-pena-community-land-trust/#outline.

11. Interview with Mario Núñez Mercado, August 2021; Hernández-Torrales et al., "Seeding the CLT," 234, note 7.

12. Hernández-Torrales et al., "Seeding the CLT," 191–192; Rafael R. Díaz Torres, "USACE, Dept. of Housing Block Environmental Justice for Caño Martín Peña Residents," Centro De Periodismo Investigativo, January 29, 2021, https://periodismoinves tigativo.com/2021/01/usace-dept-of-housing-block-environmental-justice-for-cano -martin-pena-residents/; Rivera, "Disaster Colonialism."

13. Law Number 1 of 2001, https://www.lexjuris.com/lexlex/Leyes2001/lex2001001 .htm.

14. Hernández-Torres et al., "Seeding the CLT," 195.

15. Hernández-Torres et al., "Seeding the CLT"; interview with Estrella Santiago Pérez, July 2021.

16. Hernández-Torres et al., "Seeding the CLT" 198–199.

17. Interview with Line Algoed, April 2021.

18. Lyvia N. Rodríguez Del Valle, "ENLACE Caño Martín Peña: A Restoration and Resiliency Project," Testimony before the House Subcommittee on Environment, "Response and Recovery to Environmental Concerns from the 2017 Hurricane Season," November 14, 2017.

19. Interview with Lucy Cruz Rivera, August 2021.

20. Interview with José Caraballo Pagán, August 2021; Lizzie Yarina, Miho Mazereeuw, and Larisa Ovalles, "A Retreat Critique: Deliberations on Design and Ethics in the Flood Zone," Journal of Landscape Architecture 14, no. 3 (2019): 17–20.

21. Interview with Mario Núñez Mercado, August 2021.

22. Interview with Jennifer Rivera-De Jesus, August 2021; personal correspondence with Mariolga Juliá Pacheco, May 2023.

23. Interview with Carlos Muñiz Pérez, August 2021.

24. MIT Urban Risk Lab, "Moving Together"; interview with Mario Núñez Mercado, August 2021.

25. Interview with Santiago Pérez.

26. Interview with Caraballo Pagán.

27. Laura Bachmann, "An Informal Settlement in Puerto Rico Has Become the World's First Favela Community Land Trust," RioOnWatch, April 28, 2017, https:// rioonwatch.org/?p=36081.

28. Interview with Pacheco, July 2021.

29. Interview with Muñiz.

30. Interview with Deepak Lamba-Nieves, August 2021.

31. Interview with Lamba-Nieves.

32. Interview with Rivera-De Jesus.

33. Interview with Pacheco; interview with Caraballo Pagán.

34. Interview with María Hernández-Torrales, April 2021.

35. Interview with Lamba-Nieves.

36. Bad Bunny, "El Apagón" official video, https://www.youtube.com/watch?v=1TCX _Aqzoo4&t=1s.

37. Hernández-Torrales et al., "Seeding the CLT," 197, 198, 209, note 10; Line Algoed, María E. Hernández Torrales, and Lyvia Rodríguez Del Valle, "El Fideicomiso de la Tierra del Caño Martín Peña Instrumento Notable de Regularización de Suelo en Asentamientos Informales" (Lincoln Institute of Land Policy Working Paper WP18LA1SP, June 2018), 13–17.

38. Interview with Lamba-Nieves.

39. Interview with Pacheco.

40. Interview with Pacheco.

41. Hernández-Torrales et al., "Seeding the CLT," 198, 200–201, 202–203, 209, note 9; interview with Pacheco.

42. Personal correspondence with Pacheco, May 2023.

43. Hernández-Torrales et al., "Seeding the CLT," 202–203, 209, note 11; interview with Algoed, April 2021.

44. Interview with Pacheco.

45. Interview with Cruz Rivera, August 2021.

46. Interview with Muñiz.

47. Interview with Santiago Pérez.

48. Interview with Maria del Mar Santiago, August 2021; Algoed and Hernández-Torrales, "The Land Is Ours," 38.

49. Interview with Muñiz.

50. Interview with Muñiz.

51. Marvel, "ENLACE Landscape Architecture," https://marveldesigns.com/work /enlace-landscape-architecture/340.

52. EPA, "Smart Growth Implementation Assistance: Caño Martín Peña, San Juan, Puerto Rico," https://www.epa.gov/sites/default/files/2018-04/documents/sgia_cano _martin_pena_final_report_-english.pdf, 38–44.

53. Interview with Richard Roark, April 2021.

54. "Protocolo de Protecció al Pasajero," Tutrenpr.com, December 3, 2021, https:// tutrenpr.com/index.php?lan=en.

55. Interview with Caraballo Pagán.

56. Interview with Muñiz,

57. Interview with Santiago Pérez.

58. Interview with Lamba-Nieves.

59. Interview with Lamba-Nieves.

60. Mark Rankin and Luis Deya, "USACE Signs Project Partnership Agreement for Caño Martín Peña Ecosystem Restoration Project in San Juan, Puerto Rico," US Army Corps of Engineers, July 26, 2022, http://www.saj.usace.army.mil/Media/News-Stories /Article/3106132/usace-signs-project-partnership-agreement-for-cao-martn-pea -ecosystem-restorati/.

61. Gloria Ruiz Kuilan, "Comunidades del Caño Martín Peña Celebran el Comienzo del Dragado," elnuevodia.com, January 31, 2023, https://www.elnuevodia.com /noticias/locales/notas/comunidades-del-cano-martin-pena-celebran-el-comienzo-del -dragado/.

LEARNING FROM THREE STRUGGLES FOR EQUITABLE GOVERNANCE

1. Noromitsu Onishi, "In Vancouver, Indigenous Communities Get Prime Land, and Power," *New York Times*, August 23, 2022.

CONCLUSION

1. Community organizer/activists from Marshall Ganz to Rebecca Solnit typically attribute this framing to Maimonides, without providing a more specific citation.

INDEX

Page numbers followed by the letters *f* and *t* indicate figures and tables.

Baan Mankong (cont.)
  coalitions in, 173–176, 189, 382–383
  community building and formal rec-
    ognition in, 179–182, 192, 384–385
  design approach to, 170, 387
  environmental resilience in, 171–173,
    176–179, 190f
  equitable resilience and, 176
  governance of, 184–189, 186f, 187f,
    190f, 296, 298–300
  government-led action and, 169–173,
    286, 383–384
  lessons from, 191–194
  livelihood support in, 182–184, 183f,
    190f
  location of case study sites, 23f, 24t,
    172f
  multiscalar analysis of, 192–194, 193t
  mutual support between dimensions
    of, 378–379
  overview of, 24t, 129, 169
  "people-driven" approach to, 169, 172,
    176, 182, 184–188, 187f, 193
  pride of place in, 178
  reckoning with past and present
    injustices in, 376–377
  scalar conflicts in, 379–380, 381
  security in, 177–182, 181f, 190f
  urbanization and upgrading and,
    169–176, 170f
Bad Bunny, 359
Balestra, Filipe, 240, 241
Bang Bua coalition, 173–174, 174f
Bangkok, Thailand. See Baan Mankong
Bangkok Metropolitan Administration
  (BMA), 171
Bangladesh
  climate change–driven displacement
    in, 1, 122
  environmental resilience initiatives
    in, 42–43
  livelihood disruption in, 37, 203
Barbuda, disaster capitalism in, 121

Barcelona Superilla (Superblock) initia-
  tive, 43
Barracos, 69, 77, 84, 90
Basic Services for the Urban Poor (BSUP),
  India, 238
Bechtel Corporation, 152
Bellott, Manolo, 157, 161, 166
Bengal Delta, 122
Benthemplein water square, Rotterdam,
  40
Bioswales, 39, 43, 54, 224f, 225
Blackhawk, Ned, 433n11
Black Hills, 304
Black Lives Matter movement, 117
Black-majority communities.
  Black placemaking, 124
  community land trusts, 127
  displacement of, 41, 121
  reckoning with past and present
    injustices in, 62, 376–377
  self-governance of, 296 (see also
    Northern Manhattan Climate
    Action Plan)
  structural disadvantages of, 49–51
Bloomberg Associates, 105–106
Blue and Green Streets initiative,
  Gentilly Resilience District, 54
BNIM (architecture firm), 308
Bolivia. See Comunidad María Auxili-
  adora (CMA)
Boonyabancha, Somsook, 176, 178,
  180, 185
Boston, Dudley Street Neighborhood
  Initiative, 359
"Bounce Back" website, Sri Lanka,
  202
Boyd, Emily, 292
Brazil
  Favela Verticalization Program
    (PROVER), 71–72
  Paraisópolis Condomínios project
    (see Paraisópolis Condomínios)
  2001 City Statute, 72

Elito Arquitetos, 73
ENLACE Project Corporation, formation
    of, 351, 366f. *See also* Caño Martín
    Peña
Enlightenment principles, 32–33
Environmental literacy
    in Gentilly Resilience District, 54–56,
        55f
    in Kibera Public Space Project,
        338–339
    in Paris OASIS, 101–102
    support for, 43–44
Environmental Protection Agency, 351
Environmental resilience. *See also*
        Climate change; Gentilly Resilience
        District (GRD); Paraisópolis Con-
        domínios; Paris OASIS
    in Baan Mankong, 176–179, 190f
    in Caño Martín Peña, 348, 362–364,
        363f, 366f
    at city/region scale, 41–43, 45t
    in Comunidad María Auxiliadora,
        150–153, 162–163, 163f, 166f
    in Dafen Oil Painting Village, 274–275
    equitable resilience and, 2–3, 14–16
    in Gentilly Resilience District, 45
    goals of, 1–4
    at individual/household scale, 43–44,
        45t
    inequalities in, 5–8, 14–15, 31–32
    in Kibera Public Space Project, 338–340,
        339f, 342f
    lessons from initiatives for, 111–114
    in Living Cully initiatives, 218–219
    overview of, 17–19
    in Paraisópolis Condomínios, 79, 92f
    in Paris OASIS, 98–103
    in Pasadena Trails, 140–142, 147f
    at project/community scale, 39–41,
        45t
    in Thunder Valley Community
        Development Corporation, 303–304,
        318–321, 320f, 321f

urban design and planning and, 32–34
    in Yerwada In Situ Upgrading projects,
        249–251, 250f, 254f
Equitable resilience. *See also* Environ-
        mental resilience; Governance;
        Livelihood resilience; Security; *and
        individual case studies*
    coalitions in, 381–383
    community building and formal rec-
        ognition in, 384–385
    definition of, 14–16
    design for systemic change in,
        386–388
    drivers of, 2–3
    goals of, 16–17
    government action and, 383–384
    LEGS framework for, 17–18, 23–24
    limits of project-based approaches to,
        375
    mutual support between dimensions
        of, 378–379
    productive contestation in, 385–386
    reckoning with past and present injus-
        tices in, 376–377
    requirements for, 373–375
    scales of, 24–28, 380–381
    setbacks and delays in, 377–378
    trade-offs between dimensions of,
        379–380
    transformative potential of, 14–15
Europe
    flooding and storms in, 36
    urban design and planning in, 32–33
    wildfires in, 35–36
European Regional Development Fund,
    96
Evolutionary resilience, 14
Exploitation
    of Black-majority communities, 216
    equitable resilience and, 14–15, 36
    of manufactured home park residents,
        133, 136
    of Native populations, 311, 315, 322

Urban and Industrial Environments

Series editor: Robert Gottlieb, Henry R. Luce Professor of Urban and Environmental Policy, Occidental College
Bhavna Shamasunder, Associate Professor of Urban and Environmental Policy, Occidental College

Kerry H. Whiteside, *Precautionary Politics: Principle and Practice in Confronting Environmental Risk*

Ronald Sandler and Phaedra C. Pezzullo, eds., *Environmental Justice and Environmentalism: The Social Justice Challenge to the Environmental Movement*

Julie Sze, *Noxious New York: The Racial Politics of Urban Health and Environmental Justice*

Robert D. Bullard, ed., *Growing Smarter: Achieving Livable Communities, Environmental Justice, and Regional Equity*

Ann Rappaport and Sarah Hammond Creighton, *Degrees That Matter: Climate Change and the University*

Michael Egan, *Barry Commoner and the Science of Survival: The Remaking of American Environmentalism*

David J. Hess, *Alternative Pathways in Science and Industry: Activism, Innovation, and the Environment in an Era of Globalization*

Peter F. Cannavò, *The Working Landscape: Founding, Preservation, and the Politics of Place*

Paul Stanton Kibel, ed., *Rivertown: Rethinking Urban Rivers*

Kevin P. Gallagher and Lyuba Zarsky, *The Enclave Economy: Foreign Investment and Sustainable Development in Mexico's Silicon Valley*

David N. Pellow, *Resisting Global Toxics: Transnational Movements for Environmental Justice*

Robert Gottlieb, *Reinventing Los Angeles: Nature and Community in the Global City*

David V. Carruthers, ed., *Environmental Justice in Latin America: Problems, Promise, and Practice*

Tom Angotti, *New York for Sale: Community Planning Confronts Global Real Estate*

Paloma Pavel, ed., *Breakthrough Communities: Sustainability and Justice in the Next American Metropolis*

Anastasia Loukaitou-Sideris and Renia Ehrenfeucht, *Sidewalks: Conflict and Negotiation over Public Space*

David J. Hess, *Localist Movements in a Global Economy: Sustainability, Justice, and Urban Development in the United States*

Julian Agyeman and Yelena Ogneva-Himmelberger, eds., *Environmental Justice and Sustainability in the Former Soviet Union*

Jason Corburn, *Toward the Healthy City: People, Places, and the Politics of Urban Planning*

JoAnn Carmin and Julian Agyeman, eds.,*Environmental Inequalities Beyond Borders: Local Perspectives on Global Injustices*

Louise Mozingo, *Pastoral Capitalism: A History of Suburban Corporate Landscapes*

Gwen Ottinger and Benjamin Cohen, eds., *Technoscience and Environmental Justice: Expert Cultures in a Grassroots Movement*

Samantha MacBride, *Recycling Reconsidered: The Present Failure and Future Promise of Environmental Action in the United States*

Andrew Karvonen, *Politics of Urban Runoff: Nature, Technology, and the Sustainable City*

Daniel Schneider, *Hybrid Nature: Sewage Treatment and the Contradictions of the Industrial Ecosystem*

Catherine Tumber, *Small, Gritty, and Green: The Promise of America's Smaller Industrial Cities in a Low-Carbon World*

Sam Bass Warner and Andrew H. Whittemore, *American Urban Form: A Representative History*

John Pucher and Ralph Buehler, eds., *City Cycling*

Stephanie Foote and Elizabeth Mazzolini, eds., *Histories of the Dustheap: Waste, Material Cultures, Social Justice*

David J. Hess, *Good Green Jobs in a Global Economy: Making and Keeping New Industries in the United States*

Joseph F. C. DiMento and Clifford Ellis, *Changing Lanes: Visions and Histories of Urban Freeways*

Joanna Robinson, *Contested Water: The Struggle Against Water Privatization in the United States and Canada*

William B. Meyer, *The Environmental Advantages of Cities: Countering Commonsense Antiurbanism*

Rebecca L. Henn and Andrew J. Hoffman, eds., *Constructing Green: The Social Structures of Sustainability*

Peggy F. Barlett and Geoffrey W. Chase, eds., *Sustainability in Higher Education: Stories and Strategies for Transformation*

Isabelle Anguelovski, *Neighborhood as Refuge: Community Reconstruction, Place Remaking, and Environmental Justice in the City*

Kelly Sims Gallagher, *The Globalization of Clean Energy Technology: Lessons from China*

Vinit Mukhija and Anastasia Loukaitou-Sideris, eds., *The Informal American City: Beyond Taco Trucks and Day Labor*

Roxanne Warren, *Rail and the City: Shrinking Our Carbon Footprint While Reimagining Urban Space*

Marianne E. Krasny and Keith G. Tidball, *Civic Ecology: Adaptation and Transformation from the Ground Up*

Erik Swyngedouw, *Liquid Power: Contested Hydro-Modernities in Twentieth-Century Spain*

Ken Geiser, *Chemicals without Harm: Policies for a Sustainable World*

Duncan McLaren and Julian Agyeman, *Sharing Cities: A Case for Truly Smart and Sustainable Cities*

Jessica Smartt Gullion, *Fracking the Neighborhood: Reluctant Activists and Natural Gas Drilling*

Nicholas A. Phelps, *Sequel to Suburbia: Glimpses of America's Post-Suburban Future*

Shannon Elizabeth Bell, *Fighting King Coal: The Challenges to Micromobilization in Central Appalachia*

Theresa Enright, *The Making of Grand Paris: Metropolitan Urbanism in the Twenty-first Century*

Robert Gottlieb and Simon Ng, *Global Cities: Urban Environments in Los Angeles, Hong Kong, and China*

Anna Lora-Wainwright, *Resigned Activism: Living with Pollution in Rural China*

Scott L. Cummings, *Blue and Green: The Drive for Justice at America's Port*